Water Systems towards New Future Challenges

Water Systems towards New Future Challenges

Editor

Helena M. Ramos

MDPI • Basel • Beijing • Wuhan • Barcelona • Belgrade • Manchester • Tokyo • Cluj • Tianjin

Editor
Helena M. Ramos
Department of Civil Engineering,
University of Lisbon,
IST—Tecnico Lisboa/CERIS
Portugal

Editorial Office
MDPI
St. Alban-Anlage 66
4052 Basel, Switzerland

This is a reprint of articles from the Special Issue published online in the open access journal *Water* (ISSN 2073-4441) (available at: https://www.mdpi.com/journal/water/special_issues/water-systems-Challenges?view=default&listby=date).

For citation purposes, cite each article independently as indicated on the article page online and as indicated below:

LastName, A.A.; LastName, B.B.; LastName, C.C. Article Title. *Journal Name* **Year**, *Volume Number*, Page Range.

ISBN 978-3-0365-0410-0 (Hbk)
ISBN 978-3-0365-0411-7 (PDF)

Contents

About the Editor

Helena M. Ramos is a professor at Instituto Superior Técnico (Engineering Faculty of the University of Lisbon). Her activity is mainly in the following areas: Hydropower and pumping systems, computational fluid dynamics (CFD)—Hydrodynamics, mathematical modeling, water and energy efficiency, hybrid energy solutions, hydraulic transients and surge control, safety and operation—dynamic effects, and water and energy nexus. She has participated in more than 30 scientific projects, in particular some EU research projects. She has been a member of editorial teams and a reviewer of different scientific journals. She has several publications, more than 300, including peer-reviewed international journals, and four international books.

Preface to "Water Systems towards New Future Challenges"

Water supply (WS) systems collect, store and/or treat, and distribute water among water sources and consumers. They are the transfer of drinking, irrigation, waste, storm, and industrial water from an intake to final users. This important infrastructure is becoming a dynamic environment, where new technologies and the best practices are implemented with the ambition of increasing the safety, efficiency, sustainability, and management. The monitoring system (MS), control technology (CT), management strategy (MS), energy saving (ES), eco-innovative solution (EIS), and modeling decision support system (MDSS) have to be improved to obtain technical, economic, and environmental benefits in terms of research, technology, and engineering applications. The water industry is subject to changes regarding the sustainable management of urban water systems. There are many external factors, including impacts of the climate change, drought, and population growth in urban centers, which lead to an increase in the responsibility to adopt more sustainable management of urban waters. There are many structural challenges facing the development of modern cities, from the water supply to the population and economic activities, to the improvement of urban, industrial, and rural water sectors. Population growth results in an increase in and a concentration of water needs and a consequent need for water management. Under this reality, the use of advanced studies and technologies as well as the adoption of more robust management models are necessary to better suit the critical demands in the near future. The concept of the smart water system utilizes advanced information technologies for system monitoring data to achieve greater efficiency in the resource allocation. In addition, to increase the efficiency in water loss control, the prevention and the early detection of leaks allow the development of the best practice in the asset management. A smart system uses real-time data, optimization variables, variable speed pumps, dynamic control valves, and smart meters in order to balance the demand, minimize the overpressure in aging pipelines, and save water and energy. Therefore, a sustainable water–energy nexus (WEN) arises in terms of its water and energy efficiency, reliability, and environmental integration towards smart water grids (SWGs). This is a new method for smart technology, resource management, and sustainable water infrastructure development in the near future to face climate and demand challenges.

Helena M. Ramos
Editor

Article

Modeling Irrigation Networks for the Quantification of Potential Energy Recovering: A Case Study

Modesto Pérez-Sánchez [1], Francisco Javier Sánchez-Romero [2], Helena M. Ramos [3] and P. Amparo López-Jiménez [1,*]

[1] Hydraulic and Environmental Engineering Department, Universitat Politècnica de València, Valencia 46022, Spain; mopesan1@upv.es

[2] Rural and Agrifood Engineering Department, Universitat Politècnica de València, Valencia 46022, Spain; fcosanro@agf.upv.es

[3] Civil Engineering, Architecture and Georesources Departament, CERIS, Instituto Superior Técnico, Universidade de Lisboa, Lisboa 1049-001, Portugal; hramos.ist@gmail.com

* Correspondence: palopez@upv.es; Tel.: +34-96-387700 (ext. 86106)

Academic Editor: Ashok K. Chapagain
Received: 29 February 2016; Accepted: 26 May 2016; Published: 1 June 2016

Abstract: Water irrigation systems are required to provide adequate pressure levels in any sort of network. Quite frequently, this requirement is achieved by using pressure reducing valves (PRVs). Nevertheless, the possibility of using hydraulic machines to recover energy instead of PRVs could reduce the energy footprint of the whole system. In this research, a new methodology is proposed to help water managers quantify the potential energy recovering of an irrigation water network with adequate conditions of topographies distribution. EPANET has been used to create a model based on probabilities of irrigation and flow distribution in real networks. Knowledge of the flows and pressures in the network is necessary to perform an analysis of economic viability. Using the proposed methodology, a case study has been analyzed in a typical Mediterranean region and the potential available energy has been estimated. The study quantifies the theoretical energy recoverable if hydraulic machines were installed in the network. Particularly, the maximum energy potentially recovered in the system has been estimated up to 188.23 MWh/year) with a potential saving of non-renewable energy resources (coal and gas) of CO_2 137.4 t/year.

Keywords: smart water; water-energy nexus; energy efficiency; sustainable water management; energy recovering

---

1. Introduction

Water and its management is one of the more important current and future global challenges. Its variability can cause cloudbursts, making sewers to overflow, while the scarcity of water in other components involves public services and reduces irrigation [1]. Hence, an efficient management of water irrigation networks is crucial for facing future challenges related to the energy-water nexus, considering the importance of irrigation in the whole planet [2]. The development of the modernization of irrigation systems in agriculture (replacing open channel with pressurized irrigation) has considerably increased energy consumption in recent years [3]. Nevertheless, the establishment of drip irrigation has made more efficient systems in water consumption but not in energy demand.

Spain is not an exception: The annual irrigation volume consumed in Spain is 16.344 km^3/year [4] and the global irrigation consumption in pressure systems approaches 3925 km^3/year [5]. Consequently, the theoretical energy recoverable could be a significant amount.

In Spain, the drip irrigated area (*i.e.*, 1.7 of 3.54 million of hectares are irrigated by pressure systems) [6] represents 17.56% of the world's surface irrigated by localized drip (approximately

9 million hectares) [7]. The high energy consumption and the rising cost of tariff have reduced profits or even the viability of farms [8]. The need to study strategies to decrease the energy consumption in these installations is pointed out in the consulted references. Regarding this issue, Coehlo *et al.* established the need to study the recovery in water distribution systems for increasing the energy efficiency, since the energy consumption in water networks involves 7% of the global energy consumption [9]. The objectives of this recovery are: to reduce the energy footprint of water in irrigation system and to lessen greenhouse emissions compared with other non-renewable energy sources.

Water-energy nexus analysis has become a crucial issue in recent years [3,10–13]. Baki *et al.* [10] studied water-energy interactions in water systems in Athens. Okadera *et al.* [11] and Herath *et al.* [12] analyzed water footprints of hydroelectricity. Water management improvement in irrigation networks have also been analyzed in [14], where a 40% irrigation reduction volume was achieved.

Sustainable social and economic growth based on renewable energy sources forces water networks to work as multipurpose systems [15], where power generation is not the first objective but an important complementary one [16].

Some studies and prototypes of recovering energy with small turbines can be found in the literature for power less than 100 kW [17–22]. The previous publications of Carravetta *et al.* [17,18] compare the feasible regulation systems for pump as turbines (PATs). These authors [19,20] determined performance of PATs installed in drinking systems. The efficiency oscillates between 0.4 and 0.6. Ramos *et al.* [21,22] proposed new design solutions to energy production in water pipe systems. These solutions are focused on the installing of PATs with electrical or hydraulic regulation within network. Additionally, to the previous referenced authors, the variability of the flow along time is studied as an objective in the present research. Here, a deep analysis of theoretical recovery energy in the network is proposed (*i.e.*, distinguishing values of dissipated energy, necessary energy and losses in lines and consumption points).

Particularly in irrigation networks, some studies of recovering energy in open channels flow [23–25] and preliminary studies in pressure pipe systems are described. These show the importance to analyze these networks in terms of recovery energy. An example of these studies is the network of Alqueva in Portugal [26]. In that contribution, authors analyzed the recovery energy with average steady state flows. A discretized analysis in short time intervals is proposed for determining the theoretical energy recoverable in a part of the Alqueva distribution network. This analysis was made with average consumption demands.

The present research determines the variability of flows and pressure in any point or line on the network depending on irrigation habits. This advantage (determining instant values of flows and pressure) allows performing the analysis of energy recovery in any point on the network. The methodology obtains the data pairs of flow (Q) and head (H) of the working area of the hypothetical installed machine.

Furthermore, the methodology determines the variation of flow in a network based on the habits of irrigation in order to perform energy analyses. The application of this methodology in irrigation networks aims to complement previous studies for PATs efficiency in dinking supply networks, extending its use.

The variation of flow is based on random demand of the users and the real irrigation allocations. Depending on these parameters, the proposed methodology estimates the energy dissipated by friction losses, the energy required for irrigation, and the recoverable energy in the irrigation network. The discretization of the flows leads managers to analyze power generation depending on irrigation time periods. Accordingly, the present analysis has the following objectives:

1) Proposing a new methodology for determining the flows throughout the year in an irrigation network demand, considering the need of the crop, the historic consumption and the irrigation farmers' habits

2) Estimating the flow rate and pressures with the time

3) Quantifying the energy balance in pressurized irrigation distribution systems to determine the energy footprint of water in the distribution system, and the estimated recoverable energy

4) Applying these procedures to a real case study

2. Methods and Materials

2.1. Methodology for Determining the Flow

In this section, the proposed methodology to determine the time-dependent flow throughout the year is described. In order to analyze any pressurized irrigation network from the energy point of view, the flow and pressure along pipelines are determinant variables. The requirements of the minimum pressure at any consumption point are also fundamental. Pressures are different depending on the location of irrigation points. Therefore, the spatial and timing distribution of these consumptions are important aspects to take into consideration.

The flows are variable over any irrigation campaign, depending on many factors such as distribution of crops in the irrigation area, crop maturity, weather conditions, soil characteristics, efficiency of drippers (ranging from 0.90 to 0.95), and the habits of farmers, among others.

Traditionally, the Clement methodology has been used for irrigation network sizing [27–29]. This methodology allows determining the maximum flow circulating in a network line. This maximum flow rate is calculated by assuming a binomial distribution flow. The mathematical expectation and standard deviation of the binomial probability distribution depends on the opening point of consumption. Clement assumed that this probability was uniform and equal over time. This uniform probability consideration can lead to underestimating the flows. Consequently, the Clement methodology cannot be used for analyzing potential energy recovery. Probability of irrigation at any point is non-uniform, and depends on the habits of irrigation farmers. Therefore, it varies throughout the day, week, and month. This underestimation leads to the proposal of different methodologies for estimating flows in irrigation networks. The most common are those that use statistical methods [27–29], or models based on the random opening of irrigation points by means of computer simulations [30–33]. A new methodology considering both strategies is here proposed.

Flow and energy implications are therefore separately considered and described.

The majority of water distribution networks only have water meters in each irrigation point for billing and controlling the consumed volumes. Unfortunately, it is not usual that the irrigation network has readings of flows and pressures at any time. For this reason, the proposed methodology simulates the operation of any irrigation network based on the random generation of consumption in irrigation points.

The day, start, and duration of irrigations (as function of the habits of the farmers) are considered in this research as factors for irrigation probability and flows. Furthermore, the real consumption probability weights (obtained from historical archives of the irrigation entities) can be assigned to consumptions, and the network can be very precisely simulated.

Hence, for any day of the year, consumptions can be estimated in any irrigation point by following these steps (Figure 1).

1. Estimation of cumulative volume consumed by the irrigation point

The decision to irrigate depends on the balance (V_{Na}) between the previous irrigated volume and the consumption assigned (needs) of the irrigation point (Input 1). If the volume of cumulative consumption is positive, automatically the methodology indicates that this is not an irrigation day. Only when this volume is negative, irrigation is possible. If the volume of cumulative consumption is negative, the methodology determines the irrigation probability.

2. The determination of the irrigation probability (P_I)

To randomly determine if crops are irrigated or not during a particular day, two types of weight functions are assigned. These functions are obtained from interviews with farmers. According to Figure 1, Input 2 determines the irrigation weekly pattern (w_{dj}), prioritizing the irrigation days per week. Input 3 determines the maximum days between irrigations for each month of the year (i). If in previous days no irrigation has been performed, watering is forced.

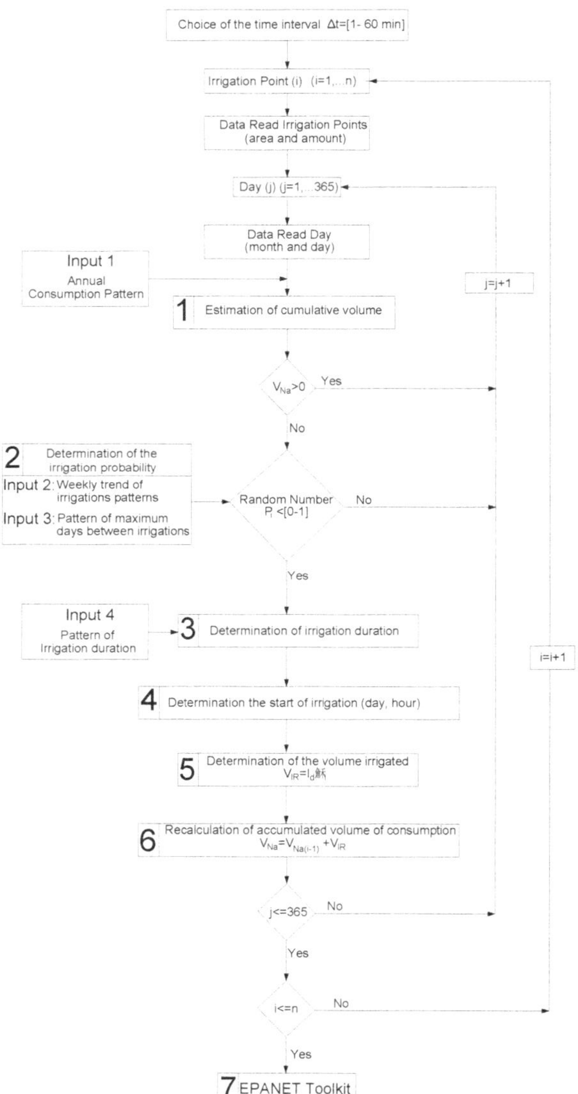

Figure 1. Schematic description of the methodology for flow estimation.

The methodology generates a random number (RN) between zero and one associated with an irrigation probability. If $RN_j \leqslant P_I$ irrigation is assigned to this consumption point.

$$P_I = \frac{w_{dj}}{\sum_{n=1}^{n=i-j+1} w_{dn}} \tag{1}$$

where:

i = numbers of days inside of interval;

j = day of decision making;

w_{dj} = pattern to irrigate one particular day inside the interval;

$\sum_{n=1}^{n=i-j+1} w_{dn}$ = total addition of patterns.

3. The determination of the irrigation duration

The methodology allows determining the estimated time based on irrigation habits of farmers to satisfy irrigation needs (Input 1). This value depends on irrigation amount and type of crop.

4. The start of irrigation

The irrigation duration randomly determines the start of irrigation as a function of the daily probability curves of irrigation time (Input 4). When the methodology determines that a consumption point has to be irrigated, the start time of irrigation is determined. Therefore, the cumulative probability must be used for starting irrigation. This curve is defined by twenty-four sections (one per hour). When no irrigation exists, the irrigate weight (w_h) in this interval is assigned to be zero.

The probability in the time interval (p_h) is:

$$p_h = \frac{w_h}{\sum_{h=0}^{h=23} w_h} \tag{2}$$

where w_h is the defined pattern (Input 2) to irrigate one particular hour inside the interval.

The cumulative probability (p_{cm}) is:

$$p_c = \sum_{h=0}^{h=m} p_h \quad (m = 0, \ldots, 23) \tag{3}$$

where m is the number of intervals in one day.

A new *RN* is generated, ranging from 0 to 1. It is compared with the values of cumulative probability (p_{cm}) and the start irrigation period is established. For this particular time period, the methodology selects within this period the start interval from zero to value equal to $\frac{60}{\Delta t}$ (where Δt is the time interval in which the simulated flow is discretized). When this step is completed, the day and hour of starting irrigation is known.

5. Determination of irrigation volume

The irrigation supply (agronomic known parameter, which depends on: framework plantation, number of dripper per plant and flow of the dripper) and the duration (Input 4) are known and the irrigation volume can be calculated for that day.

6. Calculation of cumulative consumption

When the irrigation volume is known, the methodology updates the water volume available for the plant.

7. The pressure and flow modeled for each node in the network

They are calculated for every irrigation points and each day using Epanet Toolkit. Epanet is public domain software [34] that models water distribution in pipe systems. Different elements can be represented: pipe networks composed by pipes, nodes (junctions), pumps, valves, and storage tanks or reservoirs. The model can simulate extended-period hydraulic analysis by simulating by sort of pipes systems, computing friction and minor losses, representing various types of valves, junctions, tanks and pumps, considering multiple patterns at nodes consumption with time variation, and system operation on simple tank level, timer controls or complex rule-based controls.

2.2. Balance of Energy

Once flows and pressures are estimated along the time in the whole network, the energy equation (Reynolds Theorem) must be implemented to consider the energy balance in the system [35].

According to Figure 2, a generic irrigation network with all possible elements (reservoir, pumps, turbines, and compensation tanks) is presented. The conservation of energy equation is defined as:

$$\frac{dE}{dt} = \frac{dQ}{dt} + \frac{W_{shaft}}{dt} = \frac{d}{dt}\iiint_{CV}\rho\left(gz + u + \frac{v^2}{2}\right)dV + \iint_{CS}\left(gz + u + \frac{P}{\rho} + \frac{v^2}{2}\right)\rho\left(\vec{v}\cdot d\vec{A}\right) \quad (4)$$

where:

$\frac{dE}{dt}$ = exchange of energy per unit time in the control system;

$\frac{dQ}{dt}$ = exchange of heat per unit of time (heat power);

$\frac{W_{shaft}}{dt}$ = power transmitted directly to or from the fluid (e.g., pump);

dV = differential volume of control volume for integration;

$\vec{v}$ = velocity vector of fluid;

$d\vec{A}$ = differential area of control surface for integration;

ρ = fluid density;

gz = potential energy per unit mass;

u = internal energy per unit mass;

$\frac{v^2}{2}$ = kinetic energy per unit mass;

$\frac{P}{\rho}$ = height of pressure per unit mass;

Within the control system, the following simplifications can be made:

The water density is constant.

Flow is uniform in each interval.

Exchange of heat between fluid and surroundings is negligible (adiabatic system).

The shaft work is the power transmitted directly to/from the fluid in the case that a pump or turbine exists in the network.

There is no compensation tank in the network, therefore, the time energy variation inside of the control volume as function of time is negligible.

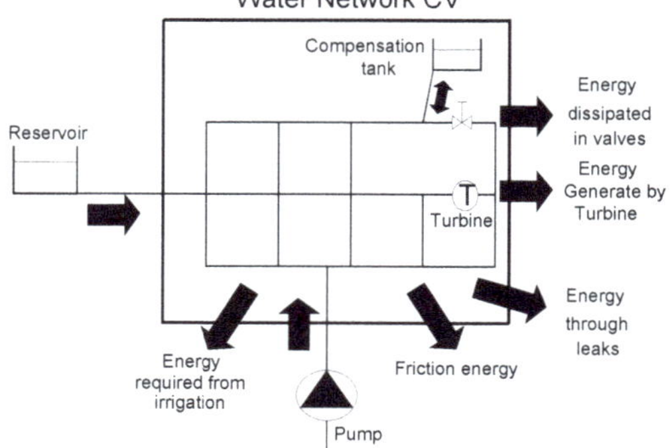

Figure 2. Energy balance in the pressurized irrigation water network adapted from [36].

If an irrigation system operates by gravity (Figure 3), the equation of energy applied to any control system along a time interval is defined by Equation (5):

$$\gamma Q_D H_D \Delta t = \sum_{i=1}^{n} \gamma Q_{oi} H_{oi} \Delta t + \rho \left(\sum_{i=1}^{n} (Q_{oi} u_{oi} - Q_{Di} u_{Di}) \right) \Delta t \tag{5}$$

where:

Δt = time interval (s);

n = total number of irrigation points;

i = individual irrigation points;

γ = specific weight of the fluid (N/m^3);

Q_D = total flow demanded by the network (m^3/s);

H_D = piezometric head of the reservoir. For a pumped system, the value is the manometric height;

Q_{oi} = flow demanded by each irrigation point (m^3/s);

H_{oi} = piezometric head of the consumption node (m);

$\gamma Q_D H_D$ = total energy (kW) supplied to the system. This term is equal to E_T, which is later defined;

$\sum_{i=1}^{n} \gamma Q_{oi} H_{oi}$ = energy consumed by all irrigation points (kW). This term will be defined as E_{RI} plus E_{TRI};

$\rho \left(\sum_{i=1}^{n} (Q_{oi} u_{oi} - Q_{Di} u_{Di}) \right)$ = Exchange of internal energy. In an adiabatic system, it is equal to friction losses. This term will be defined as E_{FR}.

Leakages are not considered in this analysis because the drip irrigation network is still new (minimum leakages), the maintenance and repair plans are usually undertaken (which reduce possible losses), and finally, these networks are not as automated as drinking systems so unmeasured volumes and leakages are difficult to discern. If an energy audit were made, this volume should be considered or estimated [6,36]. Furthermore, the installation of hydraulic machines does not affect the water quality of the final use (*i.e.*, irrigation).

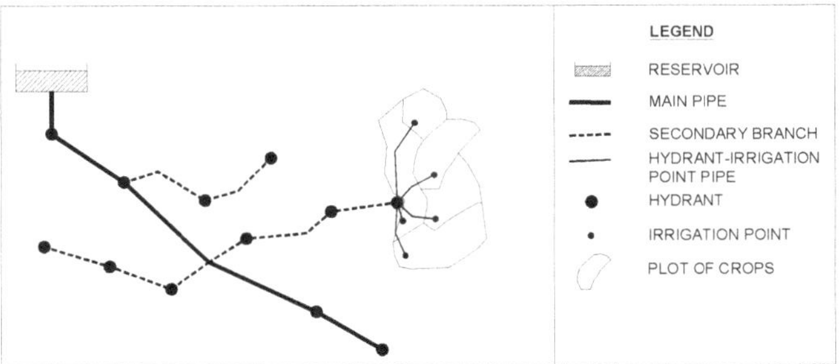

Figure 3. Scheme of irrigation network.

When a global energy balance of the network is established, it is possible to define different terms of energy. Such as lines, hydrants and irrigation points, as follow (Figure 4):

- *Total Energy (E_{Ti})*: potential total energy in an irrigation point when the consumption is null in the entire network. It corresponds to the static energy (*i.e.*, potential) of the node. For an irrigation point along a time interval, the value is:

$$E_{T_i}\ (kWh) = \frac{9.81}{3600} Q_i \left(z_0 - z_i \right) \Delta t \tag{6}$$

where:

Q_i is the flow circulating by a line that supplies to more unfavorable irrigation point (most disadvantageous consumption node in terms of need of the pressure) (m^3/s);

z_i is the geometry level above reference plane of the irrigation point. In this case, the reference is sea level (m);

z_0 is the geometry level above reference plane of the free water surface of the reservoir. In this case, the reference is sea level (m);

Δt is the time interval (s).

- *Friction Energy (E_{FRi}):* for a time interval, it is the energy dissipated in the network by the water coming from head until the irrigation point.

$$E_{FR_i} (kWh) = 2.725 \cdot 10^{-3} Q_i \left(z_0 - (z_i + P_i)\right) \Delta t \tag{7}$$

where:

P_i is the service pressure in any point of the network when consumption exists. The units are meter water column (m w.c.).

Minor losses (pressure loss in particular network components like tees, valves, and similar) are considered as a percentage of friction losses. Associated with this term, the Energy Footprint of Water (EFW) can be calculated. Energy Footprint of Water is defined as the ration between energy dissipated due to friction losses *(E_{FRi})* over the distributed volume on the network (kWh/m^3).

- *Theoretical Energy Necessary (E_{TNi}):* it is the minimum energy required in a hydrant or line to ensure the minimum pressure of irrigation in the more unfavorable point. The value is:

$$E_{TN_i} (kWh) = 2.725 \cdot 10^{-3} Q_i P_{min_i} \Delta t \tag{8}$$

where:

P_{min_i} is the minimum pressure of service of a line or hydrant to ensure the minimum pressure in the most disadvantageous consumption node. The units are meter water column (m w.c.).

- *Energy Required for Irrigation (E_{RIi}):* during an interval of time, it is the minimum energy required at an irrigation point to ensure the irrigation water evenly. The value is:

$$E_{RI_i} (kWh) = 2.725 \cdot 10^{-3} Q_i P_{min I_i} \Delta t \tag{9}$$

where:

$P_{min I_i}$ is the minimum pressure of service of an irrigation point required to ensure the irrigation water evenly. The units are meter water column (m w.c.).

- *Theoretical Available Energy (E_{TAi}):* it is the available energy for recovery in a hydrant or line. The recovery coefficient in a hydrant or line (C_{RT}) depends on losses existent between the hydrant (or pipeline) and the most disadvantageous consumption node. It is equal to the sum of the theoretical energy recoverable plus the theoretical energy unrecoverable (E_{NRT}). The value of this energy for a particular time duration, is defined as:

$$E_{TA_i} (kWh) = 2.725 \cdot 10^{-3} Q_i \left(P_i - P_{min_i}\right) \Delta t \tag{10}$$

- *Theoretical Recoverable Energy (E_{TRi}):* it is the maximum theoretical recoverable energy in an irrigation point, hydrant or line of the network, ensuring at downstream the minimum pressure of irrigation.

$$E_{TR_i}\,(kWh) = 2.725 \cdot 10^{-3} Q_i \left(P_i - \max \left(P_{min_i}; P_{minI_i} \right) \right) \Delta t = 2.725 \cdot 10^{-3} Q_i H_i \Delta t \qquad (11)$$

where

H_i is the value of head in irrigation point, hydrant or line (m w.c.), obtained as:

$$H_i = P_i - \max \left(P_{min_i}; P_{minI_i} \right) \qquad (12)$$

- *Theoretical unrecoverable Energy (E_{NTRi}):* it is the energy in a hydrant or line on the network that cannot be recovered. This energy is necessary to assume the losses from the line or hydrant to the more unfavorable irrigation point.

$$E_{NTR_i} = E_{TA_i} - E_{TR_i} \qquad (13)$$

- *Recovery coefficient in hydrant or line (C_{RTi}):* it is the quotient between E_{TRi} and E_{TAi} in an irrigation point, hydrant or line of the network. It represents the proportion of recovery energy over available energy.

$$C_{RT_i} = \frac{E_{TRi}}{E_{TAi}} \qquad (14)$$

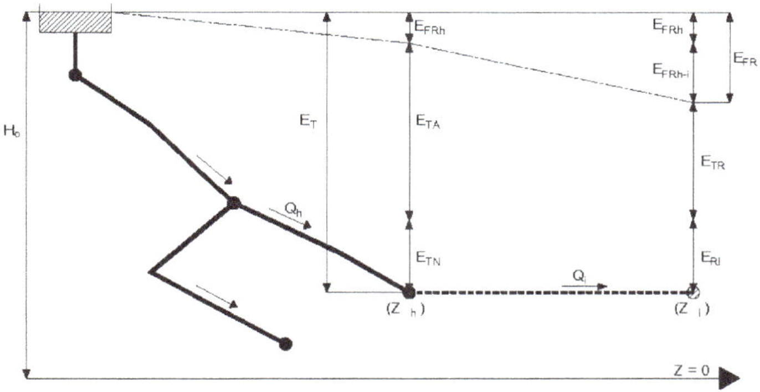

Figure 4. Scheme of hydraulic energies grade line.

Q_h (Figure 4) is the flow circulating in a line or consumed by a hydrant (m^3/s), z_h is the geometry level above reference plane of the line or hydrant. H_0 is the piezometric height of the reservoir that supplies the network. The units are meter water column (m w.c.). If reservoir is open, H_0 is equal to z_0.

When Equation (5) is applied in a point of the network, it is defined by Equation (15):

$$E_{Ti} = E_{FRi} + E_{RIi} + E_{TRi} \qquad (15)$$

When all irrigation points are considered (z_i in Figure 4), the annual balance of energy is defined by Equation (16):

$$\sum_{i=1}^{n} E_{Ti} = \sum_{i=1}^{n} \left(E_{FRi} + E_{RIi} + E_{TRi} \right) \qquad (16)$$

In the case of lines and hydrants (z_h in Figure 4), the annual balance of energy is defined by Equation (17):

$$E_T = E_{FRh} + E_{TA} + E_{TN} = E_{FRh} + E_{RI} + E_{TR} + E_{NTR} \qquad (17)$$

The energy footprint of water and the theoretical recoverable energy are crucial for the energy balance. The energy footprint on the network distribution can be obtained along the year and compared with average values analyzed in other distribution systems. Some of these values are: 0.31 kWh/m^3 in injected irrigation network [36], 0.18–0.32 kWh/m^3 according to California Energy Commission [37], 0.081 kWh/m^3 in Bangkok, 0.5 kWh/m^3 in Delhi and 0.13 kWh/m^3 in Tokyo [38].

Regarding the theoretical recoverable energy in a network, it mainly depends on the orography of the irrigation area. The networks with larger gradients between the supply and the consumption points have greater possibility to recover energy, if the appropriate machine is selected. The energy recovery can be analyzed in different parts of the network:

i) In plot of cultivation—in this case, the private user needs to reduce pressure down to 30 m w.c. to carry out drip irrigation. Generally, the user installs a pressure reducer to dissipate the excess energy. This element can be replaced by a pico-turbine to generate energy for self-consumption. This energy can be used in remote-control system, cleaning of filters, lighting and others similar consumptions.

ii) In the hydrant pipe—when the hydrant supplies to flat topography, reduction of pressure can be done. In an operating network, this reduction is carried out with a pressure reducing valve. This recovery could potentially be done if a suitable turbine could be installed.

iii) In pipe branch—in networks with large extension and irregular orography, some parts of the network can achieve higher pressure than necessary, forcing pressure to be reduced on a pipe branch. Currently, this reduction is possible by using a reducing valve installed on this branch. These valves can be replaced by turbines or pumps as turbines (PAT) [14] depending on the system characteristics to increase the energy efficiency of the network.

The presented methodology helps managers to estimate the theoretical recoverable energy in irrigation points, hydrants and branches (pipelines). According to Spadaro *et al.* [39], this recovery can contribute with a theoretical average reduction of greenhouse gases emission between 582 and 877 gCO$_2$/kWh when compared to non-renewable energy solutions (e.g., coal and gas) and 1150 gCO$_2$/kWh when compared to emissions of fossil fuel [40]. However, this reduction will depend on the water source (groundwater, superficial or residual water) and distribution (gravity or pumped) [41]. Future research should try to integrate all applications together (supply, irrigation and wastewater for better water management) in a strategy to improve system efficiency, thus reducing greenhouse gases [42].

The viability of these installations is subject to economic evaluation (incomes *vs.* costs). Zema *et al.* [43] proposed a simple method to evaluate the economic feasibility of micro-hydropower plants in irrigations systems. In preliminary studies, these methods are good indicators for taking decisions to develop more detailed projects. These decisions are focused on the selection of machine efficiency, temporal distribution of energy produced and investment analyses. Similarly, Castro [44] proposed the period simple return (PSR) and energy index (EI) as indicators of investment viability. These sorts of installations are viable if the PSR is less than six years and the energy index smaller than 0.6 €/kWh. PSR and EI are defined by the following equations:

$$PSR = \frac{IC}{I - C} \tag{18}$$

$$I = P_E E \eta \tag{19}$$

$$C = C_0 E \eta \tag{20}$$

$$EI = \frac{IC}{E} \tag{21}$$

where: *IC* is the investment cost (€); *C* is the annual operating cost (€/year); C_0 is the unit operating cost (€/kWh); *I* is the annual income (€/year); P_E is the energy price (€/kWh); *E* is the theoretical energy recovery by the turbine (kWh/year) and η is the machine efficiency.

The investment cost using PAT is 50% lower than the cost of traditional turbines [45]. Carravetta *et al.* [20] estimated *IC* of 545 €/kW if the machine is electrically regulated. Incomes depend on generated energy, which depends on the price of energy, recovery energy and the efficiency of the machine. Based on the specific speed, expert literature indicates that efficiency varies between 50% and 60% [20,46]. Castro [44] established a sales price (P_E) of 0.0842 €/kWh and C_0 of 0.0145 €/kW. These solutions, due to the smaller cost of PATs, present a lower payback period.

3. Case Study

3.1. Description

In order to apply the developed methodology, a drip irrigation network located in Vallada (Valencia, Spain) is proposed (Figure 5). The network covers 290.2 hectares, with water coming from a well. The main crop is citrus, although there is a small area of olive trees. The water is accumulated in a reservoir with a 7000 m^3 capacity. The topography varies between 378 and 248 m above sea level. The pond is located sufficiently high (399 m above sea level) to ensure the minimum pressure of 30 m w.c. in all irrigation points.

The pipelines of the network are built on asbestos cement pipes (diameters between 300 and 500 mm), polyvinyl chloride (diameters between 250 and 125 mm) and ductile iron (diameter of 150 mm). The installation has seventy multiuser hydrants. A manifold is installed in each hydrant to connect pipes of polyethylene with irrigation points. Inside the hydrant, meters are placed to read the consumption volume.

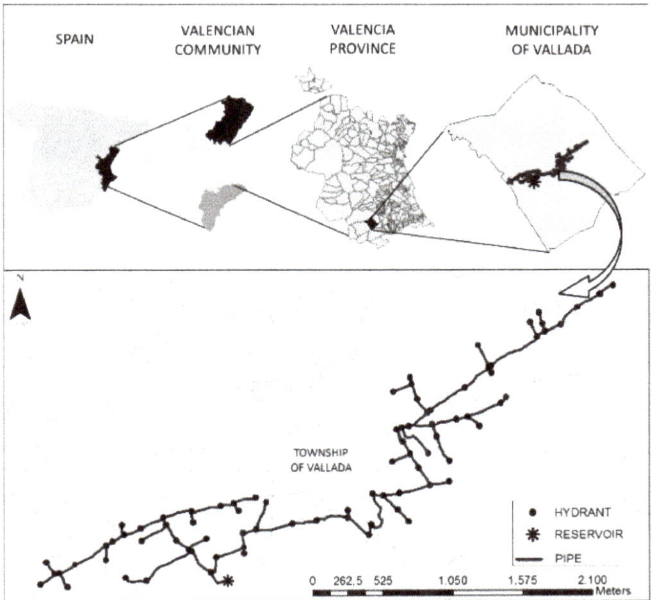

Figure 5. Study area in Vallada case study network.

The next steps are necessary to apply the proposed methodology:

a) To make use of records of the water metered in the irrigation points. There are records since 2003 (year that network began to operate). In each plot, registers were taken quarterly corresponding to the months of March, June, September and December.

b) To calculate the flow design (water requirements) for each of the considered plot, according to the crop and characteristics of the irrigation installation (distance between drippers and type). The number of sectors is established depending on the area of plots. This has allowed an allocation of irrigation according to the existing installation (Figure 1).

c) To perform interviews of users and operating staff for estimating farmer habits. The type of irrigation management at the annual, monthly, weekly and daily levels has been analyzed in this questionnaire. Based on these interviews, different consumption patterns have been established. These patterns take in to account the irrigation habits of farmers: weekly trend, maximum days between irrigations and irrigation duration (Inputs 2, 3 and 4 in Figure 1).

3.2. Methodology

3.2.1. Historical Consumption Data and Probability Function

Based on the total consumption provided by the entity, an average consumption of 3189 m^3/ha has been estimated. Therefore, the annual average consumption is 925,427 m^3 in the time series studied 2003–2014 (Figure 6).

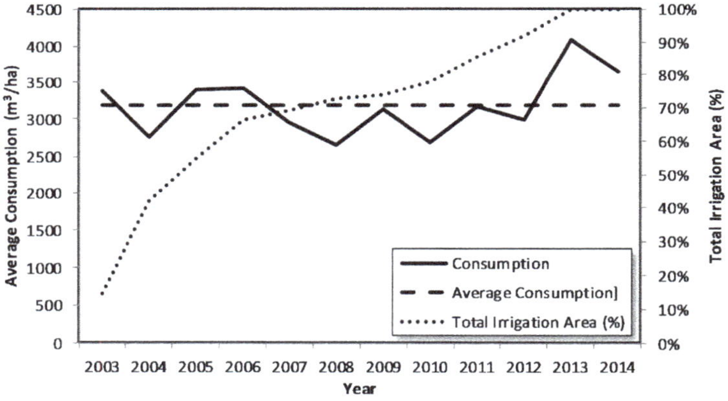

Figure 6. Measured Annual Consumption in Vallada case study network.

The next steps allow establishing consumption patterns:

1. The annual consumption is estimated for each irrigation point, grouping them in terms of similar consumptions. This classification has provided the distribution presented in Figure 7.

2. In order to determine the distribution of daily consumption, specific weights for crops of citrus and olive needs to be considered, according to the registered consumptions. The monthly pattern of irrigation needs has been set taking into account consumer groups (Figure 7). On the one hand, the irrigation points with consumptions lower than 3584 m^3/ha have been assigned the patterns of consumption under the name "Crop of Olive" (Figure 8). On the other hand, the irrigation points with consumptions higher than 4480 m^3/ha have been assigned the patterns of consumption under the name "Crop of Citrus" (Figure 8).

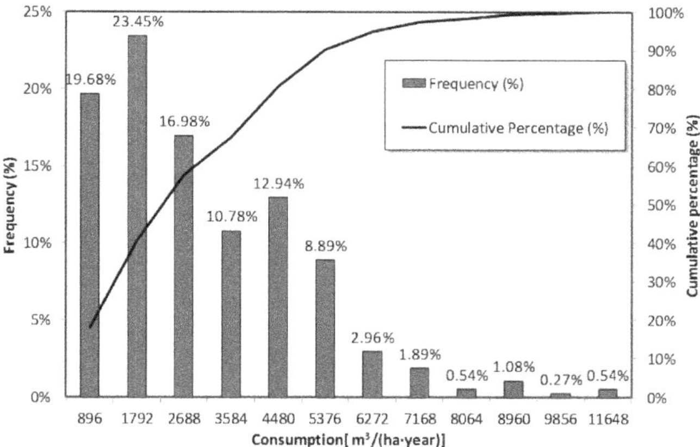

Figure 7. Distribution of annual consumptions in Vallada case study.

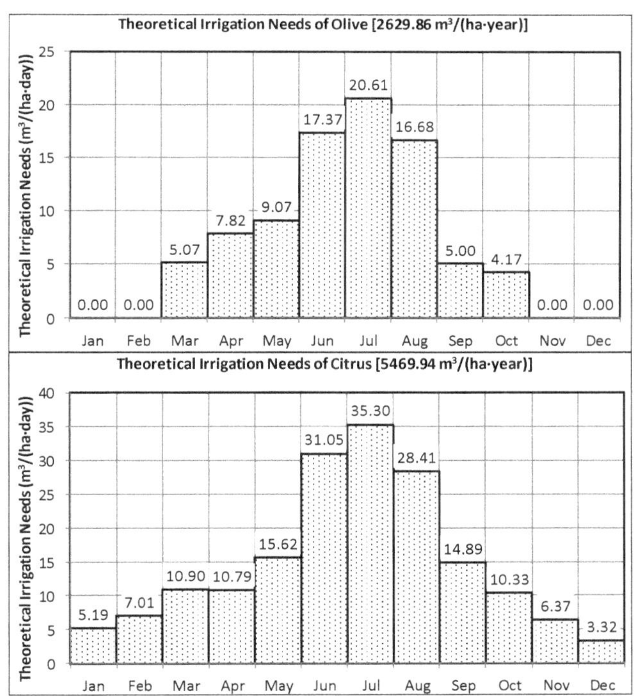

Figure 8. Pattern of irrigation needs: Top-Crop of olive. Bottom-Crop of citrus.

3.2.2. Pattern of Irrigation Habits

Habit patterns of farmers obtained by interviews, are defined as follows:

1. Analyzing the information obtained from interviews, two trends of irrigations have been depicted. Small farmers avoid Sunday as irrigation day and Saturday appears with double preference than the rest of the days (Figure 9a). Big farmers also have double preference for Saturday, but do not avoid irrigation on Sunday (Figure 9b).

2. Distribution of maximum days between irrigations: these patterns refer to the maximum interval between watering. Irrigation occurs every day during the months of higher consumption (May, June, July, August and September). In remaining months, the intervals of irrigation increase, being not a clear and well-defined pattern for all farmers. Each farmer chooses the interval according to different factors (e.g., rain, availability and soil properties). Based on the results of the interviews, four distributions have been defined. According to the results of the requested data for farmer habits across surveys, patterns have been assigned. Pattern I has been assigned approximately to 40% of the irrigation points, pattern II to 20%, pattern III to 20% and pattern IV to the other 20%. This assignment has been carried out randomly (Figure 10).

3. Patterns of irrigation duration: based on the requested information four distributions have been proposed. Again, pattern I has been assigned approximately to 40% of the irrigation points, pattern II to 20%, pattern III to 20% and pattern IV to other 20%. This assignment has been carried out randomly (Figure 11).

4. Distribution of irrigation start probability: farmers tend to irrigate in certain particular hours of the day. This aspect is considered in this methodology by using the patterns for the probability of starting irrigation in the different schedules. The watering schedule between 10 A.M. and 4 P.M. is chosen in the months of January, February, March, April, October, November and December. However, farmers irrigate in different light hours in summer months to avoid warmer hours and night. Therefore, three patterns have been developed to define the probability (see Figure 1, step 4).

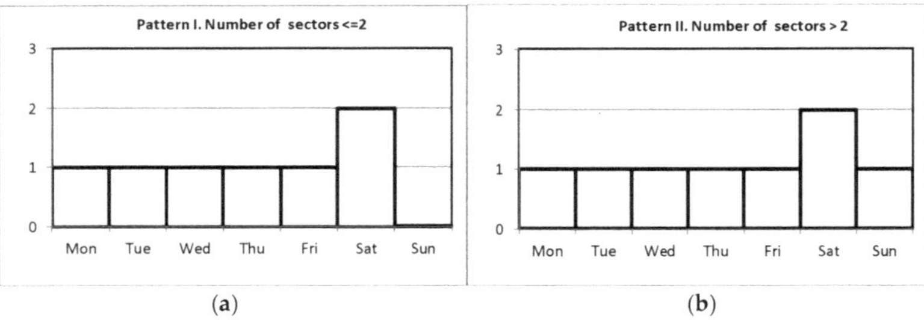

Figure 9. Weekly trend of irrigation pattern according to the number of sectors in the plot areas.

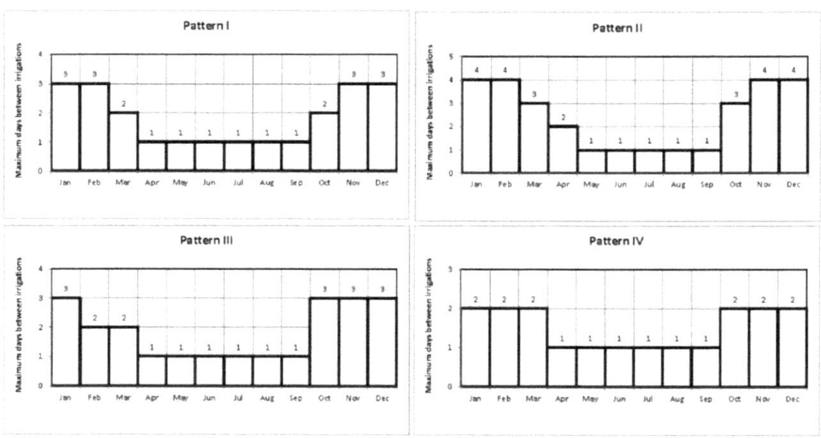

Figure 10. Patterns of maximum days between irrigations.

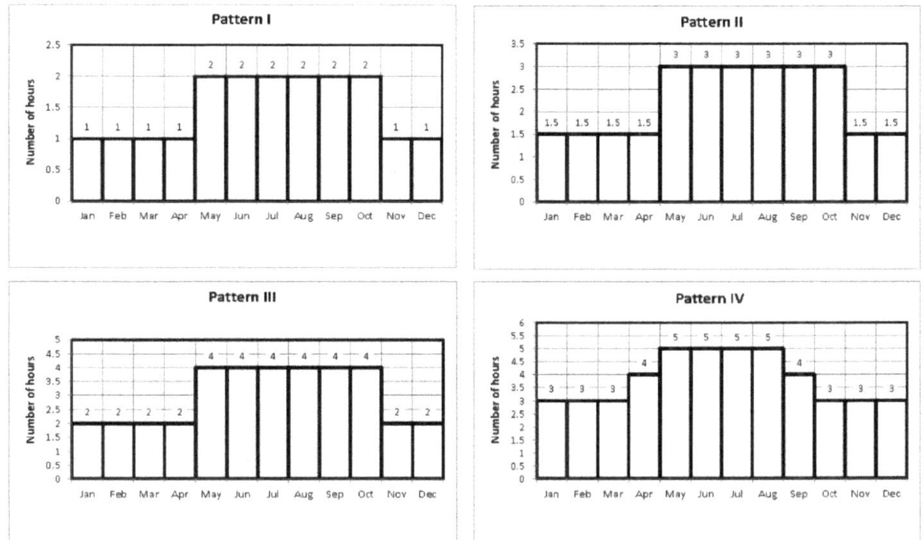

Figure 11. Patterns of irrigation duration.

A first pattern is assigned to the winter months from October to April and considers that the irrigation starts between 8 A.M. and 6 P.M. A second pattern is assigned to the summer months, from May to September, where the irrigation avoids hours of day with higher temperatures. The watering schedule starts between 4 A.M. and 12 P.M. and from 6 P.M. to 12 A.M. Finally, a third pattern function is defined for irrigating plots with more of two sectors, at any time of the day along the year.

4. Results

4.1. Basic Characteristics

Following the methodology, an analysis for flows and pressures is necessary for energy consideration in the network based on hydraulic simulations. These simulations have been run with the software EPANET.

The model of the network developed by means of the software EPANET, has been run for each day along one year. In this case, these calculations have been repeated 20, 40 and 60 times with different scenarios represented. Nevertheless, comparing these simulations, the variability obtained when the flow in the main line is compared to the average flow is smaller than 5%. As this deviation remains similar, no more repetitions are considered.

This methodology has been calibrated in other networks by the authors. The calibration has been performed with measured flow every five minutes. The results have been satisfactory with Nash-Sutcliffe index upper to 0.40, root relative squared mean error below 0.7 and percent bias below 5%, as indicated in [47].

Figures 12 and 13 depict the topology of the network to be considered and analyzed in further sections.

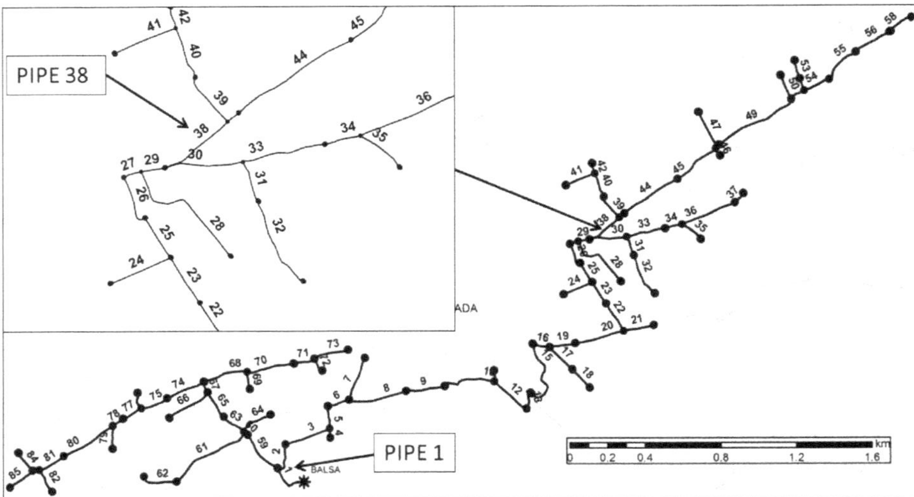

Figure 12. Identification of pipes in the irrigation network.

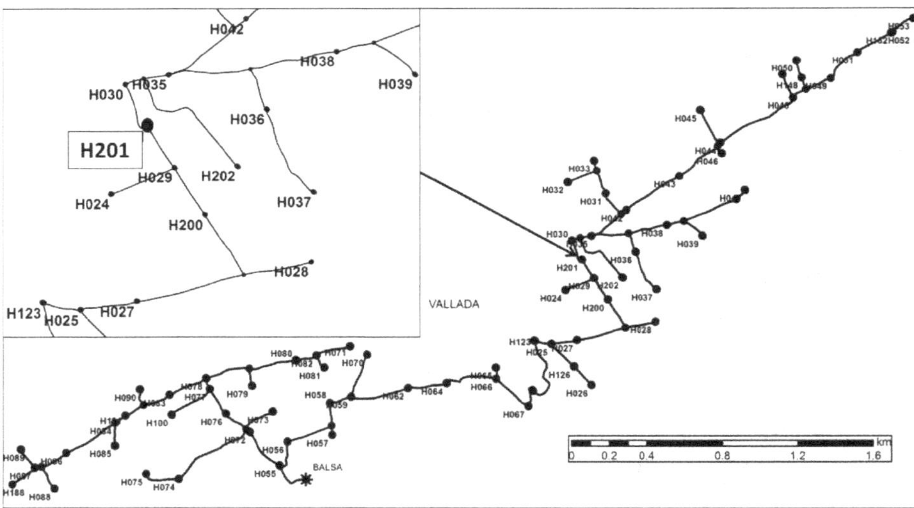

Figure 13. Identification of hydrants in the irrigation network.

4.2. Flows in the Network

Flow and pressure have been obtained along the time, based on the historical series of records registered between 2003 and 2014, in any line of the system, according to the irrigation trends. These time series of data collected in those 12 years (17,808 records), flow in the main line (see Figure 12, line 1) and hydrant 201 (see Figure 13, H201) are depicted in Figure 14.

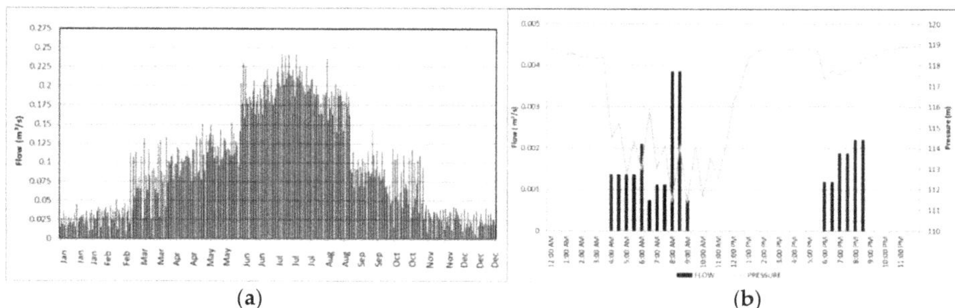

(a)   (b)

Figure 14. Flow in the main line (pipe 1) along the year (**a**); Flow and pressure variation in the hydrant (H201) over time (**b**).

The frequency histogram (Figure 15) displays the large variability of flows along the year. Hence, irrigation networks behave in a different way than drinking systems: monthly seasonal ratios range between 0.8 and 1.2 in drinking networks (except for touristic cities) and its variability of flows during the day varies between 0.7 and 1.5 of the average value. Opposite to this, the flow seasonality factor in irrigation systems is much larger than in drinking systems. In the case of citrus, seasonality factor varies between 0.14 and 2.36 times relative to the annual average consumption volume. According to this, the estimated variability in this network case study ranges between 0.1 and 2.54 times the average flow.

Additionally, there is a very high frequency of very low flows (Figure 15). Flows below 0.05 m^3/s (25% of the maximum flow rate) arise up to 80% times in the main line of the network. These small flows will become of utmost importance to be used for energy recovery.

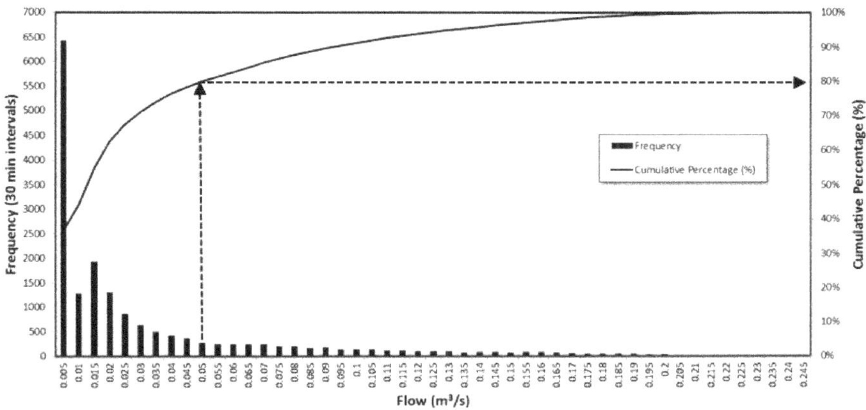

Figure 15. Histogram of flow in the main line.

4.3. Water-Energy Nexus Estimation

This section analyzes the estimation of energy dissipated in the network as a result of friction losses. Figure 16 shows the variation of the energy footprint based on time. The network is working 5943 h during of the year. Figure 16 shows the energy footprint during the distribution of flows in the water network. As shown in the histogram, 99.7% of the time the network has an energy footprint below 2 kWh/m^3. The maximum value obtained is 2.87 kWh/m^3 for a July day. However, 58.5% of the time the network has an energy footprint below 0.1 kWh/m^3.

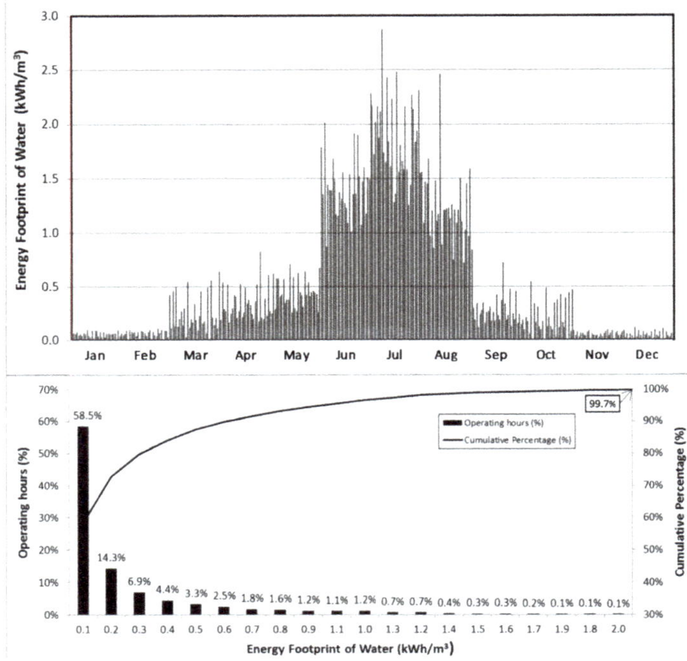

Figure 16. Network energy footprint of water.

4.4. Theoretical Recoverable Energy

If the results are analyzed in irrigation points, the total theoretical recoverable energy is 188.23 MWh/year (*i.e.*, 68.7% of the total energy supplied to the network). In these nodes, the theoretical coefficient of recovery (C_{RT}) is equal to one, as E_{RT} is equal to E_{TA}.

Figure 17 shows a detail of instantaneous power along data registered for the month of July for the analyzed time series, and the distribution of instant power frequencies over time for an irrigation point. In these points, the frequency at which the value of instantaneous power appears is practically constant because the consumption flow is uniform and only pressure varies due to the use of pressure-compensating drippers.

As an example, in irrigation point 303, the annual operating time is 2957 h. The instantaneous power oscillates between 9.96 kW and 11.64 kW. The maximum power occurs 44.2% of time, and the theoretical total energy is 33.80 MWh/year (Figure 17).

In the case of hydrants, the result is similar where the sum of the theoretical recoverable energy is 178.1 MWh/year. Figure 18 shows analogous results to those exposed in the irrigation points. Particularly, a maximum instantaneous power of 4.64 kW is achieved in hydrant H201, with an annual operating time of 1460 h and total energy of 1.59 MWh/year. The average weighted coefficient of recovery in this hydrant is 0.68 (*i.e.*, 9.68% of the total energy could be recovered if turbines had 100% efficiency). The maximum recovery occurs in the hydrant H045 with 16.12 MWh/year, with a recovery rate of 0.83 (Figure 18).

The values of theoretical energy recoverable in all hydrants are detailed in Table 1, as well as their recovery coefficients. The theoretical maximum recoverable energy is obtained in hydrant H042 with a total energy of 33.73 MWh/year, and a coefficient of recovery of 0.80. The theoretical energy ranges between 0.01 (H056) and 33.73 MWh/year. The recovery coefficient ranges between 0.14 (H055) and 0.84 (H053). The weighted average recovery coefficient is 0.75.

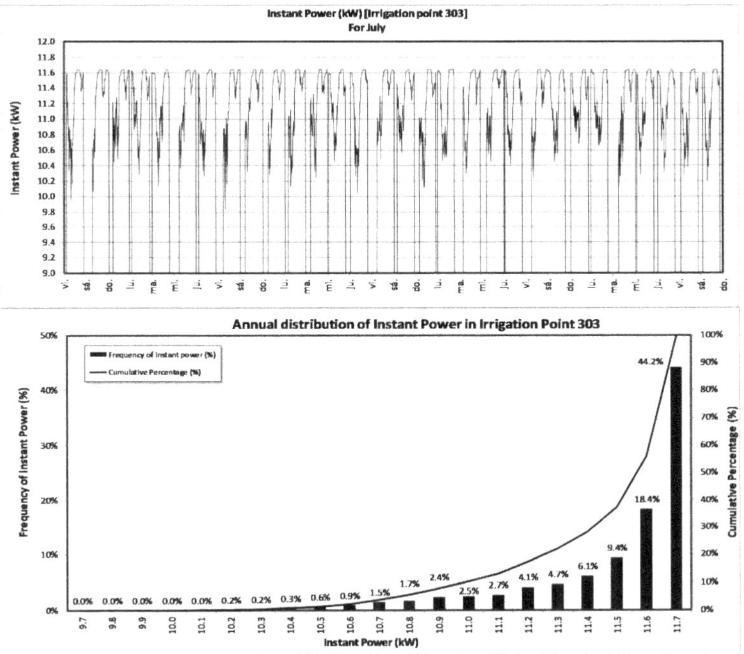

Figure 17. Potential Power in the irrigation point 303.

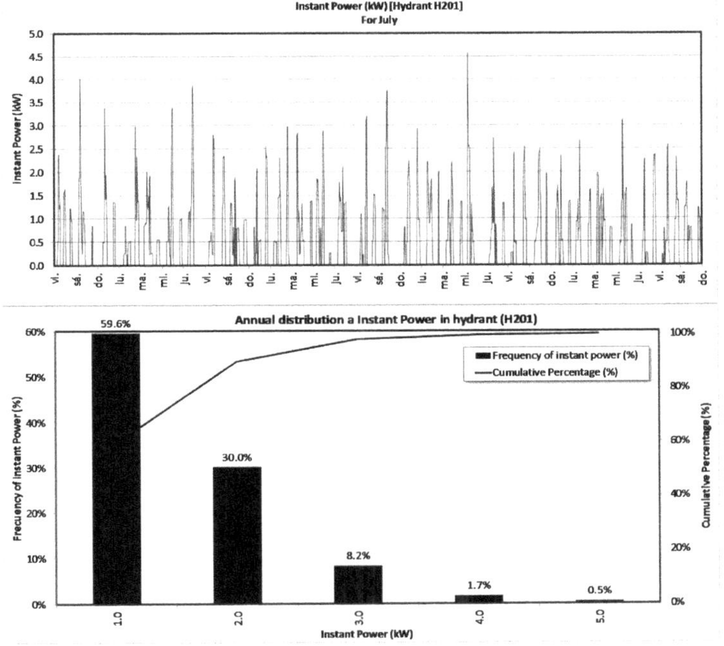

Figure 18. Potential Power in hydrant H201.

Table 1. Theoretical energy recoverable in hydrants on the irrigation network.

HYDRANT	E_{RT} MWh/Year	C_{RT}	HYDRANT	E_{RT} MWh/Year	C_{RT}
H024	1.99	0.66	H065	0.80	0.63
H025	1.56	0.76	H066	0.13	0.57
H026	0.48	0.59	H067	0.22	0.53
H027	1.92	0.67	H070	1.40	0.69
H028	0.12	0.46	H071	0.40	0.72
H029	2.27	0.72	H072	0.28	0.35
H030	0.77	0.80	H073	0.22	0.45
H031	1.36	0.81	H074	0.64	0.40
H032	3.94	0.73	H075	0.31	0.54
H033	4.19	0.81	H076	1.34	0.55
H035	3.57	0.79	H077	0.52	0.64
H036	2.50	0.76	H078	3.68	0.69
H037	1.12	0.69	H079	0.76	0.53
H038	3.54	0.71	H080	1.41	0.66
H039	0.64	0.69	H081	0.40	0.72
H040	3.57	0.65	H082	1.28	0.71
H042	33.73	0.80	H083	0.87	0.72
H043	9.13	0.81	H084	0.80	0.67
H044	5.60	0.78	H085	1.50	0.58
H045	16.12	0.83	H086	0.48	0.52
H046	5.14	0.72	H087	0.27	0.58
H047	5.03	0.80	H088	0.51	0.26
H048	13.37	0.80	H089	0.73	0.59
H049	0.26	0.82	H090	0.59	0.69
H050	1.93	0.77	H100	0.89	0.65
H051	2.67	0.59	H101	1.05	0.63
H052	0.86	0.77	H123	1.78	0.79
H053	11.05	0.84	H126	0.47	0.68
H055	0.04	0.14	H140	0.73	0.74
H056	0.01	0.28	H148	0.32	0.82
H057	0.55	0.26	H152	2.47	0.72
H058	0.99	0.60	H188	2.81	0.68
H059	0.46	0.61	H200	1.35	0.77
H062	2.32	0.57	H201	1.58	0.68
H064	0.22	0.28	H202	2.08	0.76

The line that presents the maximum recoverable energy is depicted in Figure 19 and Table 2. This condition is set on line 38, with maximum recoverable energy of 89.99 MWh/year and an average weighted recovery rate of 0.64. The maximum instantaneous power is 63.7 kW. The histogram presented in Figure 19 shows that during 918 h of the operating time (17.1%), the instantaneous power arises up to 10 kW.

Table 2 shows that the maximum recoverable energy is obtained in line 38 with total energy of 89.99 MWh/year and a coefficient of recovery 0.64. The estimated energy (E_{RT}) ranges between 0.12 (line 21) and 89.99 MWh/year, the range of recovery coefficient can be found between 0.15 (line 74) and 0.84 (line 58) and the weighted average coefficient is 0.48.

The pairs of flow, Q_i, and head, H_i, defined in Equation (11) for any point of the network are crucial to determine energetic aspects. With these data (flow and head), the estimated area of operation of the future selected machine could be determined. This cloud of point pairs is depicted in Figure 20 for an irrigation point, a hydrant and two of the lines with the maximum theoretical recoverable energy.

Figure 20 shows that not all lines have narrow operating point intervals. Line 59 has a large dispersion of operating points in the flow range. Similar circumstances occur in hydrant H201 with a wide range of flow, making the choice of a unique turbine difficult. In the case of recovery in irrigation points, the flow is constant with an interval of pressure according to the demand of the network. This becomes an additional advantage in which the performance of the chosen machine could be easily optimized.

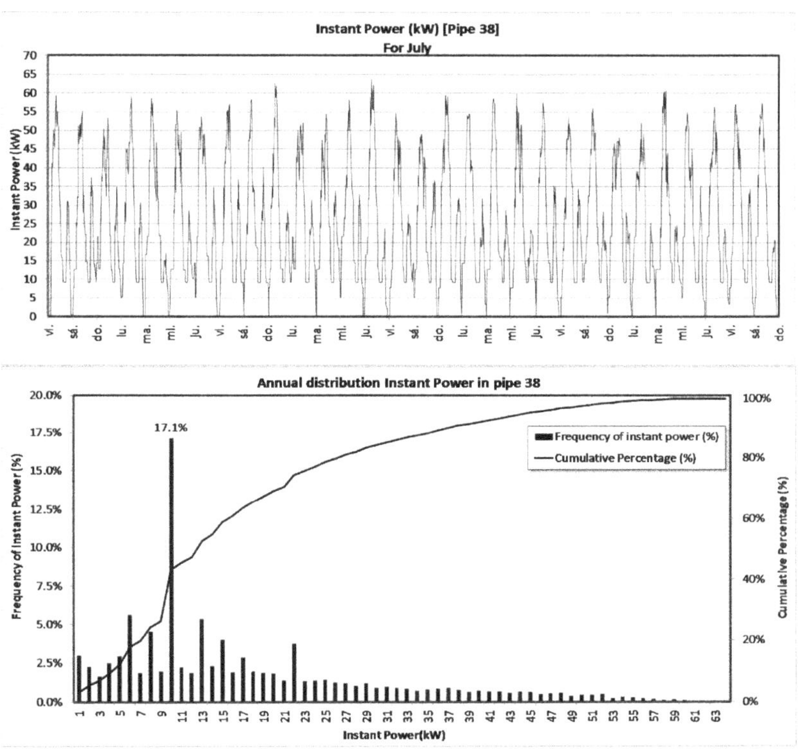

Figure 19. Potential Power in line 38.

Table 2. Estimated energy recoverable in lines on the irrigation network.

PIPE	E_{RT} MWh/year	C_{RT}	PIPE	E_{RT} MWh/year	C_{RT}
1	4.29	0.16	44	55.58	0.64
2	17.56	0.29	45	48.35	0.63
3	18.64	0.23	46	5.14	0.72
4	0.55	0.26	47	16.12	0.83
5	26.83	0.26	48	32.61	0.62
6	26.41	0.25	49	28.13	0.60
7	1.40	0.69	50	5.03	0.80
8	25.85	0.25	51	14.49	0.60
9	24.84	0.29	52	2.24	0.81
10	41.64	0.40	53	1.93	0.77
11	0.80	0.63	54	12.88	0.58
12	41.10	0.33	55	12.70	0.59
13	40.96	0.28	56	12.62	0.72
14	40.96	0.28	57	0.86	0.77
15	40.96	0.25	58	11.05	0.84
16	40.42	0.24	59	10.56	0.46

Table 2. *Cont.*

PIPE	E_{RT} MWh/year	C_{RT}	PIPE	E_{RT} MWh/year	C_{RT}
17	0.77	0.42	60	10.18	0.43
18	0.48	0.59	61	0.85	0.39
19	39.40	0.27	62	0.31	0.54
20	38.64	0.35	63	5.79	0.21
21	0.12	0.46	64	0.22	0.45
22	83.51	0.57	65	5.27	0.17
23	82.51	0.52	66	0.89	0.65
24	1.99	0.66	67	4.90	0.16
25	79.32	0.50	68	3.01	0.47
26	78.15	0.50	69	0.76	0.53
27	77.67	0.49	70	3.29	0.66
28	2.08	0.76	71	1.96	0.66
29	76.15	0.48	72	0.40	0.72
30	9.88	0.48	73	0.40	0.72
31	2.84	0.52	74	3.08	0.15
32	1.12	0.69	75	2.90	0.16
33	7.68	0.57	76	0.59	0.69
34	4.83	0.59	77	2.78	0.16
35	0.64	0.69	78	2.51	0.17
36	4.20	0.64	79	1.50	0.58
37	3.57	0.65	80	1.78	0.19
38	**89.99**	**0.64**	81	1.60	0.21
39	8.86	0.75	82	0.51	0.26
40	7.59	0.72	83	2.98	0.59
41	3.94	0.73	84	0.73	0.59
42	4.19	0.81	85	2.81	0.68
43	82.74	0.64	–	–	–

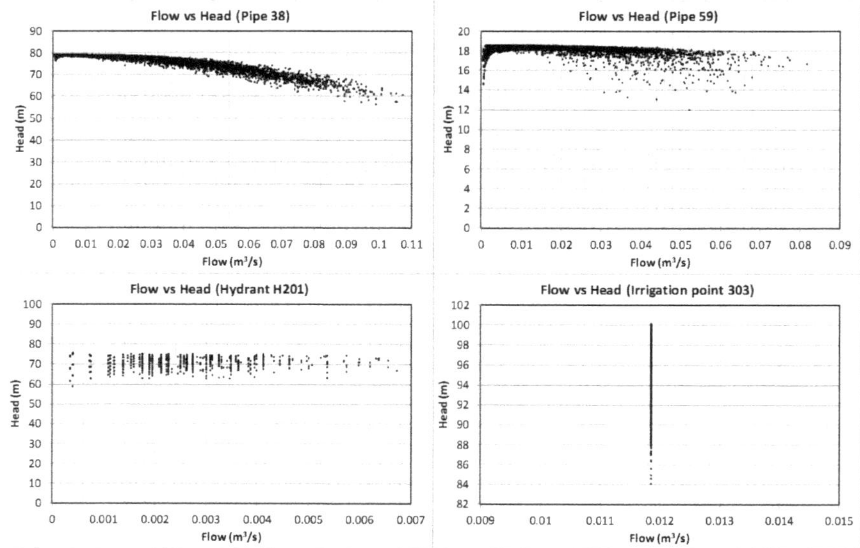

Figure 20. Representation of flow *versus* head for ideal turbine characteristics in irrigation point, hydrant and lines 38 and 55.

4.5. Global Energy Balance

The global energy network analysis in this case study, shows that with a total of 274.00 MWh/year of the network (E_T), the energy dissipated by friction (E_{FR}) is 11.25 MWh/year (4.10%), the energy required for irrigation (E_{RI}), is 74.52 MWh/year and the theoretical energy recoverable (E_{TR}) is 188.23 MWh/year (considering the sum of total individual recovery in all irrigation points).

If total energy balance in hydrants is calculated, the distribution of energy in the network is defined by Figure 21. This graphic shows that in the case of carrying out energy recovery at all hydrants, theoretical energy recoverable would be 75.2 MWh/year (27.4% compared to the total energy E_T).

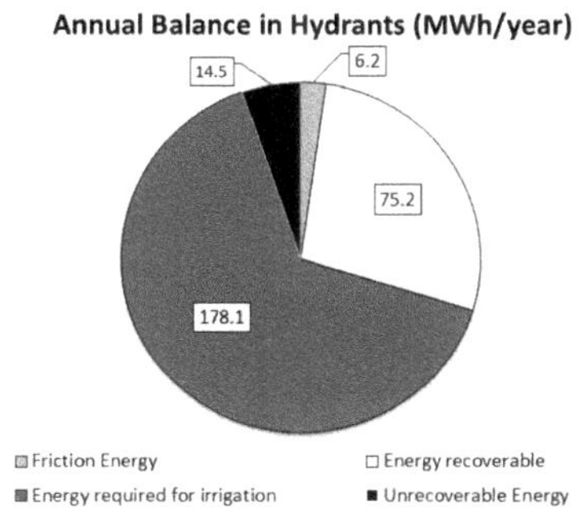

Figure 21. Annual Balance Energy in hydrants.

Finally, the virtual possibility of recovering all of the estimated energy in all irrigations points is 188.23 MWh/year. In this case, the environmental impact *versus* generation with non-renewable resources (e.g., coal and gas) would be a theoretical reduction of 137.4 t of CO_2/year or 216.2 t of CO_2/year with fuels.

4.6. Economic Feasibility

In a first economical approach, the maximum theoretical recovery line is analyzed (line 38). The corresponding PAT proposed for line 38 would have a PSR value of 5.9 years and EI of 0.22 €/kWh (Equations (18) and (21)). In this case, these results are established by considering a machine of 30 kW peak power with efficiency of 50% and the economic parameters defined in Section 2.2.

5. Conclusions

In this research, a methodology for quantifying the potential recovered energy for an average year in an irrigation network has been presented. Even nowadays, some irrigation networks do not have flowmeters. In these cases, a flow estimation method must be implemented: the flow is assigned to pipes along time depending on registered volumes, irrigation trends, and consumption patterns.

Once flows are known, the EPANET toolkit is used for estimating pressures in different scenarios along the year in order to determine pressure and energy balance for each point. Hence, it is possible to discriminate the energy needed for irrigation, friction head losses, non-recoverable energy, and potentially recoverable energy in any line or hydrant in a network.

The method was demonstrated as applied to a real case. Considering consumption records from 2003 to 2014, an irrigation network in Valencia (Spain) has been analyzed in order to determine the dissipated and recoverable energy, observing that the energy footprint achieves maximum values of 2.85 kWh/m^3, being 79.7% of time under 0.3 kWh/m^3. The potential recoverable energy, instant power, recovering coefficients relating total with recoverable energy and frequency histogram of power are studied for any irrigation point, hydrant or line.

The maximum estimated potential recoverable energy sums to 188.23 MWh/year considering all the consumption points, and 178.1 MWh/year considering all the hydrants. If only one turbine were to be installed, the more convenient location is line 38, with a potential recovery of 89.99 MWh/year.

Future works should be undertaken to analyze the performance of real turbines in lines in order to propose a method to optimize the selection of turbines and the technical and economic involvements of such installations in different irrigation networks.

Acknowledgments: This paper has been possible with the free collaboration of the "Comunidad de Regantes Virgen de Gracia". No additional funds have been received for this research. Authors thank the reviewers for their valuable comments, which have greatly contributed in the improvement of the document.

Author Contributions: All the authors have participated in any step of this research. Particularly a brief description is attached: The author Helena M. Ramos has contributed supervising the state of the art description, flow assignation and energy implications of the present study. The author Francisco Javier Sánchez-Romero has been involved in the conception of the methodology for flow assignation and computational programming for EPANET Toolkit. The author Modesto Pérez-Sánchez has analyzed the flow data for proposing the final flow distribution, has implemented the data for EPANET Toolkit and analyzed the results for energy determination. The author P. Amparo López-Jiménez has supervised the whole research and document and has been involved in final energy analysis of results and conclusions.

Conflicts of Interest: The authors declare no conflict of interest. The founding sponsors had no role in the design of the study; in the collection, analyses, or interpretation of data; in the writing of the manuscript, and in the decision to publish the results.

References

1. Kahil, M.; Albiac, J.; Dinar, A. Improving the performance of water policies: Evidence from drought in Spain. *Water* **2016**, *8*, 34. [CrossRef]
2. Llop, M.; Ponce-Alfonso, X. Water and agriculture in a Mediterranean region: The search for a sustainable water policy strategy. *Water* **2016**, *8*, 66. [CrossRef]
3. Corominas, J. Agua y Energía en el riego en la época de la sostenibilidad. *Ing. Agua* **2010**, *17*, 219–233. [CrossRef]
4. Seoane, P.; Allué, R.; Postigo, M.J.; Cordón, M.A. *Boletín Mensual de Estadística*; Ministerio de Agricultura, Alimentación y Medio Ambiente: Madrid, Spain, 2013.
5. FAO. *Agua Y Cultivos*; FAO: Rome, Italy, 2002. Availiable online: http://www.fao.org/docrep/005/y3918s/y3918s10.htm (accessed on 30 April 2016).
6. MAGRAMA. El riego Localizado Alcanza el 48.23% de la Superficie Regada en España. Minist Agric Aliment y Medio Ambient. Available online: http://www.magrama.gob.es/gl/prensa/noticias/el-riego-localizado-alcanza-el-4823--de-la-superficie-regada-en-espa%C3%B1a-/tcm7-312671-16 (accessed on 15 April 2016).
7. FAO. Superficie Equipada para el Riego. Available online: http://www.fao.org/nr/water/aquastat/infographics/Irrigation_esp.pdf (accessed on 15 April 2016).
8. Pardo, M.A.; Manzano, J.; Cabrera, E.; García-Serra, J. Energy audit of irrigation networks. *Biosyst. Eng.* **2013**, *115*, 89–101. [CrossRef]
9. Coelho, B.; Andrade-Campos, A. Efficiency achievement in water supply systems—A review. *Renew Sustain. Energy. Rev.* **2014**, *30*, 59–84. [CrossRef]
10. Baki, S.; Makropoulos, C. Tools for energy footprint assessment in urban water systems. *Procedia Eng.* **2014**, *89*, 548–556. [CrossRef]
11. Okadera, T.; Chontanawat, J.; Gheewala, S.H. Water footprint for energy production and supply in Thailand. *Energy* **2014**, *77*, 49–56. [CrossRef]
12. Herath, I.; Deurer, M.; Horne, D.; Singh, R.; Clothier, B. The water footprint of hydroelectricity: A methodological comparison from a case study in New Zealand. *J. Clean. Prod.* **2011**, *19*, 1582–1589. [CrossRef]
13. Endo, A.; Burnett, K.; Orencio, P. Methods of the water-energy-food nexus. *Water* **2015**, *7*, 5806–5830. [CrossRef]
14. Mendoza-Grimón, V.; Hernández-Moreno, J.; Palacios-Díaz, M. Improving water use in fodder production. *Water* **2015**, *7*, 2612–2621. [CrossRef]

15. Ramos, H.M.; Vieira, F.; Covas, D.I.C. Energy efficiency in a water supply system: Energy consumption and CO_2 emission. *Water Sci. Eng.* **2010**, *3*, 331–340.

16. Choulot, A. *Energy Recovery in Existing Infrastructures with Small Hydropower Plants*; FP6 Project Shapes (Work Package 5—WP5); European Directorate for Transport and Energy: Brussels, Belgium, 2010.

17. Carravetta, A.; Fecarotta, O.; Del Giudice, G.; Ramos, H. Energy recovery in water systems by PATs: A comparisons among the different installation schemes. *Procedia Eng.* **2014**, *70*, 275–284. [CrossRef]

18. Carravetta, A.; Del Giudice, G.; Fecarotta, O.; Ramos, H. Pump as turbine (PAT) design in water distribution network by system effectiveness. *Water* **2013**, *5*, 1211–1225. [CrossRef]

19. Carravetta, A.; Del Giudice, G.; Fecarotta, O.; Ramos, H.M. Energy production in water distribution networks: A PAT design strategy. *Water Resour. Manag.* **2012**, *26*, 3947–3959. [CrossRef]

20. Carravetta, A.; Del Giudice, G.; Oreste, F.; Ramos, H. PAT design strategy for energy recovery in water distribution networks by electrical regulation. *Energies* **2013**, *6*, 411–424. [CrossRef]

21. Ramos, H.; Borga, A. Pumps as turbines: An unconventional solution to energy production. *Urban Water* **1999**, *1*, 261–263. [CrossRef]

22. Ramos, H.; Borga, A.; Simão, M. New design solutions for low-power energy production in water pipe systems. *Water Sci. Eng.* **2009**, *2*, 69–84.

23. Adhau, S.P.; Moharil, R.M.; Adhau, P.G. Mini-hydro power generation on existing irrigation projects: Case study of Indian sites. *Renew. Sustain. Energy Rev.* **2012**, *16*, 4785–4795. [CrossRef]

24. Butera, I.; Balestra, R. Estimation of the hydropower potential of irrigation networks. *Renew. Sustain. Energy. Rev.* **2015**, *48*, 140–151. [CrossRef]

25. Tilmant, A.; Goor, Q.; Pinte, D. Agricultural-to-hydropower water transfers: Sharing water and benefits in hydropower-irrigation systems. *Hydrol. Earth Syst. Sci.* **2009**, *13*, 1091–1101. [CrossRef]

26. Tarragó, E.F.; Ramos, H. Micro-Hydro Solutions in Alqueva Multipurpose Project (AMP) towards Water-Energy-Environmental Efficiency Improvements. Bachelor's Thesis, Universidade de Lisboa, Lisboa, Portugal, 2015.

27. Clément, R. Calcul des débits dans le réseaux d'irrigation fonctionnant á la demande. *La Houille Blanche* **1966**, *5*, 553–575. [CrossRef]

28. Mavropoulos, T.I. Sviluppo di una nuova formula per il calcolo delle portate di punta nelle reti irrigue con esercizio alla domanda. *Riv. Irrig. Dren.* **1997**, *44*, 27–35.

29. Maidment, D.R.; Hutchinson, P.D. Modeling water demands of irrigation projects. *J. Irrig. Drain. Eng.* **1983**, *109*, 405–418. [CrossRef]

30. Alandi, P.P.; Pérez, P.C.; Álvarez, J.F.O.; Hidalgo, M.; Martín-Benito, J.M.T. Pumping selection and regulation for water-distribution networks. *J. Irrig. Drain. Eng.* **1997**, *131*, 273–281. [CrossRef]

31. Aliod, R.; Eizaguerri, A.; Estrada, C. Dimensionado y análisis hidráulico de redes de distribución a presión en riego a la demanda: Aplicación del programa GESTAR. *Riegos Dren. XXI* **1997**, *92*, 22–38.

32. Pereira, L.S.; Teixeira, J.L. *Modelling for Irrigation Delivery Scheduling: Simulation of Demand at Sector Level with Models ISAREG and IRRICEP*; FAO: Rome, Italy, 1994.

33. Lamaddalena, N.; Sagardoy, J.A. *Performance Analysis of On-Demand Pressurized Irrigation Systems*; FAO: Roma, Italy, 2007.

34. Rossman, L.A. *EPANET 2 User's Manual*; U.S. Environmental Protection Agency (EPA): Cincinnati, OH, USA, 2000.

35. White, F.M. *Fluid Mechanics*, 6th ed.; McGrau-Hill: Madrid, Spain, 2008.

36. Cabrera, E.; Cobacho, R.; Soriano, J. Towards an energy labelling of pressurized water networks. *Procedia Eng.* **2014**, *70*, 209–217. [CrossRef]

37. Klein, G.; Krebs, M.; Hall, V.; O'Brien, T.; Blevins, B.B. *California's Water–Energy Relationship*; California Energy Commission: Sacramento, CA, USA, 2005.

38. Shrestha, S.; Dhakal, S.; Shrestha, A.; Kaneko, S.; Kansal, A. Water-Energy-Carbon Nexus in Cities: Cases from Bangkok, New Delhi, Tokyo. In *Water Energy Food Nexus: International Cooperation and Technology Transfer*; Asian Institute of Technology: Paris, France, 2015.

39. Spadaro, J.V.; Langloïs, L.; Hamilton, B. Greenhouse Gas Emissions of Electricity Generation Chains: Assessing the Difference. *IAEA Bull.* **2000**, *42*, 19–28.

40. Weisser, D. A guide to life-cycle greenhouse gas (GHG) emissions from electric supply technologies. *Energy* **2007**, *32*, 1543–1559. [CrossRef]

41. Arora, M.; Aye, L.; Malano, H.; Ngo, T. Water-energy-GHG emissions accounting for urban water supply: A case study on an urban redevelopment in Melbourne. *Water Util. J.* **2013**, *6*, 9–18.
42. Nair, S.; George, B.; Malano, H.M.; Arora, M.; Nawarathna, B. Water–energy–greenhouse gas nexus of urban water systems: Review of concepts, state-of-art and methods. *Resour. Conserv. Recycl.* **2014**, *89*, 1–10. [CrossRef]
43. Zema, D.A.; Nicotra, A.; Tamburino, V.; Zimbone, S.M. A simple method to evaluate the technical and economic feasibility of micro hydro power plants in existing irrigation systems. *Renew. Energy* **2016**, *85*, 498–506. [CrossRef]
44. Castro, A. *Minicentrales Hidroeléctricas*; Instituto para la Diversificación y Ahorro de la Energía: Madrid, Spain, 2006.
45. Elbatran, A.H.; Yaakob, O.B.; Ahmed, Y.M.; Shabara, H.M. Operation, performance and economic analysis of low head micro-hydropower turbines for rural and remote areas: A review. *Renew. Sustain. Energy Rev.* **2015**, *43*, 40–50. [CrossRef]
46. Derakhshan, S.; Nourbakhsh, A. Experimental study of characteristic curves of centrifugal pumps working as turbines in different specific speeds. *Exp. Therm. Fluid Sci.* **2008**, *32*, 800–807. [CrossRef]
47. Moriasi, D.N.; Arnold, J.G.; Van Liew, M.W.; Binger, R.L.; Harmel, R.D.; Veith, T.L. Model evaluation guidelines for systematic quantification of accuracy in watershed simulations. *Trans. ASABE* **2007**, *50*, 885–900. [CrossRef]

Article

Energy Recovery in Existing Water Networks: Towards Greater Sustainability

Modesto Pérez-Sánchez [1], Francisco Javier Sánchez-Romero [2], Helena M. Ramos [3] and P. Amparo López-Jiménez [1,*

[1] Hydraulic and Environmental Engineering Department, Universitat Politècnica de València, 46022 Valencia, Spain; mopesan1@upv.es
[2] Rural and Agrifood Engineering Department, Universitat Politècnica de València, 46022 Valencia, Spain; fcosanro@agf.upv.es
[3] Civil Engineering, Architecture and Georesources Departament, CERIS, Instituto Superior Técnico, Universidade de Lisboa, 1049-001 Lisboa, Portugal; hramos.ist@gmail.com
* Correspondence: palopez@upv.es; Tel.: +34-96-387700 (ext. 86106)

Academic Editor: Arjen Y. Hoekstra
Received: 24 October 2016; Accepted: 3 February 2017; Published: 8 February 2017

Abstract: Analyses of possible synergies between energy recovery and water management are essential for achieving sustainable improvements in the performance of irrigation water networks. Improving the energy efficiency of water systems by hydraulic energy recovery is becoming an inevitable trend for energy conservation, emissions reduction, and the increase of profit margins as well as for environmental requirements. This paper presents the state of the art of hydraulic energy generation in drinking and irrigation water networks through an extensive review and by analyzing the types of machinery installed, economic and environmental implications of large and small hydropower systems, and how hydropower can be applied in water distribution networks (drinking and irrigation) where energy recovery is not the main objective. Several proposed solutions of energy recovery by using hydraulic machines increase the added value of irrigation water networks, which is an open field that needs to be explored in the near future.

Keywords: irrigation water networks; water-energy nexus; renewable energy; sustainability and efficiency; hydropower solutions; water management

---

1. Hydropower Generation

Society's energy consumption worldwide has increased by up to 600% over the last century. This increase has been a direct result of population growth since the industrial revolution, in which energy has been provided mainly by fossil fuels. Nevertheless, today and in the near future, renewable energies are expected to be more widely implemented to help maintain sustainable growth and quality of life and, by 2040, to reduce energy consumption down to the 2010 levels [1].

Sustainability must be achieved by using strategies that do not increase the overall carbon footprint, considering all levels of production (macro- and microscale) of the different supplies. These strategies' development has to be univocally linked to new technologies [2]. Special attention must be paid to those new strategies that are related to energy recovery. These new techniques have raised interesting environmental and economic advantages. Therefore, a deep knowledge of the water-energy nexus is crucial for quantifying the potential for energy recovery in any water system [3], and defining performance indicators to evaluate the potential level of energy savings is a key issue for sustainability, environmental, or even management solutions [4].

Energy recovery, with the aim of harnessing the power dissipated by valves (in pressurized flow) or hydraulic jumps (in open channels), is becoming of paramount importance in water distribution networks. Recovery will allow the energy footprint of water (i.e., the energy unit cost needed to satisfy each stage of the water cycle: catchment, pumping, treatment, and distribution) to be reduced, even considering that energy generation is not a priority for these systems [5–7], although this recovery contributes to the development of more sustainable systems. This production could also contribute to the exploitation costs reduction in these systems, increasing the feasibility of drinking and irrigation water exploitation.

Among all of the different types of renewable energy (e.g., photovoltaic, solar thermal, tidal, and wind), the hydropower plant stands out for its feasibility. Historically, large installations can be found in dams around the world to take advantage of the potential energy created by different water levels. The most important hydropower plants are located in countries such as China, the United States, Brazil, and Canada. Currently, China has the greatest installed capacity (exceeding 240 GW), with production greater than 800 TWh in 2012 and an average growth of capacity of 20 GW/year [8]. In Brazil and Canada, hydropower plants represent 84% and 56% of the total energy consumption, respectively. The production of this type of energy in these countries reaches 16% of the total consumed energy [9,10]. Figure 1 shows the technical, economic and exploited potential on each continent.

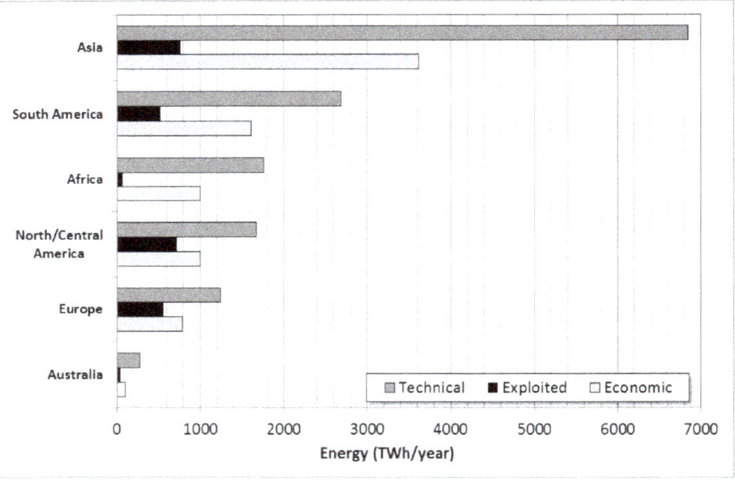

Figure 1. Worldwide hydraulic potential (adapted from [11]).

Going further, Spänhoff [12] performed a worldwide projection of the installed capacity of renewable energy for the United States Energy Information Administration. Hydropower has been the largest renewable source of energy in the period 2004–2010, and it will probably have the highest installed capacity in 2035. According to these forecasts, the installed capacity of hydropower will exceed 1400 GW. This installed capacity will be three times higher than that of wind energy and more than fifteen times greater than that of photovoltaic energy (Figure 2). The actual implementation of these renewable energy systems (solar and wind) in 2016 has been lower than the predicted values (Figure 2). The installed solar capacity is only 30% of the predicted capacity; the wind capacity is only 70% of the predicted value, and the hydropower capacity is approximately the value indicated in Figure 2 [13].

In Europe, renewable energy generation has increased by 96.17% in the period 2002–2013, with production equal to 2232.5 TWh in 2013 [14]. In this decade, the power produced by hydropower plants increased only 16.38%, but other sorts of energy (e.g., solar, wind, and biomass) have experienced greater increases. For instance, in Spain, the increase in renewable energy generation was 152% in this period, but the increase was 73% for Spanish hydropower production.

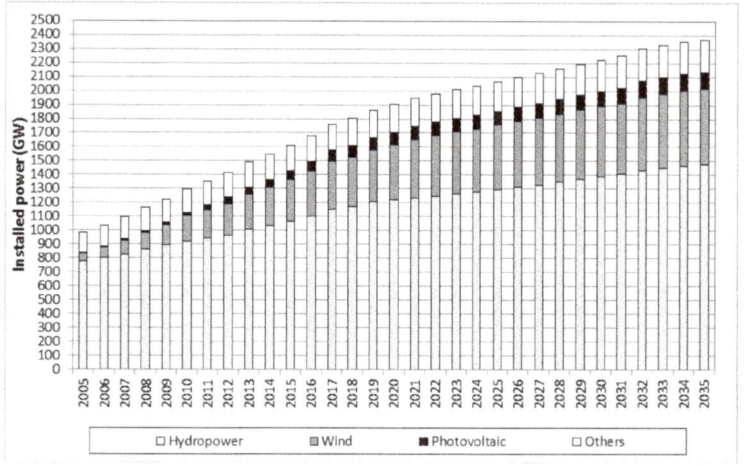

Figure 2. Trends for worldwide renewable energy (adapted from [13]).

Considering all of these renewable energy systems, wind energy has increased by 477%, with a total generation of 53.90 TWh/year. Photovoltaic energy has increased by 4300%, producing 22.85 TWh in 2013 [14]. Nevertheless, the growth of hydropower production has been uneven due to the irregularity of rainfall in the Iberian Peninsula, although the trend is upward [15]. In Spain, the Institute for Energy Diversification and Saving estimated that the untapped generation potential of small hydropower is 1000 MW [16]. Therefore, the development of renewable energy has a promising future, if the potential exploitation is considered. This promising development has positive aspects (e.g., lower environmental impact and generation of stable electrical supply) compared to other renewable energies (e.g., intermittent generation, such as solar or wind) [17]. In addition, this type of energy generation can be very important in the development of multipurpose water systems, where generation is another possible water use [18].

In the near future, part of the growth of hydropower production must come from the retrieval of potential energy embedded in water distribution networks. Considering the potential reduction of natural resources due to extensive agriculture or unsustainable water use on a global level, any investment in energy water recovery is crucial. Therefore, the whole water cycle must be included in the process of energy recovery, including both drinking and irrigation systems. This coupled water-energy nexus will allow the consideration of these systems as a new sustainable and efficient source of energy.

In this framework, the state of the art in the traditional field of hydropower (installed capacity) with a more advanced vision of the energy implications in drinking and irrigation systems is presented, considering the possibility of installing energy recovery in water distribution networks. The objective is to show the hydropower potential in water distribution networks. The installation of these systems will help increase energy efficiency. In the particular case of irrigation, improving the efficiency will allow the reduction of exploitation costs, decreasing pressure on the profit margin, as well as the environmental [19] and economic aspects [20].

2. Energy Recovery in Water Networks

Although there is not a consensus at the European level, the accepted demarcation between large and small hydro by the European Commission is whether the installed power capacity exceeds 10 MW [21]. When the installed capacity is below 100 kW, the generation system is called micro hydropower and when the generated power is below 10 kW, it is called pico hydropower.

2.1. Large Hydropower

Large hydropower in developed (20th century) and developing countries (e.g., China, Brazil, and South Africa) in the early 21st century has been very large, with this generation system providing the main energy source in those countries where topography, hydrology and climatology allow hydropower recovery.

Although China began to develop its hydropower strategic plan in the 1950s, it has quickly overtaken other countries. Its development started with the Liujiaxia dam, which was completed in 1974 and has an installed capacity of 1250 MW. In 2012, China finished the Three Gorges Dam, which has a total installed capacity of over 20,000 MW. A project is currently being developed in the Yarlon Tsagpo Canyon with an installed capacity of over 40,000 MW, double the installed power in the Three Gorges Dam [8]. China has far exceeded the milestone achieved by the Hoover Dam (Colorado River, United States) in 1936, where 2000 MW were installed; or the 12,000 MW installed in the Itaipu Dam (Parana River, Brazil and Paraguay) in 1966, now with 14,000 MW. In Spain, up to 2013, the installed capacity is approximately 18,000 MW, and the hydropower installations do not exceed 400 MW on average.

Ansar [22] made an inventory of the 245 largest hydropower sites in the world (1934–2007). Among them, 72 are located in East Asia, 50 in Latin America, 40 in North America, 29 in Africa, 29 in Europe and 25 in South Asia. Of these, 97 dams are producing electricity, 89 are multipurpose (including hydropower) and 59 are devoted to irrigation and other uses. These plants are occasionally reversible to take advantage of the available volume of water, adjusting the electric energy injected to the grid according to the energy demand. Rehman et al. [10] established that the worldwide installed capacity of reversible plants is 104 GW (presently, the total installed capacity of hydropower is 1000 GW), of which 22.2 GW are installed in North America, 44 GW in Europe (5.3 GW in Spain), 33 GW in Asia, and the remainder in Africa and Russia. These authors refer to efficiencies between 70% and 80%.

Regarding environmental performance indicators of hydropower solutions, these plants have a positive impact on global climate change [23], based on the carbon footprint, which is the parameter used to determine the environmental impact, which has taken on special importance since the 1990s. This parameter is defined as the sum of the greenhouse gases emitted by an organization, event or product, expressed in terms of CO_2 equivalent units (CO_2-e) [24]. According to Zhang and Xu [25], the influence of the carbon footprint depends on many factors, most importantly the construction and maintenance costs (because these represent more than 60% in earthen dams and 50% in concrete dams) [9]. The range of emissions for these systems is between 2 and 240 gCO_2-e/kWh [9,26], with the carbon footprint in hydropower plants being smaller than that in coal plants. These non-renewable energy plants have emissions above 890 gCO_2-e/kWh [26–28]. Considering these emissions, hydropower plants saved 3.3 billion tons of CO_2 emissions in 2014 and will help reduce emissions by over 120 billion tons between 2015 and 2050 [13] compared to coal plants.

Regarding the economic aspects of large hydropower, the investment ratio (€/kW) decreases as the installed capacity increases, reaching values from 2170 to 470 €/kW for power ranges between 200 and 1400 MW [29,30]. Civil works represent between 70% and 80% of the total investment, and the remaining costs are devoted to electro-mechanics and hydraulic equipment [10].

2.2. Small Hydropower

The expansion of these installations was due to the development of the Francis turbine (for medium heads) [31], contributing both to the reduction of greenhouse gases and to the establishment of electrical service in remote rural areas or consumption points located away from supply points. This "social contribution" should be taken into account in viability studies in developing countries. As in large hydropower, the leading countries in small hydropower energy are China, Brazil, India, Canada, and some European countries. In 2013, China had a total installed capacity above 80 GW, which supplied more than 650 rural areas [8], with a range of installed power between 0.5 and 10 MW in each plant [32]. Brazil had 397 power plants in operation in 2011, with an installed capacity of

3.5 GW. Currently, the potential capacity is equal to 25.9 GW [33]. The United States Department of Energy tabulated more than a half million sites, with an installed potential of 100 GW [34], representing 10% of the current generation in the United States.

Currently, Australia has 60 small hydropower plants with a total installed capacity of 0.15 GW, which is 10% of the potential capacity. Australian Administration has projected three new plants with an installed capacity of 20, 8.4, and 7 MW, which will be developed in the future [35]. In India, the potential capacity is 15 GW, of which 2.4 GW are currently installed in 674 plants, with an expected increase of 9.4 GW in 2017 [36]. Ushiyama [37] established an installed power capacity of 30 GW for Japan in 2010, where small hydropower was practically non-existent. However, Japan has already started to develop projects in remote areas with a rate of capacity installation greater than 300 MW per year, with an installed power and potential capacity in the near future of 3.52 and 6.82 GW, respectively [38]. Across the African continent, small hydropower is also being developed in rural zones, where these plants are generating significant social benefits with a lower installed capacity of 300 kW [39].

According to the European Small Hydropower Association (ESHA), the total installed capacity in 2005 was 12.4 GW, of which six European Union members (Italy, France, Spain, Germany, Austria, and Sweden) in addition to Switzerland and Norway possess more than 90% of the installed capacity [40]. Alonso-Tristán et al. [41] presented the distribution of the small hydropower installed in Spain using data from Red Eléctrica de España (REE). This country represented 23.1% of the hydropower generation in European small hydropower in 2008 [42]. The potential growth in installed power capacity is 10 GW, with an annual production above 38,000 GWh [30], and the installed power will reach 17.3 GW (an increment of 39.51%) [41] in the period 2005–2020, as depicted in Figure 3.

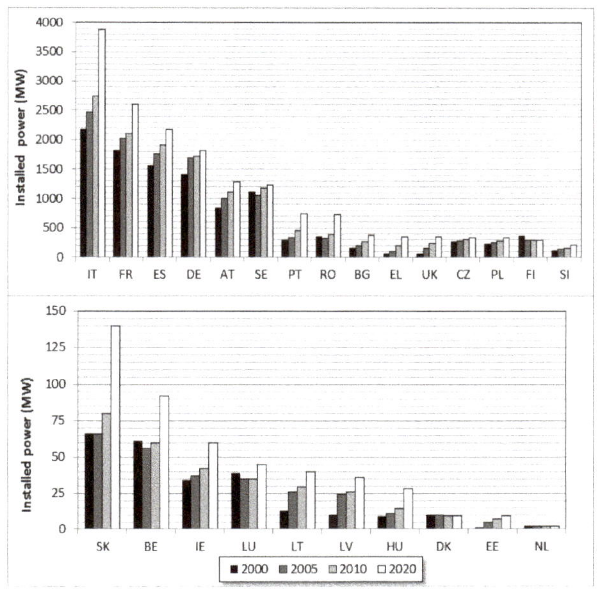

Figure 3. Planned installation of small hydropower capacity in 2005–2020 by European Union countries (adapted from [40]).

Small hydropower has several advantages: less negative impacts than large hydropower and available potential to increase renewable energy production. ESHA estimated an annual reduction of 29×10^6 tonnes of CO_2 as a result of the 13 GW installed capacity in Europe [43]. Amponsah et al. [44]

analyzed different values of the carbon footprint of small hydropower and established a range between 2 and 74.9 gCO_2-e/kWh based on the installation and the type of plant. In the particular case of micro-hydropower, Gallagher et al. [45] analyzed the carbon footprint of three plants with installed capacities of 15, 90 and 140 kW. The results of this analysis were 2.14, 4.39, and 2.78 gCO_2-e/kWh, respectively. These values emphasize the positive environmental impact of hydropower solutions.

Regarding economic aspects, Kosnik [34] developed an economic analysis based on several small plants, obtaining a non-linear relationship between the cost of implementation and installed power (small, micro or pico). Ogayar and Vidal [46] also analyzed the distribution of costs for small hydropower, which are distributed among civil work (40%), turbine (30%), electro-mechanical and regulation equipment (22%), and construction management (8%). This type of renewable energy project is viable when the required investment is below 2000 $/kW [47], although special attention should be paid to the environmental and social benefits provided by these installations. At the European level, according to the General Direction for Environment, the average cost of investment for plants with an installed capacity below 10 MW is between 2941 and 4072 €/kW, depending on the characteristics of the system (e.g., flow, head, orography) [40]. Mishra et al. [48] proposed formulas that use the turbine, installed power capacity, and net head to estimate the required investment. These expressions can be used to determine the associated costs.

Finally, the classification of these installations is referenced in European legislation [47] according to the installed power. However, other classifications have been proposed that depend on the type of plant from an operational point of view [16,49,50].

(1)	Power plant in flow or run-of-river: This system has no regulation reservoir and only takes advantage of the hydraulic head when the flow circulates. In mountain areas, with medium heads, the flow is diverted through a weir and a penstock carries the flow to the power house. If the topography does not allow it, the hydraulic head must be created by building a higher dam.
(2)	Power plant at the foot of a dam: The flow is regulated by a reservoir. In the case of small hydropower, reservoirs or dams are used to ensure project viability.
(3)	Power plant in water distribution network: The distribution network is used to take advantage of available pressure or kinetic energy, depending on the system characteristics.

2.3. Type of Hydraulic Machines

Hydraulic machines are classified according to the system (pressurized or open channels) in which they are installed (Figure 4). In open channels, all types of hydraulic wheels have been traditionally used to take advantage of waterfalls. According to the type of energy used (potential, pressure, or kinetic), the machines are classified as gravitational, hydrostatic, or kinetic. Gravitational machines take advantage of different water levels to extract energy from the flow (e.g., Archimedes screw or waterwheel). Hydrostatic machines operate by the difference of hydrostatic pressures on both faces of a blade (e.g., hydrostatic pressure wheel). Finally, if a wheel uses the velocity of the flow to move the axis of the machine, this type of machine is called a kinetic machine. There are many different types of kinetic machines (e.g., helical turbine with vertical or horizontal axis, overshot wheel, and ducted turbine) [51].

In pressurized water systems, the most frequently used machines can be grouped into traditional machines (which are categorized as action and reaction machines) and adapted machines. The last group includes hydraulic machines that normally work not as turbines but as pumps. In reaction machines, the hydraulic power is transmitted to the axis of the machine by varying the pressure flow between the inlet and outlet of the impeller, which depends on the specific speed of the machine (e.g., Francis and Kaplan). In action turbines, the energy exchange (hydraulic to mechanical) is carried out at atmospheric pressure, and the hydraulic power is due to kinetic energy of the flow (e.g., Pelton and Turgo).

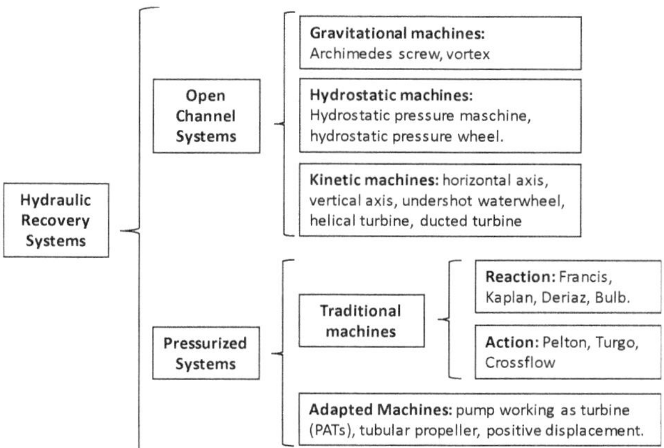

Figure 4. Classification of hydraulic machines.

These types of machines are used in large and small hydropower, depending on the nominal flow and the available head (Figure 5).

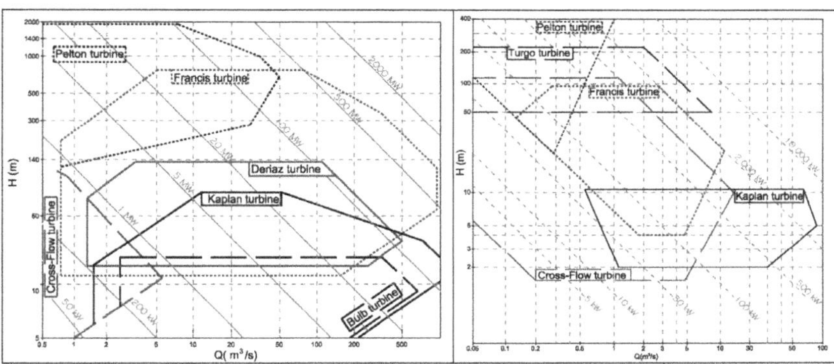

Figure 5. Selection of turbine depending on head and flow in: large (**left**); and small (**right**) hydropower [16,31].

Currently, most of the turbines installed for large hydropower are Pelton, Francis, Kaplan, or Deriaz turbines [31]. These turbines present different performance curves, which depend on their specific speed and discharge number [21,52]. Gordon [53] analyzed both the efficiency of 107 turbines that had been installed since 1908 and the increase in efficiency obtained with replacement impellers in 22 power plants, evaluating the improvement in the performance of propellers over time. Increases in the performance of the machines were obtained, rising from efficiencies lower than 50% in 1920 to above 96% in some current cases.

Regarding hydraulic machines installed for small hydropower, Paish [47] established the efficiency of these machines according to the type and head ratio. The efficiency has values of approximately 90% for Pelton turbines over a wide head ratio (0.2–0.8). In crossflow turbines, the values of efficiency are close to 80% for a head ratio (0.2–1). Francis and propeller turbines present efficiencies of approximately 85% for a head ratio (0.9–1).

The development and improvement of large and small hydropower systems have allowed the adaptation of the machines to water distribution networks, establishing the group called "adapted machines" (Figure 4). This group of machines is used in micro and pico hydropower plants. Pump as turbine (PAT) [54], tubular propeller [55], and positive displacement machine [56] are included in this group (Figure 6).

Figure 6. Hydraulic machines at IST-Universidade de Lisboa: PAT (**left**); and tubular propeller (**right**).

This group of machines can be installed in places where energy is currently dissipated for specific flow and pressure operating conditions. These conditions mainly depend on user demand and the minimum pressure required (when the machine is installed in a pressurized water network). The existing demand establishes the circulating flow over time in the line, whereas a required pressure establishes the maximum recovered head. In pressurized water networks, the excess of energy is dissipated with pressure reduction valves. In open channel flows, this dissipation of energy is carried out by means of hydraulic jumps. In micro and pico systems, conventional machines can be installed according to installed power and head characteristics (e.g., micro Francis, Pelton, Turgo, and Cross-Flow), but the high investment cost makes the installation not viable.

In 1931, Thoma [54] implemented the first pump working as a turbine (PAT). Later, other authors presented more research that presented the description, operation, performance, and theoretical model of these machines [57–64]. PATs are normally used in pressurized water networks but they can also be used in open channel flows when complementary civil works are carried out to adapt to PAT operating conditions. These machines present a high range of flow and head for installation according to the typology of the machine (Figure 7). The best efficiencies vary between 40% and 70% as a function of the specific speed [58].

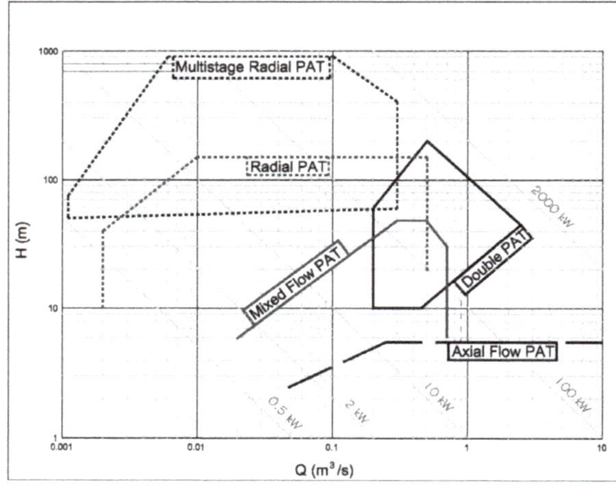

Figure 7. Range of application of PATs (adapted from [54]).

PATs become the technological solution to efficiently recover energy in water distribution networks. The main advantage of these machines is their immediate availability for installation and lower cost compared with conventional machines. Nourbakhsh and Jahangiri [65] established that the payback period of these machines is less than two years for installed capacities between 5 and 500 kW. These payback values make micro generation in water distribution networks feasible.

Elbatran et al. [64] listed the advantages of these machines in micro hydropower plants, such as a 50% reduction in the cost of the machine compared to a conventional turbine; the existence of a large availability of operating ranges depending on the hydraulic head and flow; simple management and operation; and a lifespan of twenty-five years. Furthermore, they have lower installation costs, which can improve the viability of small projects [60].

3. Micro and Pico Hydropower Solutions

3.1. Energy Recovery in Open Channel Networks

Micro hydropower can use different hydraulic heads or diversion schemes in small dams in rivers and ravines, in open irrigation channels and in drainage systems. Some examples of these systems can be found in several regions, such as the western part of the Himalayas [66], Bangladesh [67], Nigeria [68], Laos [69], Europe [41], or Lithuania [70]. Although it might seem that the development of these hydropower plants has been recent, this is not true because watermills were present in all continents many years ago, were fundamental to moving other machines in the Industrial Revolution, and can still be found operating in some countries (e.g., United Kingdom, France, Spain, USA, Africa, and part of Asia) [71,72]. This technology is currently essential to generate electrical energy in rural areas and to support the social and economic development of these isolated regions.

Water wheels can also be used in other open channel flows. For example, these solutions can be located in water treatment plants. These elements can also be installed in urban infrastructure, where energy recovery systems are established to reduce the energy footprint of urban water systems [73]. In these facilities, Ramos et al. [28] proposed the use of an urban storm-water drainage system to take advantage of storm retention ponds and to develop energy recovery systems using the rain storage volume. This solution contributes a new source of clean energy, which is involved with the water drainage system. An example of energy recovery is analyzed by Novara et al. [74] in a wastewater treatment plant in the city of Asti (Italy). The flow in this channel oscillates between 0.07 and 0.83 m^3/s, while the available head changes between 0.062 and 0.744 m. With these values, a hydrostatic pressure machine (HPM) is proposed to be installed. This wheel is an experimental waterwheel specifically designed for application in open channels with reduced head, developed and improved by Senior et al. [75], patented by Austrian inventor Adolf Brinnich, and tested under the HYLOW project as part of EU's Seventh Framework Program between 2008 and 2013 [76]. If the energy balance is carried out, the maximum electrical power is approximately 650 W with a flow equal to 0.29 m^3/s, for a daily power of 10.9 kWh. In these conditions, 48% of the hydrostatic energy is converted into electrical power, 40% is mechanical loss, and 12% is electrical loss in the generator and transmission.

Similarly, in open channel irrigation systems, energy recovery can also be implemented by installed turbines in small dams or irrigation reservoirs [77,78]. An example of these installations is the analysis made by Butera and Balestra [79], who determined the potential generation by hydropower plants for the Piedmont region (Italy). This region has an installed capacity of 46 MW, of which 45% is pico hydropower, 49% is micro and 6% is small, with an average hydraulic potential of 1.5–2 kW/ha. Tarragó et al. [26] developed a preliminary study in the Alqueva's irrigation system, where twenty-two hydrostatic pressure machines were studied in different locations with hydraulic heads below three meters. Using this assumption, the theoretical energy recovery reached 406.64 MWh/year in 67,932 ha of this region.

3.2. Energy Recovery Water Pipe and Irrigation Systems

Currently, energy recovery in pressurized water distribution networks (both urban or irrigation water supply) has great significance. Relative to urban supply systems, the energy consumption in water supply networks represents 7% of the world's consumption of energy [80]. Water distribution involves an energy footprint between 0.18 and 0.32 kWh/m^3, according to the California Energy Commission [81]. In addition to energy consumption, energy analysis of these networks has shown that an increase of pressure is correlated with increased leakage [82]. This problem justifies the installation of pressure reduction valves (PRVs) in many water distribution networks. These valves reduce pressure and, therefore, leakage volume. This directly proportional correlation between leakage and pressure caused the pioneering study of alternatives to leverage the dissipated energy by PRVs in water supply systems [57]. An unconventional solution was considered: replacing PRVs by PATs [57,59]. Ferracota et al. [60] studied leakage reduction. They presented and integrated a new technical solution with economic and system flexibility benefits, replacing pressure reduction valves by pumps used as turbines (PATs). The optimal operating point of the PATs was selected by using a variable operating strategy. Carraveta et al. [63] established a PAT operating scheme with a PRV in parallel. This operating scheme and the variability of flows over time in network pipelines due to user demand have fostered leading studies to develop variable operating strategies in these machines. These strategies allow the variation of the rotational speed of the hydraulic machine [83,84]. Ferracota et al. [85] have begun studies to improve efficiency prediction in the machine through experimental tests in semi-axial machines when the rotational speed varies. Preliminary studies in drinking water systems have been developed through computational simulations [59,60,86]. These studies considered average flows or hourly uniform patterns in all consumption joints for the development of simulations of water supply networks [87,88]. These energy recovery studies have promoted the use of water supply networks to generate clean energy, using the dissipated energy in PRVs [89]. These studies have resulted in some pilot installations emerging for evaluation (e.g., Murcia (Spain) [90], Portland (Oregon) [91], Hong Kong [89], and Kildare (Ireland) [92]).

In addition to water supply systems, water irrigation networks are very important for the improvement of energy efficiency in the water cycle. Worldwide water consumption is 3925 km^3/year [93], which is distributed such that 69.53% of water is used for irrigation, 18.70% is used for industry, and 11.77% is used for drinking water systems. In Spain, water consumption is distributed as follows: 80% for irrigation, 15% for drinking, and 5% for industry. The annual volume used for agriculture equals 16,344 hm^3 [94].

Hence, because the volume of water consumed for irrigation is higher than in urban systems, the modernization of irrigation should not only be associated with high technology and automation but also with water management that accounts for the sustainability of this infrastructure. The study of the installation of micro and pico hydropower is necessary because the irrigated surface area is huge (approximately 324 million hectares in the world are provided with irrigation installations, of which 86% are gravity irrigation, 11% sprinkler irrigation and 3% drip irrigation [95]). In Spanish economic terms, the irrigation water distribution cost was €1285 million in 2012. This value represents 20% of the total cost of the water supply service in Spain [96], considering that the irrigated surface area in Spain is 3.54 million hectares (1.09% of the worldwide irrigated surface area) [97].

Therefore, if the annual volume of water consumed in irrigation networks worldwide is measured, the development of systems to reduce the energy consumption is of the utmost importance. These new solutions should also try to improve, as much as possible, the environmental and economic sustainability of irrigation, considering that the modernization of irrigation water systems introduces an average increase of installed power equal to 2 kW/ha [98].

3.3. Strategies for Sustainability and Energy Efficiency in Pressurized Water Networks

3.3.1. Pumped Water Systems

Pumped water systems have been analyzed by different authors [99–103] whose main objective has been to minimize the energy costs. Rodriguez-Diaz et al. [99] proposed a new methodology with energy savings between 10% and 30% in real case studies, considering the most critical consumption points, which depend on needs and location. Moreno et al. [100] developed a methodology in which characteristic and efficiency curves are optimized depending on the recorded flows, obtaining a 32.33% average reduction of installed power in the studied networks.

In other research, energy reduction has been carried out using strategies to minimize energy consumption through optimal operating schedules, reducing energy footprints by 36.4% [101,102]. Costa et al. [103] presented a general optimization routine integrated with EPANET [104]. This routine allows the determination of strategic optimal rules of operation for any type of water distribution system. Cabrera et al. [105] developed a methodology to carry out an energy audit, which detects weaknesses in pressurized water networks. This methodology is applied in a real case, obtaining energy savings above 40%. In all of the cited cases, energy savings correspond to an economic reduction between 35% and 50% of the energy costs. Ferracota et al. [60] integrated a new technical solution with economic and system flexibility benefits, which replaces pressure reduction valves by pumps as turbines. In the majority of methods, when energy optimization is carried out in pumped water systems, the objective is easily defined as minimizing the energy consumption, with the solution being the establishment of optimized irrigation schedules according to the minimum necessary pressure and irrigation needs at each consumption point.

3.3.2. Gravity Water Systems

If an irrigation network is a gravity system, the best solution is not to convert an on-demand water irrigation into a scheduled network because this decision can irritate farmers, and reduce their operational freedom. Therefore, if a water manager wants to increase the system sustainability and energy efficiency of a network by the installation of energy recovery systems, the manager should know the flow distribution over time. The analysis of water distribution systems allows the establishment of some crucial aspects of the recovery system, such as the type of hydraulic machine, the best efficiency point (BEP) and the range of operating conditions for the energy converters [64].

Preliminary values of recoverable energy have been obtained using average circulating flows for both irrigation [26] and water supply networks [106] in some studies. The authors have used daily patterns in these studies. To analyze the variation of flows over time, Pérez-Sánchez et al. [107] developed a new methodology to estimate the hourly circulating flows in any line based on the opening probability of the irrigation points, which depends on farmers' habits (i.e., irrigation duration, maximum days between irrigation, weekly irrigation trend, and irrigation start). This methodology can also be used in water supply networks when behavioral patterns are known. These demand patterns allow the water network to be simulated and the energy balance to be calculated to determine the percentage of energy dissipated by friction losses, the energy necessary for irrigation, the non-recoverable energy, and the theoretically available energy. In the case of Vallada (Spain) [107], the energy dissipated by friction is approximately 4.10% of the provided energy (with a maximum energy footprint of 2.85 kWh/m^3), with the theoretical recoverable energy in the network equal to 68.70%, when all of the irrigation points are considered elective places of recovery.

The feasibility of an energy project is not guaranteed when a high number of machines is installed in the network; thus, an analysis of the water network is needed to maximize the energy recovery. Samora et al. [108] developed a methodology that uses simulated annealing to maximize recovered energy in a water supply network [109]. This methodology selects the lines depending on the recovered energy, and considering the feasibility of the facilities according to an economic criterion. For these preliminary studies of feasibility, Castro [16] proposed a simple economic balance where the payback period is only determined through the investment cost, incomes and maintenance cost, which depend on the installed power. In an advanced or existing project, other more complex and detailed methods can be used, which consider the annual interest and the inflation rate [110,111].

Therefore, if the feasibility is studied, knowledge of the performance and head curves as functions of the flow in the selected machine is necessary to determine the real recovered energy. The proposed PAT curves by Rawal and Kshirsagar [112] and Singh [113] (Figure 8) can be used in the analysis to help select a PAT. These curves allow the impellers' diameter to be selected as a function of the specific rotational speed (ns), the discharge number (ϕ), and the head number (ψ). These parameters are defined by Equations (1)–(3) [24]:

$$ns = N \frac{\sqrt{P_R}}{H_R^{1.25}} \tag{1}$$

$$\phi = \frac{Q}{ND^3} \text{ (discharge number)} \tag{2}$$

$$\psi = \frac{H}{N^2 D^2} \text{ (head number)} \tag{3}$$

where N is the rotational speed (rpm); P_R is the rated power (kW); R is the pump design point or the best efficiency condition; H_R is the rated head (m w.c.); Q is the circulating flow (m^3/s); H is the recovered head of the machine (m w.c.); and D is the impeller diameter (m).

If the previous premises are used, different studies have been developed in water systems (irrigation and drinking networks), considering strategies to maximize the energy efficiency. In the particular case study of Vallada (Spain) [107], where the annual water consumption is 930,000 m^3/year, the actual recovered energy is 26.51 MWh/year. This recovery represents 9.55% of the energy provided to the network, with a simple payback period of 5.28. In a preliminary study of Alqueva (Portugal) [26], the theoretical energy recovery is at least 2.12 MWh/year in 68 ha of pressurized irrigation, for a water consumption of 179,000 m^3/year.

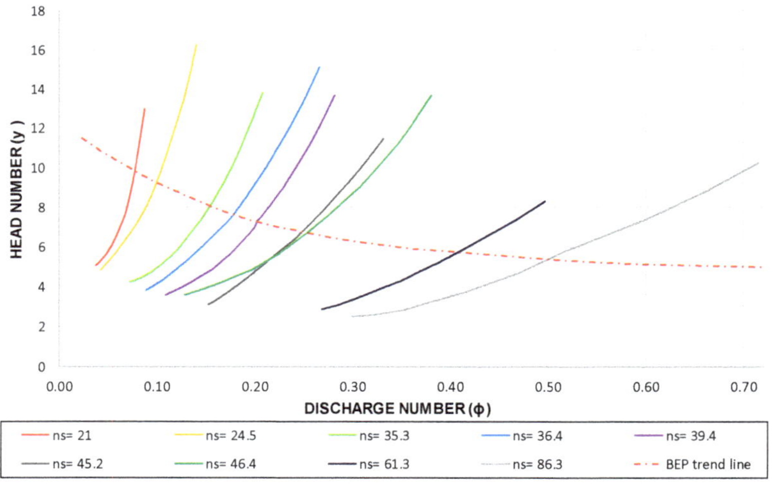

Figure 8. Head number depending on the discharge number (adapted from [112,113]).

In water supply systems, a case study of Lausanne (Switzerland) [106] finds that the real recovered energy represents up to 5% of the available energy. In a case study of Fribourg in the same country, the recovered energy reaches 10% of the available energy [114]. Another energy study developed in collaboration with the Consortium of Commons for the Monferrato Aqueduct (Italy) [74] determines an energy recovery equal to 9585 kWh/year when the pressure reducing valve is replaced by a PAT with a constant flow of 7 L/s and head of 75 m w.c. These examples show the importance of this type of solution for economic and environmental sustainability in water systems, if similar solutions are implemented.

From the economic point of view, the benefits of selling energy and generating income can be quite significant in some cases (although this generation is irregular over time because it depends on the flow, which varies as a function of consumption in water networks). Some particular analyses of these systems (and, more specifically, of PATs) present payback periods less than five years, with an installed capacity between 5 and 500 kW [8]. However, the importance of these solutions consists in the generation of energy for self-consumption by the local communities, i.e., for extracting water from their own water wells, electric supply in irrigation communities, or individual use at the irrigation points, avoiding investment in the electric grid.

From the environmental point of view, the use of these renewable energy sources reduces the emission of greenhouse gases when they are compared with non-renewable energy (e.g., fuel, usually used in electric generators installed in irrigation communities or irrigation points). Therefore, these recovery systems can supply the users' demand for low energy consumption in their facilities. Regarding the environmental added value, the theoretical reduction of CO_2 emission is 216.2 t/year in Vallada's network.

The development of studies to install energy recovery in pressurized water networks (mainly in supply systems) has been important, as previously discussed. The development of these studies has focused on the use of non-conventional hydraulic machines to generate energy and on energy analyses using different installation schemes for these machines. Table 1 summarizes the current development of small energy recovery in pressurized water supply networks.

Table 1. Analyzed topics related to recovery systems in water supply networks.

Analyzed Topic	References
First PAT	[54]
Reduction of leaks, decreasing the pressure in water supply systems and increasing the efficiency	[57,60,80,82,84,87]
Proposal to use adapted machines (PATs and tubular propeller) in water supply systems to reduce the pressure	[55,57]
Description and operation of a PAT with a review of available technologies	[56,64,67,75,76]
Performance and modeling PAT	[56,58,61,65,85]
Installation of recovery systems in water supply networks	[59,79]
Implementation of simulations to determine the theoretical recovered energy in water supply and irrigation systems	[26,74,77,86,106–108]
Design of variable operating strategies to maximize the recovered energy	[59,60,62,84]
Economic cost of implementing recovery systems in water supply and irrigation networks	[8,16,64,84]
Environmental advantages	[66,78,88]
Policies and analyses to help the development of rural areas	[41,58,68–72]
Pilot plants built in water supply networks	[89–92]
Optimization to maximize recovered energy in water supply systems	[106]

Table 1 shows different topics related to PATs that have been studied by different authors. The description, operating mode, characteristic curves (theoretical and experimental), and simulations (for hourly uniform patterns in all consumption joints) of PATs have been developed, enumerating the advantages (such as good efficiency values and low price) and limitations. The main limitations are lower efficiency when the system operates with oscillating flows and the irregular generation of energy due to variable flow, which hinders both sale and self-consumption.

The development of different research is necessary to solve the previously cited limitations. Future work should focus on obtaining better knowledge of recorded flows over time in any line; improving the variable operating strategies to adapt the rotational speed of machines in each time step; and developing sustainable and feasible electric systems (grid connection or stand-alone operation). These electric systems will increase the viability of selling the energy to the national grid or using it for self-consumption.

4. Conclusions

There are several related scales for the management of the hydraulic energy generation, including local, regional, national and international, when considering water as a resource. However, even today, energy recovery is a very attractive possibility in water networks, with small additional costs for managers and investors. The success of this novel use depends on the experience acquired in hydropower plants with higher installed power.

A deep review has been presented, analyzing the different alternatives for hydropower production from large to pico facilities, according to the production levels, the economic and the environmental points of view, as well as the classification of the hydraulic machines. Considering the evolution of these energy solutions, energy recovery in water distribution networks is an alternative for the development of systems towards more sustainable and efficient solutions. Technological, economic and environmental implications of hydropower systems for energy recovery around the world have to be considered. Despite the large list of references, only a few can be related to agricultural water network energy recovery because this subject has not yet been explored.

Analysis of the references cited in this document establishes that recovery systems in pressurized water networks are:

(1) Recovery systems with less installed power, called mini and pico hydropower plants. These energy recovery systems appeared due to the need to replace waste or non-renewable energy devices with renewable energy solutions. The building of large hydropower plants has been maximized in different developed countries and the development of new large hydropower plants is currently limited due to environmental and social factors. However, the experience in these facilities (i.e., large and small hydropower) has contributed to the development of new recovery systems in pressurized water networks. The most important transfer has been advances in possible recovery machines and improvement in the efficiency of impellers and in water pipe systems as a whole.

(2) The description of machines used in different hydropower plants (i.e., pressurized and open channel flows) has shown that on the one hand, classical machines cannot directly be used or scaled to pico hydropower plants because the adaptation of flow and head presents some difficulties in terms of viability. In contrast, similar or adapted machines can be developed based on classical machines (e.g., Francis turbine vs. radial and mixed PATs; axial turbine vs. tubular propeller or axial PATs). The development of new adapted machines and improvement in the efficiency of the current ones are fundamental challenges for increasing the installation of recovery energy systems in water pipe networks in the near future.

(3) For energy recovery in pico hydropower plants, the PAT is currently the most successful machine to be adapted to these systems, according to previous studies and installed pilot plants. The main positive aspects of these machines are that: (i) the installation of a PAT allows the replacement of a PRV to dissipate excess flow energy; (ii) the PATs' efficiency values vary between 0.40 and 0.70, operating in reverse mode; (iii) theoretical studies can be developed with the current technology (e.g., computational fluid dynamics (CFD)) based on the classical theory of hydraulic machines (i.e., Euler's Theorem), for comparison with existing experimental tests; and (iv) they have low investment costs and a high number of available machines. These advantages allow the

installation of these machines in water pipe systems to be promoted. The main aspects negative of PATs are related to their low efficiency when operating outside their best efficiency point. Operation with different flows can be solved by the development of new regulation techniques (e.g., variable operation strategies (VOS)) with electronic regulation. The positive resolution of this aspect is a crucial point for expanding use of PATs in water distribution networks. Issues related to the use of the generated energy for self-consumption may include storage in batteries and integrating this renewable energy in a similar manner as other supplementary sources (e.g., solar and wind).

(4)   Different case studies have been developed using specific software (e.g., EPANET and WaterGEMS), which have been combined with optimization methodologies to maximize the recovered energy. Future simulations should take into account the integration of VOS as well as the variation of the machine efficiency with the rotational speed. These simulations should consider discretized demand over time to improve the analyzed energy values because the majority of studies only consider the mean demand value or modulation curves. The development of a specific methodology to determine this variation of flow over time in water supply networks is crucial to improve the fit between theoretical and real values of recovered energy. Regarding the software used, it is necessary to implement operation rules for these machines in specific algorithms. This implementation is the key point in the development of optimized techniques, making possible studies similar to those with water pump systems. The primary need is for correct machine selection and establishment of the rotational speed as a function of the flow, maintaining the maximum efficiency at each operation point of the machine.

Therefore, hydraulic recovery in water networks is a real and necessary alternative to improve the energy efficiency of the whole system. By means of implementing energy converters, the energy efficiency will be increased and operating costs will be reduced (i.e., the energy footprint of water). The implementation of these systems will essentially depend on the physical characteristics of the systems. The orography, topology, and volumes of water consumed establish the economic viability of these recovery strategies in water distribution networks. When the investment analyses are developed, recovery systems have acceptable values of economic feasibility indexes (e.g., payback value and internal rate of return). A better understanding of the operation of each recovery system in terms of water-energy management is needed that considers the high global volume distributed in pressurized water networks (i.e., drinking and irrigation) each year. This understanding will positively contribute to the sustainability and efficiency of near future recovery system applications.

Acknowledgments: No additional funds have been received for this research.

Author Contributions: All the authors have participated in any step of this research. Particularly a brief description is attached: Helena M. Ramos has contributed by supervising the state of the art and revision of the whole paper. Francisco Javier Sánchez-Romero developed the study of large and small hydropower. Modesto Pérez-Sánchez wrote and analyzed the state of the art in water networks taking account the energy recovery. P. Amparo López-Jiménez supervised the whole research and she was involved in conclusions determination.

Conflicts of Interest: The authors declare no conflict of interest. The founding sponsors had no role in the design of the study, in the collection, analyses, or interpretation of data, in the writing of the manuscript, and in the decision to publish the results.

References

1.   Pasten, C.; Santamarina, J.C. Energy and quality of life. *Energy Policy* **2012**, *49*, 468–476. [CrossRef]
2.   Huesemann, M.H. The limits of technological solutions to sustainable development. *Clean Technol. Environ. Policy* **2003**, *5*, 21–34.
3.   Gilron, J. Water-energy nexus: Matching sources and uses. *Clean Technol. Environ. Policy* **2014**, *16*, 1471–1479. [CrossRef]
4.   Emec, S.; Bilge, P.; Seliger, G. Design of production systems with hybrid energy and water generation for sustainable value creation. *Clean Technol. Environ. Policy* **2015**, *17*, 1807–1829. [CrossRef]

5. Fontanazza, C.; Freni, G.; Loggia, G.; Notaro, V.; Puleo, V. Evaluation of the Water Scarcity Energy Cost for Users. *Energies* **2013**, *6*, 220–234. [CrossRef]

6. Herath, I.; Deurer, M.; Horne, D.; Singh, R.; Clothier, B. The water footprint of hydroelectricity: A methodological comparison from a case study in New Zealand. *J. Clean. Prod.* **2011**, *19*, 1582–1589. [CrossRef]

7. Mushtaq, S.; Maraseni, T.N.; Reardon-Smith, K.; Bundschuh, J.; Jackson, T. Integrated assessment of water–energy–GHG emissions tradeoffs in an irrigated lucerne production system in eastern Australia. *J. Clean. Prod.* **2015**, *103*, 491–498. [CrossRef]

8. Hennig, T.; Wang, W.; Feng, Y.; Ou, X.; He, D. Review of Yunnan's hydropower development. Comparing small and large hydropower projects regarding their environmental implications and socio-economic consequences. *Renew. Sustain. Energy Rev.* **2013**, *27*, 585–595. [CrossRef]

9. Zhang, S.; Pang, B.; Zhang, Z. Carbon footprint analysis of two different types of hydropower schemes: Comparing earth-rockfill dams and concrete gravity dams using hybrid life cycle assessment. *J. Clean. Prod.* **2014**, *103*, 854–862. [CrossRef]

10. Rehman, S.; Al-Hadhrami, L.M.; Alam, M.M. Pumped hydro energy storage system: A technological review. *Renew. Sustain. Energy Rev.* **2015**, *44*, 586–598. [CrossRef]

11. Paish, O. Micro-hydropower: Status and prospects. *Proc. Inst. Mech. Eng. A J. Power Energy* **2005**, *216*, 31–40. [CrossRef]

12. Spänhoff, B. Current status and future prospects of hydropower in Saxony (Germany) compared to trends in Germany, the European Union and the World. *Renew. Sustain. Energy Rev.* **2014**, *30*, 518–525. [CrossRef]

13. Hans-Wilhelm, S. *World Energy Scenarios—The Role of Hydropower in Composing Energy Futures to 2050*; World Hydropower Congress: Beijing, China, 2015; pp. 40–46.

14. Eurostat. 2015. Available online: http://ec.europa.eu/eurostat/tgm/refreshTableAction.do?tab=table&plugin=1&pcode=ten00081&language=en (accessed on 31 May 2015).

15. López-González, L.M.; Sala-Lizarraga, J.M.; Míguez-Tabarés, J.L.; López-Ochoa, L.M. Contribution of renewable energy sources to electricity production in the autonomous community of Navarre (Spain): A review. *Renew. Sustain. Energy Rev.* **2007**, *11*, 1776–1793. [CrossRef]

16. Castro, A. *Minicentrales Hidroeléctricas*; Instituto para la Diversificación y Ahorro de la Energía: Madrid, Spain, 2006. Available online: http://www.energiasrenovables.ciemat.es/adjuntos_documentos/Minicentrales_hidroelectricas.pdf (accessed on 10 September 2016).

17. Gaudard, L.; Romerio, F. Reprint of "The future of hydropower in Europe: Interconnecting climate, markets and policies". *Environ. Sci. Policy* **2014**, *43*, 5–14. [CrossRef]

18. Choulot, A. *Energy Recovery in Existing Infrastructures with Small Hydropower Plants*; FP6 Project Shapes (Work Package 5—WP5); European Directorate for Transport and Energy: Brussels, Belgium, 2010.

19. Huisingh, D.; Zhang, Z.; Moore, J.C.; Qiao, Q.; Li, Q. Recent advances in carbon emissions reduction: Policies, technologies, monitoring, assessment and modeling. *J. Clean. Prod.* **2015**, *103*, 1–12. [CrossRef]

20. Ihle, C.F. The need to extend the study of greenhouse impacts of mining and mineral processing to hydraulic streams: Long distance pipelines count. *J. Clean. Prod.* **2014**, *84*, 597. [CrossRef]

21. Ramos, H.M.; Almeida, A. *Small Hydro as One of the Oldest Renewable Energy Sources*; Water Power and Dam Construction; Small Hydro: Lisbon, Portugal, 2000.

22. Ansar, A.; Flyvbjerg, B.; Budzier, A.; Lunn, D. Should we build more large dams? The actual costs of hydropower megaproject development. *Energy Policy* **2014**, *69*, 43–56. [CrossRef]

23. Hamududu, B.; Killingtveit, A. Assessing Climate Change Impacts on Global Hydropower. *Energies* **2012**, *5*, 305–322. [CrossRef]

24. Chakraborty, D.; Roy, J. Energy and carbon footprint: Numbers matter in low energy and low carbon choices. *Curr. Opin. Environ. Sustain.* **2013**, *5*, 237–243. [CrossRef]

25. Zhang, J.; Xu, L. Embodied carbon budget accounting system for calculating carbon footprint of large hydropower project. *J. Clean. Prod.* **2015**, *96*, 444–451. [CrossRef]

26. Tarragó, E.F.; Ramos, H. Micro-Hydro Solutions in Alqueva Multipurpose Project (AMP) towards Water-Energy-Environmental Efficiency Improvements. Bachelor's Thesis, Universidade de Lisboa, Lisboa, Portugal, 2015.

27. Pacca, S. Impacts from decommissioning of hydroelectric dams: A life cycle perspective. *Clim. Chang.* **2007**, *84*, 281–294. [CrossRef]

28. Ramos, H.M.; Teyssier, C.; López-Jiménez, P.A. Optimization of Retention Ponds to Improve the Drainage System Elasticity for Water-Energy Nexus. *Water Resour. Manag.* **2013**, *27*, 2889–2901. [CrossRef]

29. Steffen, B. Prospects for pumped-hydro storage in Germany. *Energy Policy* **2012**, *45*, 420–429. [CrossRef]

30. Deane, J.P.; Gallachóir, B.P.; McKeogh, E.J. Techno-economic review of existing and new pumped hydro energy storage plant. *Renew. Sustain. Energy Rev.* **2010**, *14*, 1293–1302. [CrossRef]

31. Mataix, C. *Turbomáquinas Hidráulicas*; Universidad Pontificia Comillas: Madrid, Spain, 2009.

32. Cheng, C.; Liu, B.; Chau, K.W.; Li, G.; Liao, S. China's small hydropower and its dispatching management. *Renew. Sustain. Energy Rev.* **2015**, *42*, 43–55. [CrossRef]

33. Pereira, A.O.; Cunha, R.; Costa, V.; Marreco, J.; Rovere, E.L. Perspectives for the expansion of new renewable energy sources in Brazil. *Renew. Sustain. Energy Rev.* **2013**, *23*, 49–59. [CrossRef]

34. Kosnik, L. The potential for small scale hydropower development in the US. *Energy Policy* **2010**, *38*, 5512–5519. [CrossRef]

35. Bahadori, A.; Zahedi, G.; Zendehboudi, S. An overview of Australia's hydropower energy: Status and future prospects. *Renew. Sustain. Energy Rev.* **2013**, *20*, 565–569. [CrossRef]

36. Nautiyal, H.; Singal, S.K.; Sharma, A. Small hydropower for sustainable energy development in India. *Renew. Sustain. Energy Rev.* **2011**, *15*, 2021–2027. [CrossRef]

37. Ushiyama, I. Renewable energy strategy in Japan. *Renew. Energy* **1999**, *16*, 1174–1179. [CrossRef]

38. Liu, H.; Masera, D.; Esser, L. World Small Hydropower Development Report 2013. Available online: www.smallhydroworld.org (accessed on 13 September 2016).

39. Miller, C.A.; Altamirano-Allende, C.; Johnson, N.; Agyemang, M. The social value of mid-scale energy in Africa: Redefining value and redesigning energy to reduce poverty. *Energy Res. Soc. Sci.* **2015**, *5*, 67–69. [CrossRef]

40. ESHA. Statistical Releases. 2012. Available online: http://streammap.esha.be/ (accessed on 13 September 2016).

41. Alonso-Tristán, C.; González-Peña, D.; Díez-Mediavilla, M.; Rodríguez-Amigo, M.; García-Calderón, T. Small hydropower plants in Spain: A case study. *Renew. Sustain. Energy Rev.* **2011**, *15*, 2729–2735. [CrossRef]

42. Instituto para la Diversificación y Ahorro de la Energía. *National Action Plan for Renewable Energy in Spain (PANER) 2011–2020*; Ministerio Industria, Turismo y Comercio: Madrid, Spain, 2010. Available online: http://www.idae.es/uploads/documentos/documentos_11227_per_2011-2020_def_93c624ab.pdf (accessed on 8 February 2017).

43. European Small Hydropower Association. Current Status of Small Hydropower Development in the EU-27. Available online: http://www.streammap.esha.be/fileadmin/documents/Raising_awareness_doc___press_release/FINAL_SHP_Awareness_2011.pdf (accessed on 15 September 2016).

44. Amponsah, N.Y.; Troldborg, M.; Kington, B.; Aalders, I.; Hough, R.L. Greenhouse gas emissions from renewable energy sources: A review of lifecycle considerations. *Renew. Sustain. Energy Rev.* **2014**, *39*, 461–475. [CrossRef]

45. Gallagher, J.; Styles, D.; McNabola, A.; Williams, A.P. Life cycle environmental balance and greenhouse gas mitigation potential of micro-hydropower energy recovery in the water industry. *J. Clean. Prod.* **2015**, *99*, 152–159. [CrossRef]

46. Ogayar, B.; Vidal, P.G. Cost determination of the electro-mechanical equipment of a small hydro-power plant. *Renew. Energy* **2009**, *34*, 6–13. [CrossRef]

47. Paish, O. Small hydro power: Technology and current status. *Renew. Sustain. Energy Rev.* **2002**, *6*, 537–556. [CrossRef]

48. Mishra, S.; Singal, S.K.; Khatod, D.K. Optimal installation of small hydropower plant—A review. *Renew. Sustain. Energy Rev.* **2011**, *15*, 3862–3869. [CrossRef]

49. European Small Hydropower Association. Guía Para El Desarrollo de Una Pequeña Central Hidroeléctrica. European Small Hydropower Association. 2006. Available online: www.esha.be/fileadmin/esha_files/.../GUIDE_SHP_ES_01.pdf (accessed on 15 September 2016). (In Spanish)

50. Ramos, H. *Guidelines for Design of Small Hydropower Plants*; WREAN (Western Regional Energy Agency &Network) and DED (Department of Economic Development): Belfast, North-Ireland, UK, 2000.

51. Yuce, M.I.; Muratoglu, A. Hydrokinetic energy conversion systems: A technology status review. *Renew. Sustain. Energy Rev.* **2015**, *43*, 72–82. [CrossRef]

52. Ramos, H.M.; Almeida, A. *Caracterização Dinâmica Global do Funcionamento de Aproveitamentos Hidroeléctricos*; IV SILUSBA–Simpósio de Hidráulica e Recursos Hídricos dos Países de Língua Oficial Portuguesa: Lisboa, Portugal, 1999. (In Portuguese)

53. Gordon, J.L. Hydraulic turbine efficiency. *Can. J. Civ. Eng.* **2001**, *28*, 238–253. [CrossRef]
54. Chapallaz, J.M. *Manual on Pumps Used as Turbines*; Vieweg: Braunschweig, Germany, 1992.
55. Caxaria, G.; Mesquita, D.; Ramos, H.M. *Small Scale Hydropower: Generator Analysis and Optimization for Water Supply Systems*; World Renewable Energy Congress: Linköping, Sweden, 2011; pp. 1386–1393.
56. Simão, M.; Ramos, H.M. Hydrodynamic and performance of low power turbines: Conception, modelling and experimental tests. *Int. J. Energy Environ.* **2010**, *1*, 431–444.
57. Ramos, H.; Borga, A. Pumps as turbines: An unconventional solution to energy production. *Urban Water* **1999**, *1*, 261–263. [CrossRef]
58. Arriaga, M. Pump as turbine—A pico-hydro alternative in Lao People's Democratic Republic. *Renew. Energy* **2010**, *35*, 1109–1115. [CrossRef]
59. Carravetta, A.; Fecarotta, O.; Del Giudice, G.; Ramos, H. Energy Recovery in Water Systems by PATs: A Comparisons among the Different Installation Schemes. *Procedia Eng.* **2014**, *70*, 275–284. [CrossRef]
60. Fecarotta, O.; Aricò, C.; Carravetta, A.; Martino, R.; Ramos, H.M. Hydropower Potential in Water Distribution Networks: Pressure Control by PATs. *Water Resour. Manag.* **2014**, *29*, 699–714. [CrossRef]
61. Derakhshan, S.; Nourbakhsh, A. Experimental study of characteristic curves of centrifugal pumps working as turbines in different specific speeds. *Exp. Therm. Fluid Sci.* **2008**, *32*, 800–807. [CrossRef]
62. Carravetta, A.; Del Giudice, G.; Fecarotta, O.; Ramos, H. Pump as Turbine (PAT) Design in Water Distribution Network by System Effectiveness. *Water* **2013**, *5*, 1211–1225. [CrossRef]
63. Carravetta, A.; Del Giudice, G.; Fecarotta, O.; Ramos, H.M. Energy Production in Water Distribution Networks: A PAT Design Strategy. *Water Resour. Manag.* **2012**, *26*, 3947–3959. [CrossRef]
64. Elbatran, A.H.; Yaakob, O.B.; Ahmed, Y.M.; Shabara, H.M. Operation, performance and economic analysis of low head micro-hydropower turbines for rural and remote areas: A review. *Renew. Sustain. Energy Rev.* **2015**, *43*, 40–50. [CrossRef]
65. Nourbakhsh, A.; Jahangiri, G. Inexpensive small hydropower stations for small areas of developing countries. In Proceedings of the Conference on Advanced in Planning-Design and Management of Irrigation Systems as Related to Sustainable Land Use, Louvain, Belgium, 14–17 September 1992; pp. 313–319.
66. Kumar, D.; Katoch, S.S. Small hydropower development in western Himalayas: Strategy for faster implementation. *Renew. Energy* **2015**, *77*, 571–578. [CrossRef]
67. Razan, J.I.; Islam, R.S.; Hasan, R.; Hasan, S.; Islam, F. A Comprehensive Study of Micro-Hydropower Plant and Its Potential in Bangladesh. *Renew. Energy* **2012**, *2012*, 635396. [CrossRef]
68. Ohunakin, O.S.; Ojolo, S.J.; Ajayi, O.O. Small hydropower (SHP) development in Nigeria: An assessment. *Renew. Sustain. Energy Rev.* **2011**, *15*, 2006–2013. [CrossRef]
69. Vicente, S.; Bludszuweit, H. Flexible design of a pico-hydropower system for Laos communities. *Renew. Energy* **2012**, *44*, 406–413. [CrossRef]
70. Punys, P.; Dumbrauskas, A.; Kasiulis, E.; Vyčienė, G.; Šilinis, L. Flow Regime Changes: From Impounding a Temperate Lowland River to Small Hydropower Operations. *Energies* **2015**, *8*, 7478–7501. [CrossRef]
71. Abbasi, T.; Abbasi, S.A. Small hydro and the environmental implications of its extensive utilization. *Renew. Sustain. Energy Rev.* **2011**, *15*, 2134–2143. [CrossRef]
72. Punys, P.; Dumbrauskas, A.; Kvaraciejus, A.; Vyciene, G. Tools for Small Hydropower Plant Resource Planning and Development: A Review of Technology and Applications. *Energies* **2011**, *4*, 1258–1277. [CrossRef]
73. Vilanova, M.R.; Balestieri, J.A. Hydropower recovery in water supply systems: Models and case study. *Energy Convers. Manag.* **2014**, *84*, 414–426. [CrossRef]
74. Novara, D.; Stanek, W.; Ramos, H. Energy Harvesting from Municipal Water Management Systems: From Storage and Distribution to Wastewater Treatment. Master's Thesis, Universidade de Lisboa, Lisboa, Portugal, 2016.
75. Senior, J.A.; Muller, G.; Wiemann, P. The development of the rotary hydraulic pressure machine. In Proceedings of the Congress IAHR, Venice, Italy, 1–6 July 2007.
76. Senior, J.A.; Saenger, N.; Müller, G. New hydropower converters for very low-head differences. *J. Hydraul. Res.* **2010**, *48*, 703–714. [CrossRef]
77. Adhau, S.P.; Moharil, R.M.; Adhau, P.G. Mini-hydro power generation on existing irrigation projects: Case study of Indian sites. *Renew. Sustain. Energy Rev.* **2012**, *16*, 4785–4795. [CrossRef]
78. Tilmant, A.; Goor, Q.; Pinte, D. Agricultural-to-hydropower water transfers: Sharing water and benefits in hydropower-irrigation systems. *Hydrol. Earth Syst. Sci.* **2009**, *13*, 1091–1101. [CrossRef]

79. Butera, I.; Balestra, R. Estimation of the hydropower potential of irrigation networks. *Renew. Sustain. Energy Rev.* **2015**, *48*, 140–151. [CrossRef]

80. Coelho, B.; Andrade-Campos, A. Efficiency achievement in water supply systems—A review. *Renew. Sustain. Energy Rev.* **2014**, *30*, 59–84. [CrossRef]

81. Klein, G.; Krebs, M.; Hall, V.; O'Brien, T.; Blevins, B.B. *California's Water—Energy Relationship*; California Energy Commission: Sacramento, CA, USA, 2005.

82. Almandoz, J.; Cabrera, E.; Arregui, F.; Cobacho, R. Leakage Assessment through Water Distribution Network Simulation. *J. Water Resour. Plan. Manag.* **2005**, *131*, 458–466. [CrossRef]

83. Carravetta, A.; Del Giudice, G.; Oreste, F.; Ramos, H. PAT design strategy for energy recovery in water distribution networks by electrical regulation. *Energies* **2013**, *6*, 411–424. [CrossRef]

84. Colombo, A.F.; Karney, B.W. Energy and Costs of Leaky Pipes: Toward Comprehensive Picture. *J. Water Resour. Plan. Manag.* **2002**, *128*, 441–450. [CrossRef]

85. Fecarotta, O.; Carravetta, A.; Ramos, H.M.; Martino, R. An improved affinity model to enhance variable operating strategy for pumps used as turbines. *J. Hydraul. Res.* **2016**, *54*, 332–341. [CrossRef]

86. Sitzenfrei, R.; von Leon, J. Long-time simulation of water distribution systems for the design of small hydropower systems. *Renew Energy* **2014**, *72*, 182–187. [CrossRef]

87. Fontana, N.; Giugni, M.; Portolano, D. Losses Reduction and Energy Production in Water-Distribution Networks. *J. Water Resour. Plan. Manag.* **2012**, *138*, 237–244. [CrossRef]

88. Ramos, H.; Mello, M.; De, P.K. Clean power in water supply systems as a sustainable solution: From planning to practical implementation. *Water Sci. Technol. Water Supply* **2010**, *10*, 39–49. [CrossRef]

89. Hong Kong Polytechnic University, Novel Inline Hydropower System for Power Generation from Water Pipelines. Available online: http://phys.org/news/2012-12-inline-hydropower-power-pipelines.html (accessed on 16 August 2016).

90. Imbernón, J.A.; Usquin, B. Sistemas de generación hidráulica. Una nueva forma de entender la energía. In Proceedings of the II Congreso Smart Grid, Madrid, Spain, 27–28 October 2014. (In Spanish)

91. Lisk, B.; Greenberg, E.; Bloetscher, F. *Implementing Renewable Energy at Water Utilities*; Case Studies; Water Research Foundation: Denver, CO, USA, 2012.

92. McNabola, A.; Coughlan, P.; Williams, A.P. Energy recovery in the water industry: An assessment of the potential of micro hydropower. *Water Environ. J.* **2014**, *28*, 294–304. [CrossRef]

93. Food and Agriculture Organization (FAO). Aquastat. 2015. Available online: http://www.fao.org/nr/water/aquastat/data/query/results.html?regionQuery=true&yearGrouping=SURVEY&showCodes=false&yearRange.fromYear=1958&yearRange.toYear=2017&varGrpIds=4250%2C4251%2C4252%2C4253%2C4257&cntIds=®Ids=9805%2C9806%2C9807%2C9808%2C9809&edit (accessed on 9 June 2015).

94. Seoane, P.; Allué, R.; Postigo, M.J.; Cordón, M.A. *Boletín Mensual de Estadística*; Instituto Nacional de Estadística: Madrid, Spain, 2013. (In Spanish)

95. Food and Agriculture Organization (FAO). Agua Y Cultivos, 2002. Available online: http://www.fao.org/docrep/005/y3918s/y3918s10.htm (accessed on 19 September 2016). (In Spanish)

96. Maestu, J.; Villar, A. Precios Y Costes de Los Servicios Del Agua En España, Madrid, Spain, 2007. Available online: http://hispagua.cedex.es/sites/default/files/especiales/Tarifas_agua/precios_costes_servicios_agua.pdf (accessed on 19 September 2016). (In Spanish)

97. MAGRAMA. El Riego Localizado Alcanza el 48.23% de la Superficie Regada en España. Minist Agric Aliment y Medio Ambient. 2014. Available online: http://www.magrama.gob.es/gl/prensa/noticias/el-riego-localizado-alcanza-el-4823--de-la-superficie-regada-en-espa%C3%B1a-/tcm7-312671-16 (accessed on 9 June 2015). (In Spanish)

98. Instituto para la Diversificación y Ahorro de la Energía. *Ahorro Y Eficiencia Energética en Agricultura de Regadío*; Ministerio Industria, Turismo y Comercio: Madrid, Spain, 2005. (In Spanish)

99. Rodríguez-Díaz, J.A.; Montesinos, P.; Poyato, E.C. Detecting Critical Points in On-Demand Irrigation Pressurized Networks—A New Methodology. *Water Resour. Manag.* **2012**, *26*, 1693–1713. [CrossRef]

100. Moreno, M.A.; Planells, P.; Córcoles, J.I.; Tarjuelo, J.M.; Carrión, P.A. Development of a new methodology to obtain the characteristic pump curves that minimize the total cost at pumping stations. *Biosyst. Eng.* **2009**, *102*, 95–105. [CrossRef]

101. Jiménez-Bello, M.A.; Royuela, A.; Manzano, J.; Prats, A.G.; Martínez-Alzamora, F. Methodology to improve water and energy use by proper irrigation scheduling in pressurised networks. *Agric. Water Manag.* **2015**, *149*, 91–101. [CrossRef]

102. Prats, A.G.; Picó, S.G.; Alzamora, F.M.; Bello, M.A. Random Scenarios Generation with Minimum Energy Consumption Model for Sectoring Optimization in Pressurized Irrigation Networks Using a Simulated Annealing Approach. *J. Irrig. Drain. Eng.* **2012**, *138*, 613–624. [CrossRef]

103. Costa, L.; de Athayde-Prata, B.; Ramos, H.; de Castro, M. A Branch-and-Bound Algorithm for Optimal Pump Scheduling in Water Distribution Networks. *Water Resour. Manag.* **2015**, *30*, 1037–1052. [CrossRef]

104. Rossman, L.A. *EPANET 2 User's Manual*; U.S. Environmental Protection Agency (EPA): Cincinnati, OH, USA, 2000.

105. Cabrera, E.; Cobacho, R.; Soriano, J. Towards an Energy Labelling of Pressurized Water Networks. *Procedia Eng.* **2014**, *70*, 209–217. [CrossRef]

106. Samora, I.; Franca, M.; Schleiss, A.; Ramos, H.M. Simulated Annealing in Optimization of Energy Production in a Water Supply Network. *Water Resour. Manag.* **2016**, *30*, 1533–1547. [CrossRef]

107. Pérez-Sánchez, M.; Sánchez-Romero, F.; Ramos, H.; López-Jiménez, P. Modeling Irrigation Networks for the Quantification of Potential Energy Recovering: A Case Study. *Water* **2016**, *8*, 234. [CrossRef]

108. Samora, I.; Manso, P.; Franca, M.J.; Schleiss, A.J.; Ramos, H.M. Opportunity and Economic Feasibility of Inline Microhydropower Units in Water Supply Networks. *J. Water Resour. Plan. Manag.* **2016**, *142*, 04016052. [CrossRef]

109. Kirkpatrick, S.; Gelatt, C.; Vecchi, M. Optimization by simulated annealing. *Science* **1983**, *220*, 671–680. [CrossRef] [PubMed]

110. Forouzbakhsh, F.; Hosseini, S.M.H.; Vakilian, M. An approach to the investment analysis of small and medium hydro-power plants. *Energy Policy* **2007**, *35*, 1013–1024. [CrossRef]

111. Zema, D.A.; Nicotra, A.; Tamburino, V.; Zimbone, S.M. A simple method to evaluate the technical and economic feasibility of micro hydro power plants in existing irrigation systems. *Renew. Energy* **2016**, *85*, 498–506. [CrossRef]

112. Rawal, S.; Kshirsagar, J. Simulation on a pump operating in a turbine mode. In Proceedings of the 23rd International Pump Users Symposium, Houston, TX, 5–8 March 2007; Texas A&M University: College Station, TX, USA, 2007; pp. 21–27.

113. Singh, P. Optimization of the Internal Hydraulic and of System Design in Pumps as Turbines with Field Implementation and Evaluation. 2005. Available online: http://docplayer.net/18607634-Optimization-of-internal-hydraulics-and-of-system-design-for-pumps-as-turbines-with-field-implementation-and-evaluation.html (accessed on 19 September 2016).

114. Samora, I.; Manso, P.; Franca, M.; Schleiss, A.; Ramos, H. Energy Recovery Using Micro-Hydropower Technology in Water Supply Systems: The Case Study of the City of Fribourg. *Water* **2016**, *8*, 344. [CrossRef]

Article

Optimal Node Grouping for Water Distribution System Demand Estimation

Donghwi Jung [1], Young Hwan Choi [2] and Joong Hoon Kim [2,*]

[1] Research Center for Disaster Prevention Science and Technology, Korea University, Seoul 136-713, Korea; donghwiku@gmail.com

[2] School of Civil, Environmental and Architectural Engineering, Korea University, Anam-ro 145, Seongbuk-gu, Seoul 136-713, Korea; younghwan87@korea.ac.kr

* Correspondence: jaykim@korea.ac.kr; Tel.: +82-2-3290-3316; Fax: +82-2-928-7656

Academic Editor: Helena Ramos
Received: 29 February 2016; Accepted: 15 April 2016; Published: 20 April 2016

Abstract: Real-time state estimation is defined as the process of calculating the state variable of interest in real time not being directly measured. In a water distribution system (WDS), nodal demands are often considered as the state variable (*i.e.*, unknown variable) and can be estimated using nodal pressures and pipe flow rates measured at sensors installed throughout the system. Nodes are often grouped for aggregation to decrease the number of unknowns (demands) in the WDS demand estimation problem. This study proposes an optimal node grouping model to maximize the real-time WDS demand estimation accuracy. This Kalman filter-based demand estimation method is linked with a genetic algorithm for node group optimization. The modified Austin network demand is estimated to demonstrate the proposed model. True demands and field measurements are synthetically generated using a hydraulic model of the study network. Accordingly, the optimal node groups identified by the proposed model reduce the total root-mean-square error of the estimated node group demand by 24% compared to that determined by engineering knowledge. Based on the results, more pipe flow sensors should be installed to measure small flows and to further enhance the demand estimation accuracy.

Keywords: water distribution system; demand estimation; Kalman filter; node grouping; genetic algorithm

---

1. Introduction

A water distribution system (WDS) comprises various components (e.g., nodes, pipes and pumps), each of which has its own purpose and function. For example, a pumping unit elevates the total head of water to supply demand at high elevation, whereas a pipe transports water from one location to another. Hydraulic WDS models have been developed for many reasons, including design and rehabilitation, operation and management and system surveillance. The real-world system is simplified for modeling purposes during the development process of a hydraulic model. As shown in Figure 1, a group of households (h1–h5) can be simplified to node N1, because they are the same type of users (*i.e.*, residential) and in proximity to each other. Similarly, households h6–h9 can be modeled as node N2. Node N3 is the sum of the commercial demands c1–c5.

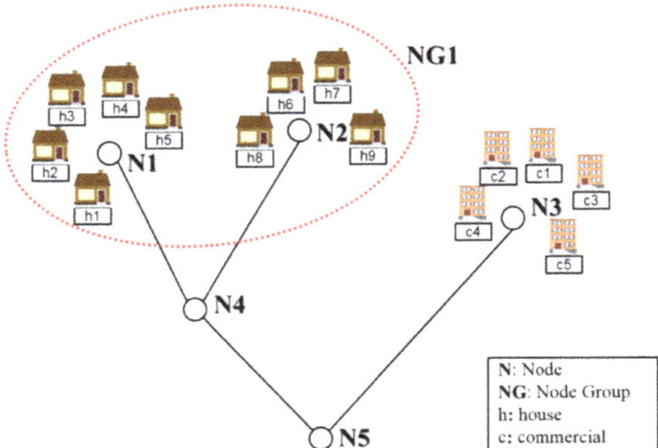

Figure 1. Residential and commercial demands represented as nodes in a hydraulic WDS model. The dashed ellipse shows a potential node group.

State estimation is defined as the process of calculating a state variable of interest that cannot be directly measured [1,2]. In a WDS, nodal demands are often considered as state variables (*i.e.*, unknown variables) and can be estimated using nodal pressures and pipe flow rates measured in the field. Previous demand forecasting/estimation studies have focused on estimating hourly to monthly macro-scale demand (*i.e.*, total system demand) by using time series models [3–6]. A few studies have recently proposed micro-scale demand (*i.e.*, nodal demands) estimation methodologies with a more precise time step (e.g., 15 min) based on coupling the Kalman filter (KF) with a hydraulic simulator (e.g., EPANET [7]) [2,8]. Advances in sensor technology, such as increased battery lifetime and communication frequency, have encouraged the development of demand estimation techniques compatible with field measurements from advanced sensors (e.g., advanced metering infrastructure (AMI)).

The WDS demand estimation can produce accurate results when sufficient field measurements are available. However, a sufficient number of meters cannot be installed in a system because of budget constraints. In demand estimation, the demand is the unknown variable, whereas the pipe flow/nodal pressure is the known variable. Therefore, the number of unknowns should be reduced to make the demand estimation problem even-determined. An alternative is to aggregate (*i.e.*, group) the nodes. For example, nodes N1 and N2 can be aggregated to Node Group 1 (NG1) because they have the same residential usage pattern (Figure 1).

Many methodologies have been proposed for the WDS component grouping, clustering and aggregation in the last two decades. However, most do not target demand estimation. Mallick *et al.* [9] investigated the tradeoff between the WDS model error caused by simplification (*i.e.*, pipe grouping) and the model prediction error. They subsequently proposed a method to identify the best number of pipe groups. Deuerlein [10] proposed a water distribution network decomposition method that classifies network components into forest, bridge and biconnected-block components. The latter two are called "core components". The three component types are used to augment the network graph. Perelman and Ostfeld [11] developed a multilayered clustering method as an extension of their previous clustering algorithm [12] based on depth-first search [13] and breadth-first search [14].

Diao *et al.* [15] proposed a community structure identifier that uses modularity (*M*) as an indicator to "quantify the quality of the graph division into communities" [16]. This approach merges vertices so that communities are aggregated with the greatest increment of modularity (Δ*M*). In their follow-up paper [17], they proposed a methodology to decompose a WDS into a twin-simplified pipeline structure comprising backbone mains and community feedlines.

Di Nardo *et al.* [18] used graph theory principles and a heuristic procedure for WDS sectorization based on minimizing the amount of dissipated power in a WDS. Giustolisi and Ridolfi [19] developed a multi-objective strategy for optimal WDS segmentation that maximizes a modified modularity index and minimizes the cost of newly-installed devices to obtain network segments. In summary, little effort has been devoted to the application of an optimization technique to the WDS component grouping, clustering and aggregation.

Few studies have considered node grouping as a method of decreasing the unknown dimensions for the WDS demand estimation. Kang and Lansey [2] compared the two following real-time demand estimation methods with respect to their WDS demand estimation accuracy: the tracking state estimator and the KF. They aggregated nodes in a study network into multiple groups to reduce the number of unknowns and to make the demand estimation problem overdetermined or even-determined, which was also suggested in [20]. Jung and Lansey [8] employed the nodal demand aggregation approach of Kang and Lansey in their KF-based WDS pipe burst detection model. They then demonstrated that using the same number of node groups as the number of flowmeters yielded the best accuracy with the KF-based demand estimation scheme.

Note that the KF-based estimation method has been used for real-time WDS demand estimation [2,8], whereas optimization techniques have been widely used for demand and its hourly pattern calibration, which estimates base demands and leakage (non-real-time estimation) [21–24]. Less estimation time is generally allowed for the former (e.g., 5 min), which results in using different methodologies for the two different estimation purposes. The base demand and its hourly pattern are not to be determined in the real-time demand estimation. The commonality is in the use of a hydraulic model (e.g., EPANET). Accordingly, the demand calibration methodologies solve a non-linear system equation to quantify the accuracy of possible solutions, whereas the KF-based method solves a network equation to obtain pipe flow rates compared to field measurements.

Node grouping is one of the most important factors affecting the accuracy of the WDS demand estimation. However, previous studies have determined node groups based on engineering knowledge/sense (e.g., grouping nodes with the same demand pattern or in proximity) or assumed that they are given. Identifying optimal node groups using such approaches is very difficult considering the complex hydraulic relationship between the pipe flows/nodal pressure at sensor locations and the demand of node groups, especially in actual large networks (mostly loop-dominated). Note that the task is more difficult when sensors are not located at the best points for demand estimation, which is generally not the main concern during the sensor network design. Therefore, an optimization-based approach to find the optimal node groups for a highly accurate WDS demand estimation is required.

This study proposes an optimal node grouping (ONG) model to maximize the accuracy of the real-time WDS demand estimation. The KF-based demand estimation method is linked with a genetic algorithm (GA) for node group optimization. The modified Austin network demand is estimated to demonstrate the proposed model's validity. True demands and field measurements are synthetically generated using a hydraulic model of the study network.

2. Methodology

This research proposes an ONG approach that minimizes the sum of the root-mean-square error (RMSE) for the estimated demands of a given number of node groups. The nodes should be aggregated to decrease the number of unknowns in the demand estimation problem. Figure 2 shows the structure of the proposed ONG model, which comprises three main submodules. Each submodule represents an important factor that affects the demand estimation accuracy.

The WDS demand estimation accuracy is affected by the available information, estimation method and node grouping. The amount of information from field measurements on the estimated demand is a function of the number and types of sensors and their locations. The sensor network layout (*i.e.*, number and location of sensors) is assumed to be provided ("sensor network" in Figure 2) because the demand estimation is not a primary concern for decisions on the number and type of sensors and their locations. The KF-based WDS demand estimation methodology proposed by Jung and Lansey [8] is employed in this study because of its high accuracy. Based on the potential node groupings provided from the optimization algorithm (GA) submodule and field measurements obtained from the sensor network, the KF-based method estimates the node group demands. The accuracy is then quantified using the RMSE. The RMSE serves as the fitness value of the node grouping optimization and is minimized to increase the WDS demand estimation accuracy.

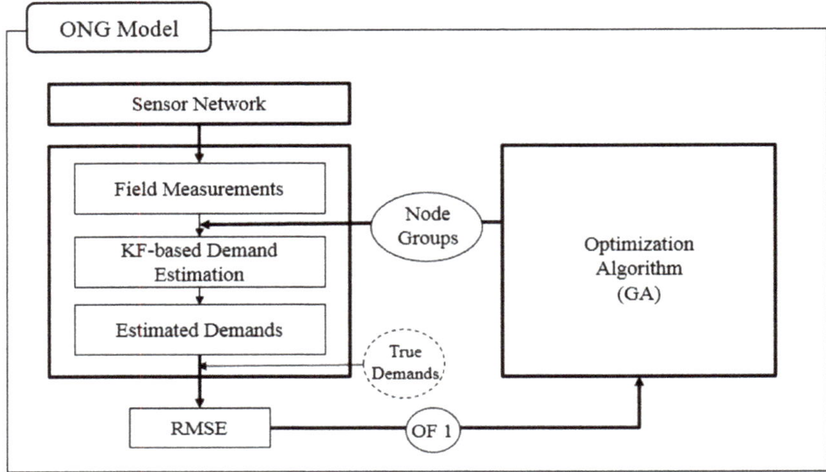

Figure 2. Structure of the proposed optimal node grouping (ONG) model. OF indicates objective function.

The following subsections detail the node aggregation, field measurements, KF-based demand estimation method, ONG model and GA. The blocks corresponding to each subsection are shown in Figure 2.

2.1. Node Groups and Demand Patterns

Each node can be classified into one of the node groups. However, the total number of groups is predefined. Each node is labeled with an integer number from 1 to the total number of node groups during optimization. For example, each node is indexed among integers 1–14 if the WDS has a total of 14 node groups. This number is set equal to the total number of flowmeters. Therefore, the decision variables of the proposed ONG problem are all integer values (Figure 3).

Figure 4 shows the five following general diurnal demand patterns considered by Kang and Lansey [2]: three residential, one industrial and one commercial. The three residential users are an apartment (Residential 1 in Figure 4), houses with half-acre lots and large home lots (Residential 2 and 3, respectively). The residential apartment demand is characterized by higher peaks early in the morning (6–7 a.m.) and evening (6–7 p.m.) than those of the other two residential demands. Note that a residential user with a large lot has the lowest peak factor and attenuated demand change during the day. The industrial water usage is relatively constant throughout the day, while the commercial demand sharply rises and falls at the start and end of the workday and is constant during the workday.

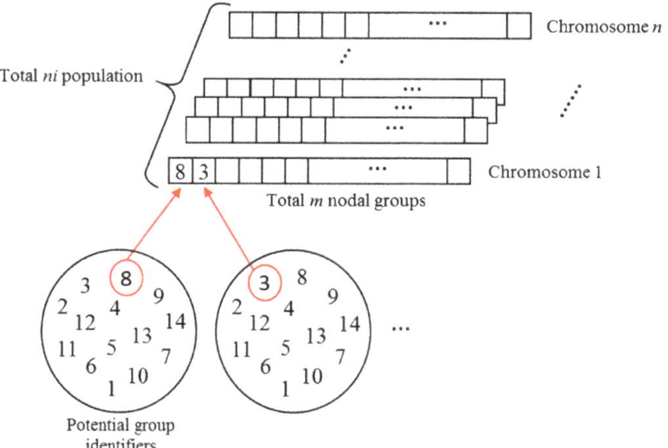

Figure 3. Decision variables of the proposed ONG model.

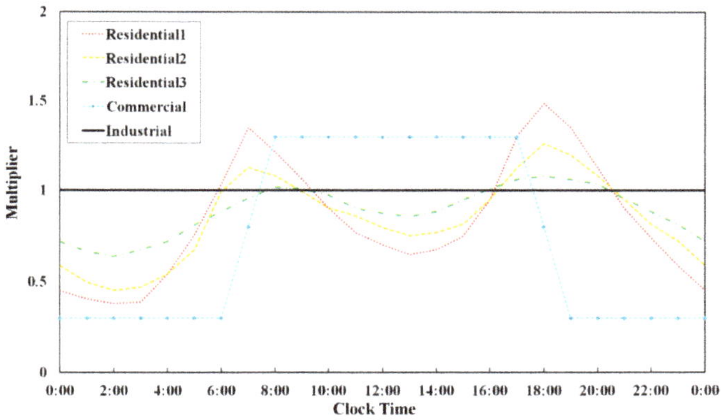

Figure 4. Diurnal demand curves for five user types: Residential 1 (apartments), Residential 2 (houses with 1/2-acre lots), Residential 3 (houses with large home lots), commercial and industrial.

All users are assumed to be in apartments (Residential 1), which is easily valid for highly dense urban areas. More complicated demand estimation problems can be formulated by randomly setting the user type for a node from the above five types, assuming that there is no spatial correlation in the demand patterns. However, this is not realistic. In the real world, demand patterns show strong spatial correlations (e.g., commercial districts are in the downtown area of a city).

2.2. Field Measurements

The KF-based demand estimation method estimates real-time node group demands using field measurements (e.g., pipe flow rate and nodal pressure) received from a supervisory control and data acquisition system. The pipe flow rate and nodal pressure are the output variables, whereas the node group demands are the input variable of the WDS. Note that the nodal demands can easily be calculated from the node group demands. Therefore, the KF-based method reverse-engineers the input variable, which cannot be directly measured, using the output variables, which can be measured by meters installed throughout the system.

In this study, the field datasets of the pipe flow rate are synthetically generated to assess the accuracy of the KF-based demand estimation method given the potential node groupings (Figure 2). The field measurements for the nodal pressure are assumed to not be utilized for the WDS demand estimation because low accuracy is observed when nodal pressure measurements are included [8].

The synthetic pipe flow rates are generated following these procedures: (1) an identical demand pattern (Residential 1 in Figure 4) is assigned to each node in the system (Figure 5a); (2) random deviations $N\left(0,\sigma_q\right)$ (*i.e.*, white noise) are added to each demand to consider the randomness and heterogeneous nature of true nodal demands (Figure 5b); (3) true node group demands are calculated by summing the demand of nodes in a group (Figure 5c); (4) the nodal demands generated in Step (2) are entered into an EPANET hydraulic model of the study network to generate true pipe flow rates; and (5) another white noise is added to each pipe flow rate to introduce measurement errors $N\left(0,\sigma_Q\right)$ (Figure 5d).

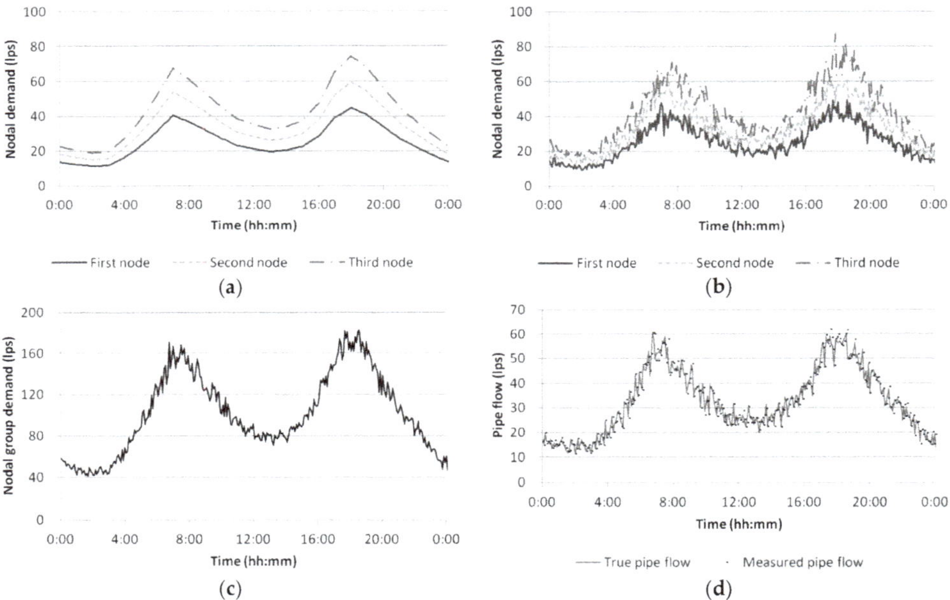

Figure 5. Synthetic field measurement generation steps: (**a**) an identical demand pattern (Residential 1) is assigned to each node (*i.e.*, time-varying demand factors are multiplied by the base demand); (**b**) random variability is added to the nodal demands; (**c**) the true node group demand (*i.e.*, the sum of the three nodes' demand generated in the previous step); and (**d**) the true pipe flow and the addition of measurement error.

While the assumption of the normal distribution for the nodal demand is common in WDS design [8,25,26], Surendran and Tota-Maharaj [27] have recently confirmed that normal and lognormal distributions are appropriate for WDS demand based on analyzing real daily water consumption data with a 15-min interval for four years in the U.K. Auto-correlated noise (e.g., a sports event lasting for 2 h) can be considered for Step (2) to simulate more realistic demand fluctuations [28,29]. Note that the effect of considering different noise types on the accuracy of the KF-based demand estimation method would be minimal because it estimates the state variable, such that the error covariance is minimized.

2.3. KF-Based Demand Estimation

The KF-based demand estimation method links the field measurements with the hydraulic simulation model (EPANET) that solves the non-linear system equation comprising mass and energy conservations [2]. Before demand estimation, the nodes are grouped into a single demand (*i.e.*, nodal group demand) following the grouping from the optimization submodule (Figure 2) to overcome the limitation of low measurement redundancy. The field pipe flow data are measured in real time. First, aggregated demands are estimated using the final demand estimates of the previous time step. The estimated group demands are disaggregated to individual nodes, which are entered into the hydraulic simulation model to calculate the pipe flow rate estimates. The demand estimates are updated such that the error covariance is minimized using the gap between the field measurement and the estimate of the pipe flow rate.

The KF comprises the recursive implementation of forecast and update steps, such that the *a posteriori* error covariance is minimized. The forecast step estimates the state at the current time step using the state estimate from the previous time step. The update step refines the forecast based on measurements at the current time step for a more accurate state estimate in the current time step. When calculating the state estimate in the update stage, more weight is given to the state estimate with the smaller uncertainty between those from the forecast stage and those from the current measurement [30].

The KF has the advantage of embedding the system dynamics in the estimates, which enables it to consider system operational changes when estimating the demand for a set of nodes (node group demands) based on the measured pipe flows.

The two types of KF are linear (LKF) and non-linear (NKF). The NKF represents the full non-linear relationship between measurements and states, whereas the LKF employs the linearization of these functions. The NKF is used herein because of its higher accuracy for the state estimation of a WDS (*i.e.*, a non-linear system) than the LKF [8].

The state forecast $\mathbf{x}_k^-$ (*i.e.*, the *a priori* state estimate of the node group demands) is computed using the state equation as follows:

$$\mathbf{x}_k^- = \mathbf{A}_k \mathbf{x}_{k-1}^+ + \mathbf{w}_k, \quad \mathbf{w}_k \sim N\left[0, \mathbf{Q}_k\right] \tag{1}$$

where $\mathbf{A}_k$ relates the state at the previous time step $k-1$ to the state at the current step k. This matrix is updated at each time step and calculated from the historical mean node group demand values; $\mathbf{w}_k$ is a random variable representing the process noise; and $\mathbf{Q}_k$ is the process noise covariance. Note that node grouping is provided from the optimization algorithm submodule (Figure 2) based on the methodology described in Section 2.1.

In the NKF, the non-linear system function h in the measurement equation relates the *a priori* state estimate ($\mathbf{x}_k^-$) and exogenous variable ($\mathbf{u}_k$, operational information) to the measurements (pipe flows) as follows:

$$\mathbf{z}_k = h\left(\mathbf{x}_k^-, \mathbf{u}_k\right) + \mathbf{v}_k, \quad \mathbf{v}_k \sim N\left[0, \mathbf{R}_k\right] \tag{2}$$

where $\mathbf{z}_k$ denotes the measurement variables; $\mathbf{v}_k$ is a random variable representing the measurement noise; and $\mathbf{R}_k$ is the measurement noise covariance.

The updated state estimate (*i.e.*, the *a posteriori* state estimate of the demand) is given as:

$$\mathbf{x}_k^+ = \mathbf{x}_k^- + \mathbf{K}_k\left(\mathbf{z}_k - h\left(\mathbf{x}_k^-, \mathbf{u}_k\right)\right) \tag{3}$$

where $\mathbf{K}_k$ is the Kalman gain matrix expressed as follows:

$$\mathbf{K}_k = \mathbf{P}_k^- \mathbf{H}_{uk}^T \left(\mathbf{H}_{uk} \mathbf{P}_k^- \mathbf{H}_{uk}^T + \mathbf{R}_k\right)^{-1} \tag{4}$$

where $\mathbf{H}_{uk}$ is the Jacobian matrix of the partial derivatives of h with respect to x and u, and a unique $\mathbf{H}_{uk}$ is computed for each system operational state $(\mathbf{u}_k)$ around the *a priori* state estimate $\mathbf{x}_k^-$; and $\mathbf{P}_k^-$ is the *a priori* estimate error covariance calculated as follows:

$$\mathbf{P}_k^- = \mathbf{A}_k \mathbf{P}_{k-1}^+ \mathbf{A}_k^T + \mathbf{Q}_k \tag{5}$$

Therefore, the *a posteriori* state estimate $\mathbf{x}_k^+$ is computed as a linear combination of the *a priori* estimate $\mathbf{x}_k^-$ and the weighted difference between the actual $\mathbf{z}_k$ and predicted $h\left(\mathbf{x}_k^-, \mathbf{u}_k\right)$ measurements. For example, a large measurement error covariance $\mathbf{R}_k$ results in a small update correction to the forecast state vector $\mathbf{x}_k^-$.

The *a posteriori* estimate error covariance is finally calculated as follows:

$$\mathbf{P}_k^+ = \left(\mathbf{I} - \mathbf{K}_k \mathbf{H}_{uk}\right) \mathbf{P}_k^- \tag{6}$$

The LKF uses the linear transition matrix $\mathbf{H}$ in Equations 2 ($\mathbf{z}_k = \mathbf{H}\mathbf{x}_k^- + \mathbf{v}_k$) and 3 ($\mathbf{x}_k^+ = \mathbf{x}_k^- + \mathbf{K}_k \left(\mathbf{z}_k - \mathbf{H}\mathbf{x}_k^-\right)$). The LKF can be used for non-linear systems $h(x)$ with weak non-linearity, but may perform poorly as the non-linearities increase. Both the LKF and NKF use first-order approximations for the error covariance propagation ($\mathbf{H}_{uk}\mathbf{P}_k^-\mathbf{H}_{uk}^T$ in Equation 4 and $\mathbf{A}_k\mathbf{P}_{k-1}^+\mathbf{A}_k^T$ in Equation 5). A perturbation method, wherein the derivatives are approximated by the numerical forward finite differences, is used to calculate $\mathbf{H}_{uk}$.

Note that a KF is classified as linear or non-linear based on the type of underlying state and measurement equation and not on the first-order approximation. The extended KF can only be applied to an NKF that estimates $\mathbf{w}_k$ and $\mathbf{v}_k$ in Equations 1 and 2, respectively, along with the state.

2.4. Demand Estimation Accuracy Indicator: RMSE

The state estimate includes errors because of measurement and model errors and the state variable randomness. Measurement errors can originate from deterioration and imperfect calibration of meters and delays in data communication (e.g., missing data). Model errors originate from model parameter uncertainties and a lack of system knowledge (e.g., erroneous system structure). This study aims to find the optimal node grouping that minimizes the model errors when estimating the WDS node group demands. The RMSE is used as an indicator of the WDS demand estimation accuracy (Figure 2). The RMSE measures the difference between values predicted by a model or estimator (KF) and the true values and is calculated as follows:

$$RMSE = \sqrt{\frac{\sum_{k=1}^{nt} \left(E_k - O_k\right)^2}{nt}} \tag{7}$$

where nt is the total number of time steps; E_k is the estimated group demand at time step k (obtained from the KF-based demand estimation method described in Section 2.3); and O_k is the true group demand at the k-th time step (synthetic measurements are obtained by the methodology described in Section 2.2).

2.5. Elitism-Based GA

A GA [31] is a metaheuristic optimization algorithm that mimics the process of natural selection (*i.e.*, "the survival of the fittest"). A standard GA (SGA) generally begins with a randomly-generated initial population (*i.e.*, group of potential solutions). A pair of solutions is selected with a high probability to have high fitness (selection process), which is then subjected to crossover and mutation processes that share their genetic traits (*i.e.*, decision variables) and modify a few of them, respectively. This series of processes (selection–crossover–mutation) is called an "iteration." This series is continued until the child population is filled. Better fitness solutions tend to appear in the population over iterations.

Various versions of the GA have been released after Holland [31] first introduced the SGA. These versions include the elitism GA, greedy GA, adaptive GA and refined GA [32–35]. One of the reasons for such a massive number of releases is that SGA is not capable of keeping the current best solution for the next generation. In other words, SGA is inefficient at considering new and good solutions found in the iterations. Therefore, an elitism-based GA (eGA) is proposed to optimize the WDS node grouping problem.

The proposed eGA selects two solutions from the parent population through roulette wheel selection, where the probability of a solution to be selected is proportional to its fitness. Similar to the general GA, a high fitness solution has a better chance to be selected. The general crossover and mutation processes are then performed at frequencies defined by the crossover and mutation rates. Crossover generally occurs with a probability of 70%–90%, while mutation happens with a probability of 2%–10%. The resulting two solutions are called children solutions. The two new children solutions compete with the selected parent solutions in a tournament with respect to fitness to determine whether any of the two children solutions can replace the parent solution(s). Therefore, one iteration of eGA requires only two functional evaluations, whereas the SGA requires ni number of functional evaluations, where ni is the total number of individuals (*i.e.*, solutions) in the population. The prompt inclusion of newly-found good solutions in the eGA improves the search efficiency of the SGA because the solution can be considered as a parent solution for selection in the immediately following iteration. Chromosomes are integer-coded (*i.e.*, decision variables are in integers) in the eGA (Figure 3) for the node grouping optimization in the WDS demand estimation.

2.6. Optimal Node Grouping Model

While the number of unknown variables (*i.e.*, demands) should be reduced because of the lack of available information, an accurate demand estimation can be achieved when each pipe flowmeter is linked to an appropriate node group, which is usually the group of nodes in proximity to the meter, whose demand affects the pipe flow rate. Therefore, given the layout of pipe flowmeters, finding the optimal node groups plays a very important role in determining the WDS demand estimation accuracy. Provided that there is no spatial proximity information between the meters and the nodes, the proposed optimal node group model finds the optimal node groups that minimize the total RMSE of the estimated node group demand using the KF-based method given the total number of node groups:

$$\text{Minimize } F = \sum_{i=1}^{ng} RMSE_i \tag{8}$$

$$\text{subject to } ng = m \tag{9}$$

where $RMSE_i$ is the RMSE of the i-th node group; ng is the total number of node groups ($i = 1, 2, ..., ng$); and m is the total number of node groups predefined by the user generally equal to the total number of pipe flowmeters in the WDS.

Note that pressure constraints are not included in the proposed ONG problem, which differs from other optimization problems (e.g., WDS design problem) developed in the WDS field. However, the synthetic demand generated by the methodologies described in Sections 2.1 and 2.2 results in nodal pressures in a normal operation range of 21–28 m (30–40 psi).

3. Application

The proposed ONG model is applied to estimate the Austin network demand [36] with modifications [8]. The Austin network is a non-district metering area (DMA) structured loop-dominated network. As shown in Figure 6, the modified system comprises 126 nodes and 90 pipes supplied by two reservoirs and one pumping station. Similar to the original network, the modified Austin network is solely supplied by a single reservoir (Source 1) with a fixed source head

(306 m). The second reservoir (total head = 366 m) in the middle of the loop in the east corner of the system is not operated.

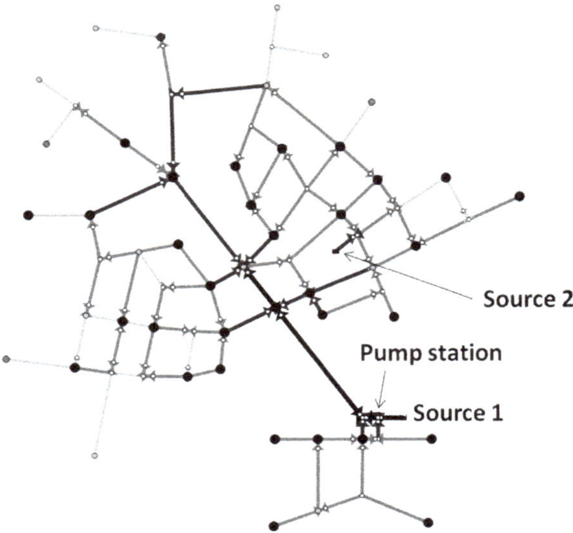

Figure 6. Study network layout. A thicker pipe represents a larger diameter, while a larger node represents a larger demand.

Only 47 nodes have external demands. Therefore, the total number of unknown demand nodes is 47. The rest of the nodes (*i.e.*, 79 nodes) are pipe connections with no external demands. A total of 14 pipe flowmeters are assumed to be installed throughout the system. The meters are assumed to be installed at the same locations selected by Jung and Lansey [8]. The number of node groups is set equal to the number of meters to realize high demand estimation accuracy. Therefore, the 47 demand nodes are assumed to be aggregated into 14 node groups (*i.e.*, $m = 14$).

A two-day series of node group demands and pipe flow rates is generated at 5-min time steps using the network's hydraulic model based on the methodologies described in Section 2.1. Similar to Jung and Lansey [8], the Kang and Lansey approach [2] is applied to demand aggregation and disaggregation. The method is briefly described here. The forecast node group demands at the current time (aggregated demands) are estimated from the transition of the updated demand of the previous time step (Equation (1)). The forecast node group demands are then disaggregated to individual nodes under the assumption that the nodal demands (disaggregated demands) in the same group are perfectly correlated. During the disaggregation, the ratio of the individual nodes' base demands to the total node group base demand is multiplied by the forecast node group demands to obtain the individual nodal demands.

The demands are normally distributed with a coefficient of variation (CV) of 0.1 ($\sigma_q = 0.1\mu_q$). The measurement errors are random variables $N(0,\sigma)$, where $\sigma_Q = 0.1\mu_Q/3.27$ (corresponding to a measurement error of 10%) for pipe flow measurements. All nodes are considered to be residential users in apartments.

eGA is used to find an optimal solution for the ONG problem, wherein the number of possible solutions is 7.379×10^{53} (= 14^{47}). Note that the well-known Hanoi network design problem has 2.865×10^{26} (6^{34}) possible solutions [37]. The crossover and mutation processes are conducted with probabilities of 85% and 5%, respectively. The genetic traits of the chromosomes are shared at multiple scattered points, while the standard mutation is employed. The node grouping determined by the engineering decision of Jung and Lansey [8] (*i.e.*, a node group comprising nodes of the same user type) is seeded as an initial solution, whereas the other initial solutions, where the population is 100 (ni = 100), are randomly generated by uniform sampling among 1–14 integer values. The eGA returns the best solution when the solution is not changed over 300 iterations.

4. Application Results

The KF-based method was used to consider a two-day series of pipe flow rates measured at 14 meter locations for node group demand estimation. The RMSE for each group (Equation (7)) were calculated by comparing the synthetically-generated node group demands for two days (both the pipe flow rates and node group demands were in 5-min time steps) with the estimated values. The erroneous node groups resulted in the divergence of the state estimates, which mostly originated from a badly-scaled or (nearly) singular matrix in Equation (4) (*i.e.*, the determinant of the matrix $(\mathbf{H}_{uk}\mathbf{P}_k^-\mathbf{H}_{uk}^T + \mathbf{R}_k)$ to be inversed was close to zero). In the optimization of the proposed ONG model, a high penalty value (e.g., 10,000 × 14 = 140,000) was added to the objective function value (*i.e.*, RMSE) in Equation 8 if a matrix was ill-conditioned. The demand estimation was terminated at the time step to speed up the optimization when the ill-conditioned matrix was identified.

Note that the KF-based demand estimation method estimated the nodal group demands only using pipe flow measurements without being given any information/values of the true nodal group demands. In this study, the true nodal (group) demands were synthetically generated and entered into the WDS system equations to obtain pipe flow rates at locations, to which the measurement error was added to finally produce the pipe flow measurements. Processed using the non-linear governing equations and added noises, the pipe flow measurements did not contain clear clues for tracking the true nodal group demands. Therefore, estimating the nodal group demands was not a circular numerical calculation.

First, the differences between the node groupings of the initial and final optimal solutions were investigated. Then, the spatial distributions of the nodes in a group and node groups were examined along with whether or not there was a one-to-one relationship between a meter and a node group. The RMSE values of the individual node group demand estimates were determined. Conclusions were drawn on the demand estimation error, base demand of the node group and meter locations.

4.1. Optimal Node Grouping Results

Figure 7 shows the trajectory of the best fitness value (*i.e.*, sum of node group RMSEs) over the iterations. Note that the sum of the RMSEs reported by Jung and Lansey [8] was 74.1 L/s. They determined the node grouping based on engineering judgment. The best RMSE with the proposed ONG model was 56.2 L/s, which was about 76% of the RMSE obtained by Jung and Lansey [8] (Figure 7). Jung and Lansey [8] included seven nodes (out of 47 demand nodes) with industrial and commercial demands, which slightly complicated their demand estimation. Other conditions (e.g., meter locations and measurement time interval) were the same as in this study.

Most solutions found in the early stage of the optimization had diverging demand estimates (*i.e.*, infeasible solutions with a penalty value), which resulted in the solution's fitness value of 140,000 (not included in Figure 7). The main reason for the divergence was the scattered distribution of node groups (Figure 8). Generally, the highest accuracy is achieved when a group of gathered nodes had a one-to-one relationship with a meter in close proximity (opposite condition to the node grouping in the infeasible solutions). A feasible solution was found after a few iterations, and step decreases in the fitness value were observed until the best fitness was reached at the optimal value of 56.2 L/s. The results for the individual node group demands were discussed in Section 4.2.

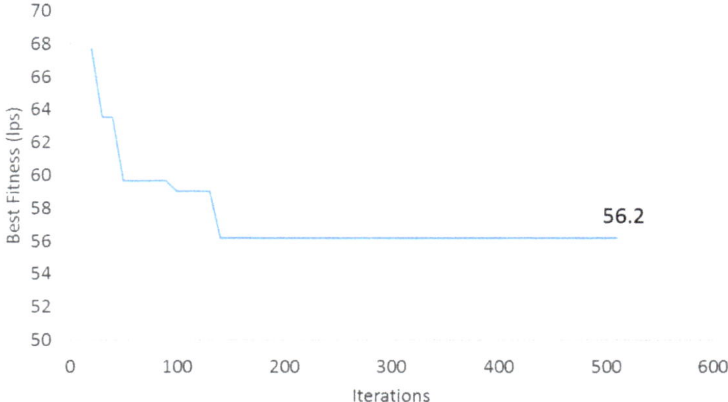

Figure 7. Trajectory of the fitness value of the best solution.

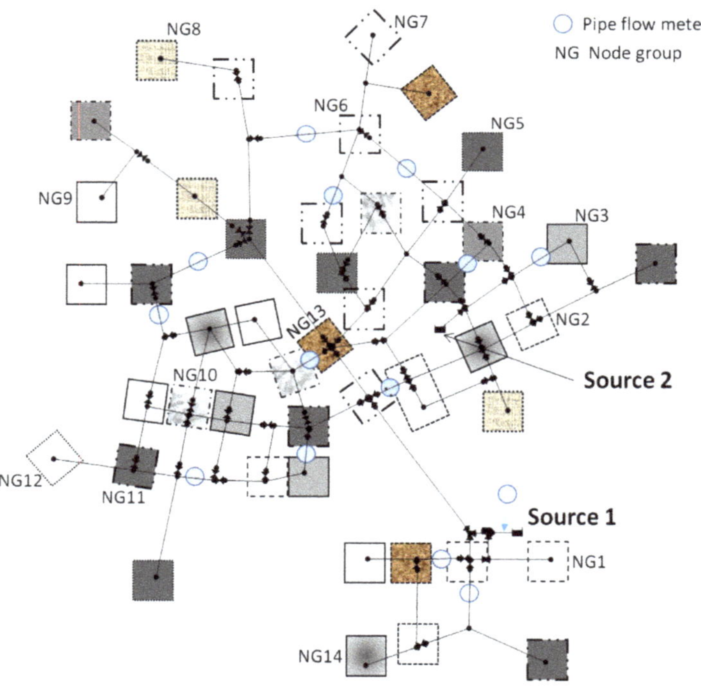

Figure 8. Node groups of a representative initial random solution and meter locations. Each node group is delineated by either a different shape outline or filled by different colors/styles.

Figure 9 shows the optimal node grouping layout found by the proposed node grouping model. In contrast to the representative initial solution shown in Figure 8, the optimal solution spatially gathered nodes in each group. For example, Node Group (NG) 14 comprised three nodes in proximity at the south end of the study network, whereas the five nodes at the north end composed NG8. NG4 only had one node. Some node groups (e.g., NG3, NG6 and N10) comprised nodes that were not close to each other (*i.e.*, slightly scattered), which was mostly caused by the existence of node(s) with no external demand between the nodes. For example, there was a zero-demand node between the leftmost and middle nodes in NG3. In addition, the three zero-demand nodes near NG11 were also worthy of being seen (Figure 9).

Figure 9. Optimal node groups identified by the proposed model. A node that is not grouped (boxed) has no external demand. Therefore, demand estimation is not required.

Figure 9 also shows the location of the 14 meters installed in the study network. Note that not only was a meter located at the transmission mains (*e.g.*, the source pipe linked to Source 1 and two pipes on the right and left sides of the NG13 box), but one was also installed in the distribution pipes within the loops at the northeast and southwest parts of the network. Finding an apparent one-to-one relationship between a node group and a meter was very difficult because of the complex hydraulic relationship of a looped network. Compared to a simple branched network, the response of the pipe flow rate at the meter location to a perturbation in the node group demand, which was information required for demand estimation, cannot be easily identified by visual inspection or engineering knowledge/sense in a looped network. Therefore, this result highlighted the need for the proposed model when trying to find optimal node groups that result in a highly accurate WDS demand estimation.

Multiple optimal (or near-optimal) solutions for the ONG problem of real large networks could be employed. More preference would be given to the solution with a high tendency of classifying nodes in proximity or of the same user type into the same group.

4.2. Accuracy of Individual Nodal Group Demand Estimation

Table 1 summarizes the RMSE of the individual node group demand estimates. Figure 10 plots the actual and estimated demands of representative multiple node groups (*i.e.*, NG1, 4, 7, 8, 12 and 13). The non-linear KF was used as the unbiased minimum-variance estimator to minimize the error between two values. While accurate demand estimation was achieved in most node groups, the highest RMSE was obtained in NG13 followed by NG1 (Table 1 and Figure 10a,f, respectively). The latter mainly originated from the fact that NG1 had the largest base demand among the node groups (see the range of the *y*-axis in Figure 10a compared to the other plots in Figure 10). The node group with the largest base demand should have the largest error value if the proportion of error to the base demand value was assumed to stay similar or the same. In contrast, the former was caused by the lack of available information/signals for the demand estimation.

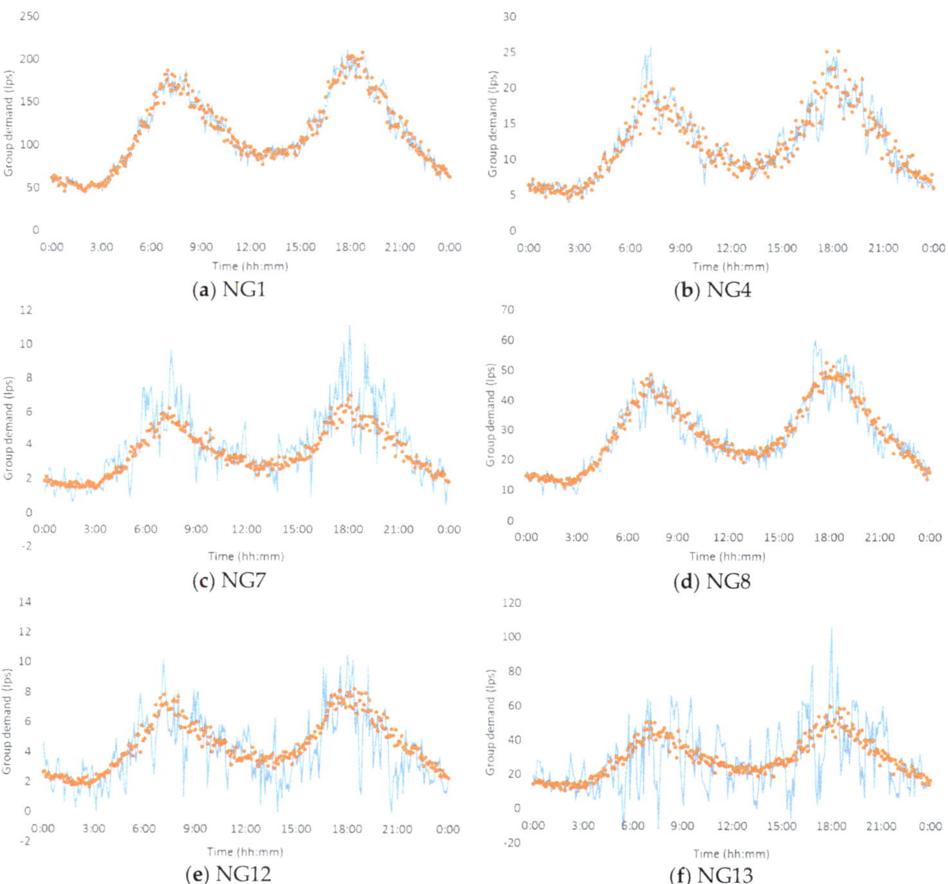

Figure 10. Actual (circle) and estimated (line) group demands at 5-min time steps for the first 24 h: (**a**) NG1; (**b**) NG4; (**c**) NG7; (**d**) NG8; (**e**) NG12; and (**f**) NG13 (NG = node group).

Table 1. Node group demand estimation RMSE (sum of node group demands' RMSE = 56.2 L/s).

Node Group (NG)	1	2	3	4	5	6	7
RMSE (L/s)	7.7	4.9	2.4	1.8	4.4	2.0	1.2
Node group (NG)	8	9	10	11	12	13	14
RMSE (L/s)	4.3	1.9	3.8	4.2	1.6	15.0	1.0

The nodes in NG13 were located along the transmission pipes that delivered bulk flows to supply the north, northeast and southwest parts of the study network (Figures 6 and 9). While only a meter located at the source pipe linked to Source 1 could measure NG13's demand (with other demands), the proportion of the group demand was very small compared to the total pipe flow (*i.e.*, the total system demand because Source 2 was not operated) measured at the meter. NG13's demand was less than the natural randomness of the total system demand (its base value was 726 L/s), which made it difficult to extract useful information from the measurements. This result led to the fluctuating demand with large deviation from the actual demand value (Figure 10f).

Similarly, fluctuating demand estimates with large deviations from the actual values were observed in node groups with a small base demand (e.g., NG7 and 12 in Figure 10c and e, respectively). In other words, the proportion of error seemed higher in these groups than the others. This result implied that the demand estimation was not reliable and difficult to utilize for management purposes. Such cases were avoided by placing meters at transmission pipes and distributing them among small pipes to capture the pipe flow information of various magnitudes, which was required for WDS demand estimation. The examples included a conventional meter located at the end of a branched network (delivering demand to one or two pendant nodes) or an AMI installed at the inlet pipe to each household.

5. Summary and Conclusions

State estimation is defined as the process of calculating a state variable of interest that is not directly measured. In a WDS, nodal demands are considered as the state variable (*i.e.*, an unknown variable that cannot be directly measured) and can be estimated using nodal pressures and pipe flow rates measured at sensors installed throughout the system. The number of unknown variables should be minimized given the lack of available information for demand estimation. An alternative way is to group a set of nodes with a correlation (e.g., the same demand pattern and proximity). Finding the optimal node grouping that results in the best demand estimation accuracy is a challenging task given the number and location of meters and complex hydraulic conditions.

This study proposes an ONG model that minimizes the sum of the RMSEs of the node groups' estimated demand given the number of node groups. The KF-based demand estimation method is linked with a genetic algorithm for node group optimization. The proposed model is applied to estimate the modified Austin network demand as a demonstration. The true demands and field measurements are synthetically generated using a hydraulic model of the study network. Note that the proposed model is the first to combine a WDS demand estimation tool (*i.e.*, the KF-based model) and a node clustering/aggregation optimization model (*i.e.*, eGA).

The sum of the RMSEs of the final best solution found with the proposed model is 56.2 L/s, which is about 76% of the value obtained by Jung and Lansey [8] based on their engineering judgment. In contrast to the randomly-generated initial solutions, the nodes are spatially grouped in the final solution. However, no apparent one-to-one linkage (mapping) is observed between a meter and a node group. This result indicates that the proposed ONG model should be applied to identify the best node grouping for the demand estimation of the loop-dominated networks. Among individual node groups, a high RMSE is obtained for the node group with the largest base demand. In addition, the proportion of the error to the true demand is large for node groups with a small base demand. Therefore, micrometers (*i.e.*, meters that can measure a small flow, such as a conventional meter installed at pipes to dead-end nodes or AMIs) should be installed to further increase the demand estimation accuracy.

The work in this study has several limitations that future research must address. First, this model finds the optimal node grouping given that the number of node groups is predefined as equal to the number of meters. The model can be extended to include the number of node groups as a decision variable. Assuming that the same number of meters should be installed for an accurate demand estimation, the demand estimation problem can be formulated as a multi-objective problem that minimizes the sum of RMSEs and minimizes the cost of meters (*i.e.*, meter instrument and installation costs). Many meters should be installed to achieve high demand estimation accuracy, which requires a large investment. Such a tradeoff relationship can be explored by solving the multi-objective demand estimation and meter placement problem.

Second, the proposed model can be verified using real demand data measured in a real large network fully equipped with AMI/automatic water meter reading.

Third, an advanced warm-start approach to the initial solution should be developed to shorten the time of finding a feasible solution in the early optimization phase. The Euclidean distance among nodes can be considered for the initial node grouping to avoid spatially-distributed nodes in a group. This study uses a loop-dominated network to demonstrate the proposed model. Different network types (e.g., branched or DMA-structured) can be used to check (1) whether the conclusions of this study are still valid or not and (2) whether such network layouts affect the demand estimation accuracy and optimal node grouping or not. Finally, the proposed model can serve as a submodule for WDS operation and management tools (e.g., a real-time WDS operation model that determines the status of pumping units given the estimated future demand).

Acknowledgments: This work was supported by a grant from the National Research Foundation (NRF) of Korea funded by the Korean government (MSIP) (No. 2013R1A2A1A01013886).

Author Contributions: Donghwi Jung and Young Hwan Choi carried out the survey of previous studies and wrote the draft of the manuscript. Donghwi Jung made Matlab codes of the proposed ONG model and performed the optimization runs. Joong Hoon Kim and Donghwi Jung conceived the original idea of the proposed model and revised the draft to the final manuscript.

Conflicts of Interest: The authors declare no conflict of interest.

References

1. Bargiela, A.; Hainsworth, G.D. Pressure and flow uncertainty in water systems. *J. Water Resour. Plan. Manag.* **1989**, *115*, 212–229. [CrossRef]
2. Kang, D.; Lansey, K. Real-time demand estimation and confidence limit analysis for water distribution systems. *J. Hydraul. Eng.* **2009**, *135*, 825–837. [CrossRef]
3. Shvartser, L.; Shamir, U.; Feldman, M. Forecasting hourly water demands by pattern recognition approach. *J. Water Resour. Plan. Manag.* **1993**, *119*, 611–627. [CrossRef]
4. Homwongs, C.; Sastri, T.; Foster, J.W., III. Adaptive forecasting of hourly municipal water consumption. *J. Water Resour. Plan. Manag.* **1994**, *120*, 888–905. [CrossRef]
5. Zhou, S.L.; McMahon, T.A.; Walton, A.; Lewis, J. Forecasting daily urban water demand: A case study of Melbourne. *J. Hydrol.* **2000**, *236*, 153–164. [CrossRef]
6. Ghiassi, M.; Zimbra, D.K.; Saidane, H. Urban water demand forecasting with a dynamic artificial neural network model. *J. Water Resour. Plan. Manag.* **2008**, *134*, 138–146. [CrossRef]
7. Rossman, L.A. *EPANET 2 User's Manual*; US Environmental Protection Agency (USEPA): Cincinnati, OH, USA, 2000.
8. Jung, D.; Lansey, K. Water distribution system burst detection using a nonlinear Kalman filter. *J. Water Resour. Plan. Manag.* **2014**, *141*. [CrossRef]
9. Mallick, K.N.; Ahmed, I.; Tickle, K.S.; Lansey, K.E. Determining pipe groupings for water distribution networks. *J. Water Resour. Plan. Manag.* **2002**, *128*, 130–139. [CrossRef]
10. Deuerlein, J.W. Decomposition model of a general water supply network graph. *J. Hydraul. Eng.* **2008**, *134*, 822–832. [CrossRef]
11. Perelman, L.; Ostfeld, A. Water-distribution systems simplifications through clustering. *J. Water Resour. Plan. Manag.* **2012**, *138*, 218–229. [CrossRef]
12. Perelman, L.; Ostfeld, A. Topological clustering for water distribution systems analysis. *Environ. Model. Softw.* **2011**, *26*, 969–972. [CrossRef]
13. Tarjan, R. Depth-first search and linear graph algorithms. *SIAM J. Comput.* **1972**, *1*, 146–160. [CrossRef]
14. Pohl, I. Bi-directional and Heuristic Search in Path Problems. Ph.D. Thesis, Department of Computer Science, Stanford University, Stanford, CA, USA, 1969.

15. Diao, K.; Zhou, Y.; Rauch, W. Automated creation of district metered area boundaries in water distribution systems. *J. Water Resour. Plan. Manag.* **2013**, *139*, 184–190. [CrossRef]

16. Novák, P.; Neumann, P.; Macas, J. Graph-based clustering and characterization of repetitive sequences in next-generation sequencing data. *BMC Bioinform.* **2010**, *11*. [CrossRef] [PubMed]

17. Diao, K.; Fu, G.; Farmani, R.; Guidolin, M.; Butler, D. Twin-hierarchy decomposition for optimal design of water distribution systems. *J. Water Resour. Plan. Manag.* **2015**, *142*. [CrossRef]

18. Di Nardo, A.; Di Natale, M.; Santonastaso, G.F. A comparison between different techniques for water network sectorization. *Water Sci. Technol. Water Supply* **2014**, *14*, 961–970. [CrossRef]

19. Giustolisi, O.; Ridolfi, L. New modularity-based approach to segmentation of water distribution networks. *J. Hydraul. Eng.* **2014**, *140*. [CrossRef]

20. Savic, D.A.; Kapelan, Z.S.; Jonkergouw, P.M. Quo vadis water distribution model calibration? *Urban Water J.* **2009**, *6*, 3–22. [CrossRef]

21. Lansey, K.E.; Basnet, C. Parameter estimation for water distribution networks. *J. Water Resour. Plan. Manag.* **1991**, *117*, 126–144. [CrossRef]

22. Vítkovský, J.P.; Simpson, A.R.; Lambert, M.F. Leak detection and calibration using transients and genetic algorithms. *J. Water Resour. Plan. Manag.* **2000**, *126*, 262–265. [CrossRef]

23. Dini, M.; Tabesh, M. A new method for simultaneous calibration of demand pattern and Hazen-Williams coefficients in water distribution systems. *Water Resour. Manag.* **2014**, *28*, 2021–2034. [CrossRef]

24. Di Nardo, A.; Di Natale, M.; Gisonni, C.; Iervolino, M. A genetic algorithm for demand pattern and leakage estimation in a water distribution network. *J. Water Supply Res. Technol. AQUA* **2015**, *64*, 35–46. [CrossRef]

25. Kapelan, Z.S.; Savic, D.A.; Walters, G.A. Multiobjective design of water distribution systems under uncertainty. *Water Resour. Res.* **2005**, *41*. [CrossRef]

26. Giustolisi, O.; Laucelli, D.; Colombo, A.F. Deterministic *versus* stochastic design of water distribution networks. *J. Water Resour. Plan. Manag.* **2009**, *135*, 117–127. [CrossRef]

27. Surendran, S.; Tota-Maharaj, K. Log logistic distribution to model water demand data. *Procedia Eng.* **2015**, *119*, 798–802. [CrossRef]

28. Aly, A.H.; Wanakule, N. Short-term forecasting for urban water consumption. *J. Water Resour. Plan. Manag.* **2004**, *130*, 405–410. [CrossRef]

29. van Zyl, J.E.; le Gat, Y.; Piller, O.; Walski, T.M. Impact of water demand parameters on the reliability of municipal storage tanks. *J. Water Resour. Plan. Manag.* **2012**, *138*, 553–561. [CrossRef]

30. Welch, G.; Bishop, G. *An Introduction to the Kalman Filter*; Department of Computer Science, University of North Carolina: Chapel Hill, NC, USA, 2006.

31. Holland, J.H. *Adaptation in Natural and Artificial Systems: An Introductory Analysis with Applications to Biology, Control, and Artificial Intelligence*; U Michigan Press: Ann Arbor, MI, USA, 1975.

32. Yang, S. Genetic algorithms with elitism-based immigrants for changing optimization problems. In *Applications of Evolutionary Computing*; Springer: Berlin, Germany, 2007; pp. 627–636.

33. Ahuja, R.K.; Orlin, J.B.; Tiwari, A. A greedy genetic algorithm for the quadratic assignment problem. *Comput. Oper. Res.* **2000**, *27*, 917–934. [CrossRef]

34. Wu, Q.H.; Cao, Y.J.; Wen, J.Y. Optimal reactive power dispatch using an adaptive genetic algorithm. *Int. J. Electr. Power Energy Syst.* **1998**, *20*, 563–569. [CrossRef]

35. Sheblé, G.B.; Brittig, K. Refined genetic algorithm-economic dispatch example. *IEEE Trans. Power Syst.* **1995**, *10*, 117–124. [CrossRef]

36. Brion, L.M.; Mays, L.W. Methodology for optimal operation of pumping stations in water distribution systems. *J. Hydraul. Eng.* **1991**, *117*, 1551–1569. [CrossRef]

37. Fujiwara, O.; Khang, D.B. A two-phase decomposition method for optimal design of looped water distribution networks. *Water Resour. Res.* **1990**, *26*, 539–549. [CrossRef]

Article

Energy Saving in a Water Supply Network by Coupling a Pump and a Pump As Turbine (PAT) in a Turbopump

Armando Carravetta [1], Lauro Antipodi [2], Umberto Golia [3] and Oreste Fecarotta [1,*]

[1] Department of Civil, Structure and Environmental Engineering, Università di Napoli Federico II, via Claudio, 21, 80125 Napoli, Italy; arcarrav@unina.it

[2] Caprari s.p.a., Via Emilia Ovest, 900, 41123 Modena, Italy; l.antipodi@caprari.it

[3] Department of Civil Engineering, Industrial Design, Environment and History, Second University of Naples, Real Casa dell'Annunziata, Via Roma, 9, 81031 Aversa, Italy; umbgolia@unina.t

* Correspondence: oreste.fecarotta@unina.it; Tel.: +39-81-768-3462

Academic Editor: Helena Ramos and Arjen Y. Hoekstra

Received: 28 October 2016; Accepted: 9 January 2017; Published: 20 January 2017

Abstract: The management of a water distribution network (WDN) is performed by valve and pump control, to regulate both the pressure and the discharge between certain limits. The energy that is usually merely dissipated by valves can instead be converted and used to partially supply the pumping stations. Pumps used as turbines (PAT) can be used in order to both reduce pressure and recover energy, with proven economic benefits. The direct coupling of the PAT shaft with the pump shaft in a PAT-pump turbocharger (P&P plant) allows us to transfer energy from the pressure control system to the pumping system without any electrical device. Based on experimental PAT and pump performance curves, P&P equations are given and P&P working conditions are simulated with reference to the operating conditions of a real water supply network. The annual energy saving demonstrates the economic relevance of the P&P plant.

Keywords: energy saving; Pump As Turbine (PAT); PAT and pump system (P&P); energy recovery; pumping

---

1. Introduction

The sustainable management of the water supply is becoming a major challenge within the framework of the water–energy–food nexus [1,2]. A reduction in water leakage and an efficient use of energy are considered essential in order to contain greenhouse gas emissions and to limit the impacts of climate change [3–5]. In the last few years, a number of research papers in literature have focused on new methodologies for the efficient management of a WDN [6]. Mini hydro power plants are a common practice in water transmission lines [7,8], where a stable discharge and pressure drop are present and can be exploited to produce energy. Pressure reducing valves (PRV) were first proposed to obtain optimal pressure values in water distribution lines, with the benefit of reducing the amount of water leakage [9,10]. Several methods for the optimal location of PRVs have been developed [11,12]. Recently, the substitution of PRVs with micro hydro power plants has been proposed [13–15]. In the presence of pumping systems, the strategy for energy cost savings and energy efficiency could be based on the increase in the hydraulic efficiency of the pumps and the electrical efficiency of motors; on the introduction of performance standards to which the pumps on the market must comply; on the use of variable frequency drivers to increase the efficiency performance of pumping groups working in variable conditions; and on the energetic assessment of energy use in the network [16,17].

Compensation tanks, which fix the pressure in a particular water district, are commonly inserted between water transmission and water distribution systems. At the tank intake, a residual pressure of the flow can be present and it is usually merely dissipated by a valve. In the presence of a large variability of ground elevation between the parts of the system, a second tank positioned at a higher elevation may be necessary in order to guarantee sufficient pressure to the users of this higher district. The hydraulic energy dissipated at the lower tank could be comparable with the energy to be supplied to the pump to deliver water to the higher tank. Therefore, a turbine could be placed at the tank inflow in order to both reduce pressure and recover energy to supply the pump. Three main problems limit the convenience of the installation of a turbine governed generator to power the pumping system, namely (i) the small amount of power available; (ii) the high cost of traditional turbines; and (iii) the low combined efficiency of the two electromechanical devices (the turbine generator and the pump motor) [18–20]. PATs could be employed instead of classic turbines, since they are considered a viable and flexible solution for energy production in a WDN, due to their lower cost and their generally acceptable efficiency [21–24]. In order to reduce the investment cost and increase the efficiency of the system, the PAT and a pump can be directly coupled and mounted on the same shaft without any intermediate generator or motor. The resulting PAT-pump turbocharger should be a viable solution. However, an appropriate P&P theory for water systems has, to the best of the authors' knowledge, not yet been developed. The main difficulty in the design of such a device derives from the variability of the rotational speed. While an asynchronous generator defines the rotational speed of the turbine, according to the electric grid frequency, the rotational speed of a turbocharger is unknown and influences the hydraulic behaviour of the whole system. In this paper, a complete P&P formulation is developed, and a complete mathematical model is given for the design of the P&P plant when the boundary conditions are assigned. In addition, a non-dimensional solution is presented to discuss the general behaviour of the P&P plant. The solutions show that the performance of the system depends on the number of stages of both the pump and the PAT and on the pump–PAT discharge ratio. Furthermore, the model has been applied to four different scenarios based on the experimental discharge pattern of a pressure control station in a water distribution network in southern Italy. The results are investigated in terms of pressure reduction and pumping discharge and head, to show the hydraulic potential of the P&P device. The annual energy saving is also calculated for each scenario.

2. State-of-the-Art

2.1. Machine Behaviour

Experimental tests can provide the performance curves of a pump or a PAT at a certain rotational speed value or, alternatively, CFD techniques can be employed. Such curves express the relations of head (H), mechanical power (P) and efficiency (η) with the discharge (Q), and can be expressed as follows:

$$H = H(Q)$$
$$P = P(Q) \tag{1}$$
$$\eta = \eta(Q)$$

Usually, a quadratic polynomial can be successfully used to fit the head data, while a cubic polynomial best explains the power data. The efficiency curve presents a maximum value η_B for the Q_B discharge. The working point (Q_B, H_B, P_B, η_B) is called the best efficiency point (BEP). Affinity laws can be used to calculate the performance of two similar machines, having D^I and D^{II} diameters, respectively, and rotating at different speeds. The performances of two homologous working points, (Q^I, H^I, P^I) and (Q^{II}, H^{II}, P^{II}), respectively, are related to both the ratios of the respective speeds, N^I and N^{II}, and the scale ratios, D^I/D^{II}, as follows:

$$\frac{Q^I}{Q^{II}} = \left(\frac{N^I}{N^{II}}\right)^1 \left(\frac{D^I}{D^{II}}\right)^3$$
$$\frac{H^I}{H^{II}} = \left(\frac{N^I}{N^{II}}\right)^2 \left(\frac{D^I}{D^{II}}\right)^2 \qquad (2)$$
$$\frac{P^I}{P^{II}} = \left(\frac{N^I}{N^{II}}\right)^3 \left(\frac{D^I}{D^{II}}\right)^5$$

The Equation (2) is generally used by pump manufacturers to predict the behaviour of their machines for different rotational speeds, even if some studies demonstrate that Equation (2) does not predict the real behaviour of turbomachines with a high accuracy in the whole range of rotational speeds [25,26]. Principally, according to affinity laws, the efficiency value of two homologous points is constant with speed. However, this result is not in agreement with the real behaviour of pumps. Simpson and Marchi [27] showed that the efficiency value at the BEP is attained only at a given optimal N^{max} speed value, while a decrease is observed as the speed diverges. Furthermore, a recent experimental study [28] on the behaviour of PATs has demonstrated that the use of affinity laws could lead to significant errors in the prediction of machine performance, up to 30% in the prediction of head and 100% in the prediction of efficiency. The authors proposed a new semi-theoretical model for semi axial machines for a better prediction of characteristic PAT performance curves. N^{max} was identified experimentally and related to geometrical pump parameters. Then, the BEP at a N^{II} rotational speed could be related to the BEP at N^{max} by a relaxation of the affinity equations (RAE):

$$\frac{Q_B^{II}}{Q_B^{max}} = f_1 \frac{N^{II}}{N^{max}}$$
$$\frac{H_B^{II}}{H_B^{max}} = f_2 \frac{N^{II}}{N^{max}} \qquad (3)$$
$$\frac{P_B^{II}}{P_B^{max}} = f_3 \frac{N^{II}}{N^{max}}$$

By experimental estimation of the functions f_1, f_2 and f_3 [28], a better agreement was found between the theoretical and experimental characteristic curves with a significant reduction in the prediction error.

2.2. Energy Recovery in Water Systems

Hydropower production in a water supply system is a viable option where the available hydraulic power is fairly large and constant, with a great economic benefit and a small environmental impact. Examples of such scenarios are found along water transmission lines, from the water source to the distribution centre, where several small hydropower projects are already active [29]. Hydropower plants could also be used in water distribution networks, as a replacement of a PRV. Nevertheless, such a plant would face a large variability of both discharge and head, due to the hourly variability of water users' demand, as well as the small amounts of available power [30,31]. Furthermore, the turbine head curve exhibits a behaviour which is completely different from the head drop curve of a regulating valve: the former increases with the discharge, while the latter decreases as the flow rate increases [32]. All these difficulties, together with the small amount of power available, hinder the use of classic turbines, due to their high price and their long payback period. Some studies demonstrate that the use of PATs, instead of classic turbines, to substitute the dissipation valves could be a convenient practice even if the available power is low, due to the low price and the mechanical simplicity of such devices [33]. Several studies have shown that the best efficiency of a PAT could be greater than 70% [21–23]. Ramos et al. [34] showed that the cost of a PAT is much smaller if compared with the cost of a classic turbine, with unit costs ranging between 200 €/kw and 400 €/kw, where the turbine cost ranges between 300 €/kw and

800 €/kW. Carravetta et al. [35] showed that the payback period of a PAT energy recovery system could be really short, ranging between 6 months and 3 years, while Fecarotta et al. [36] showed that the coupling of the pressure control strategy with an energy recovery strategy within a water network could be convenient if the valves are replaced by PATs, with high 10 year net present values. Other studies propose different solutions for the regulation of the PAT plant to address the large variability of hydraulic characteristics and to match the turbine head and discharge with the needs of the network. Carravetta et al. [37] propose regulating the PAT either by a hydraulic or electric regulation system. In the former case, two hydraulic valves are placed in a series–parallel circuit to dissipate the excess head when the turbine head is too low or bypasses a part of the discharge when the flow rate is high. In the latter case, an electronic speed driver adjusts the rotational speed of the PAT to modify its performance. In 2014 , Carravetta et al. [38] compared the effectiveness of the different installation schemes of a PAT plant. Other preliminary studies have included an analysis of the optimal location of a PAT within a WDN [36,39,40] to maximize the power production.

3. Pump and PAT Characteristic Curves

Experiments were performed in the Hydro Energy Laboratory (HELab) of the CESMA—University of Naples Federico II and in the qualification laboratory of Caprari Pumps Ltd. (Peterborough, UK). Pictures of the two laboratories are shown in Figure 1.

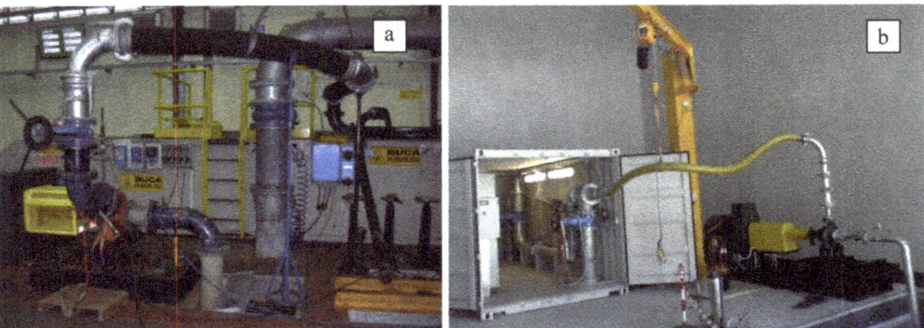

Figure 1. Pictures of the qualification laboratory of Caprari ltd (**a**); and of HELab of University of Naples (**b**).

Both laboratories are equipped with digital magnetic flow meters, piezometric pressure transducers and digital power meters. A fixed level water tank is located below ground level to feed the hydraulic circuit and receive the circulated water. A gate valve is used to regulate the flow. The testing conditions in both laboratories comply with ISO 9906 (Rotodynamic pumps—Hydraulic performance acceptance test) level 1, IEC 60034-2-1 (Rotating electrical machines) and EN 50598-2 (Efficiency classes of converters and drive systems). The maximum measurement uncertainties are reported in Table 1.

Table 1. Maximum measurement device uncertainties.

Measured Quantity	Maximum Uncertainty Value
Q	2.0%
H	1.5%
P	2.0%
η	3.2%

A two-stage pump unit, the Caprari model HMU, has been tested in the HELab. The specifications of the pump are reported in Table 2.

Table 2. Specification of the Caprari HMU pump, as tested in HELab.

Specification	Value	Unit
Nominal speed	2900	rpm
Q_B	14.2	L/s
H_B	78.9	m
η_B	73.0	%

The working conditions for different rotation speeds have been tested and the main hydraulic and electric parameters have been measured. Head and power curves have been determined experimentally. Test results in the normalized parameter are shown in Figure 2.

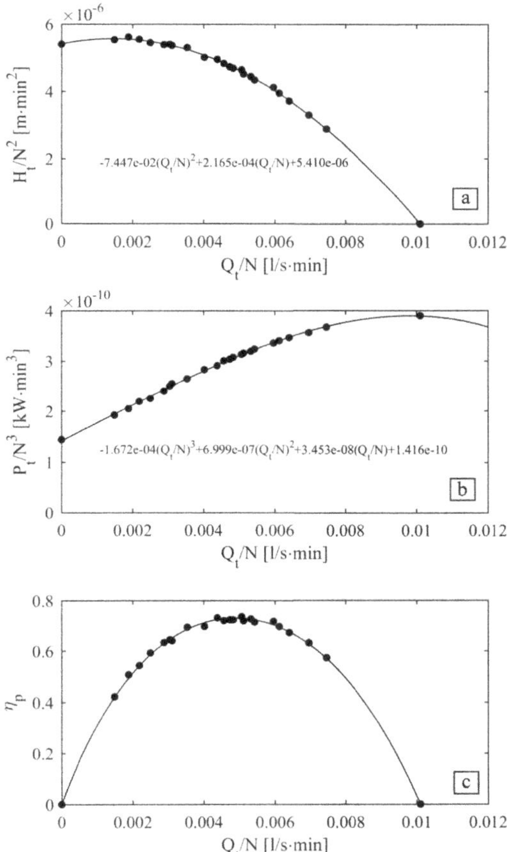

Figure 2. Experimental normalized head (**a**), power (**b**) and efficiency (**c**) of the HMU pump and regression curves.

Another pump, the Caprari model NC80, has been tested in inverse mode in the Caprari qualification laboratory. The head and efficiency as determined experimentally are shown in the normalized plot of Figure 3, while the specifications are reported in Table 3.

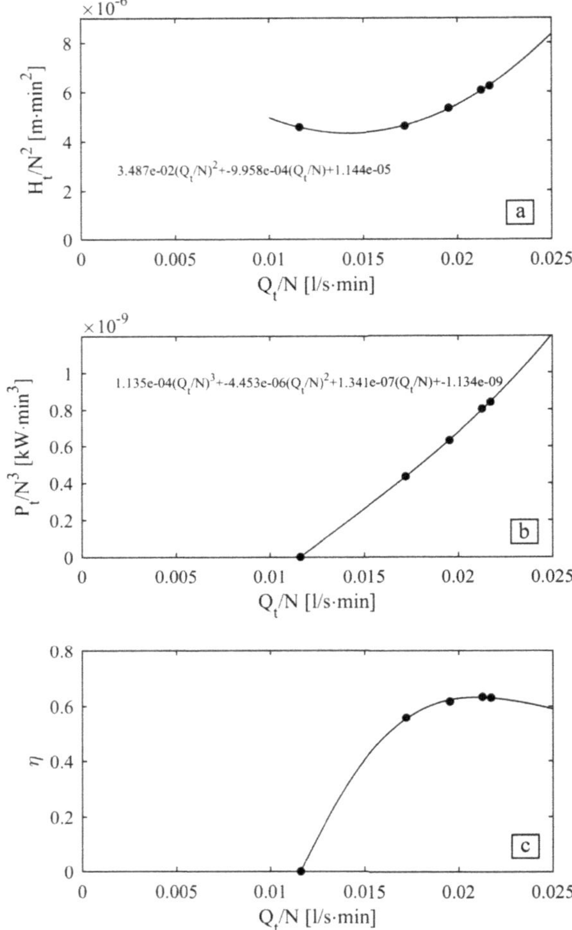

Figure 3. Experimental normalized head (**a**), power (**b**) and efficiency (**c**) of NC80 PAT and regression curves.

Table 3. Specification of Caprari NC80 PAT, as tested in Caprari laboratory.

Specification	Value	Unit
Nominal speed	1550	rpm
Q_B	32.6	L/s
H_B	14.2	m
η_B	63.2	%

4. PAT and Pump System (P&P) Modelling

In the design of a pumping system or a hydro power plant, the rotational speed of the device is imposed by the grid frequency or by a variable frequency driver. Instead, in a P&P, the group is free to achieve any rotational speed. The PAT provides the power for the pump and the rotational speed is set by the combination of the performance curves of the two devices with the network characteristics.

Figure 4 shows a simplified scheme of a P&P plant. The whole water supply system can be considered as two separated network districts which are connected by the P&P plant. The residual head at the end of district 1 that can be turbined within the PAT is represented by the difference $H_T = H_1^u - H_1^d - \Delta H_1^r$, where H_1^u is the head measured at the end point of district 1, ΔH_1^r is the head loss in the pipeline approaching the PAT and H_1^d is the head downstream of the PAT. The lengths of the pipelines are not representative of the real system and, if the values of H_1^u and H_2^d are measured near the P&P system, the head losses ΔH_1^r and ΔH_2^r can be neglected. As a general case, the two values H_1^d and H_2^u are considered different, but the outlet tank of the PAT and the inlet tank of the pump are often the same. The presence of four tanks, however, could even be unnecessary if the P&P system is inserted in a fully pressurized network, and the values of the four variables, H_1^u, H_1^d, H_2^u and H_2^d depend on time according to the network behaviour. Q_T is the flow rate available at the PAT inlet. The power produced by the PAT is $\gamma Q_T H_T \eta_T$, η_T being the PAT efficiency. Such power is transmitted to the pump by the shaft which connects the two machines. Q_P is the pumped flow rate while the total head required at the pump outlet is $H_P = H_2^d - H_2^u + \Delta H_2^r$, where the meaning of the symbols is evident. The efficiency of the plant can be calculated as the ratio between the output hydraulic power at pump side and the input hydraulic power at PAT side. Thus, the plant efficiency η can be calculated as:

$$\eta = \eta_T \cdot \eta_P \tag{4}$$

η_P being the efficiency of the pump. If such a simplified scheme is used to model the water system, the design problem is defined by the following equations:

$$\begin{cases} H_T = H_1^u - H_1^d - \Delta H_1^r \\[4pt] H_P = H_2^d - H_2^u + \Delta H_2^r \\[4pt] \dfrac{H_T}{N^2} = \left(a_T \dfrac{Q_T}{N}^2 + b_T \dfrac{Q_T}{N} + c_T \right) n_t \\[10pt] \dfrac{P_T}{N^3} = \left(\alpha_T \dfrac{Q_T}{N}^3 + \beta_T \dfrac{Q_T}{N}^2 + \gamma_T \dfrac{Q_T}{N} + \delta_T \right) n_t \\[10pt] \dfrac{H_P}{N^2} = \left(a_P \dfrac{Q_P}{N}^2 + b_P \dfrac{Q_P}{N} + c_P \right) n_p \\[10pt] \dfrac{P_P}{N^3} = \left(\alpha_P \dfrac{Q_P}{N}^3 + \beta_P \dfrac{Q_P}{N}^2 + \gamma_P \dfrac{Q_P}{N} + \delta_P \right) n_p \\[10pt] P_T = P_P \end{cases} \tag{5}$$

in the seven variables H_T, H_P, Q_T, Q_P, P_T, P_P and N, P_T and P_P being the mechanical power of the PAT and pump respectively, n_T and n_P the number of stages of the PAT and pump respectively and a_T, b_T, c_T, α_T, β_T, γ_T, δ_T, a_P, b_P, c_P, α_P, β_P, γ_P, and δ_P the experimental regression coefficients of the head curve and the power curve of the PAT and pump, respectively. A P&P node in the water system is described by the last five equations of Equation (5). The energy equations at the PAT intake and pump outlet balance the number of unknowns and equations. More generally, along the pipeline of the two districts, derivations, dissipation points, or additional supply points could be present. In this case, the design of the P&P plant is linked to the operating conditions of the two networks and the complexity of the design solution is connected to the complexity of the water system.

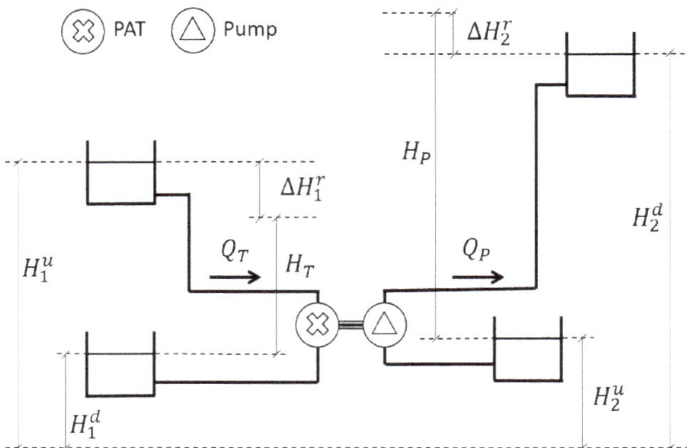

Figure 4. Simplified scheme of the P&P plant.

Behaviour of the P&P Plant

The last five equations of Equation (5) can be rewritten if $q_T = Q_T/N$, $q_P = Q_P/N$, $h_T = H_T/N^2$ and $h_P = H_P/N^2$:

$$\begin{cases} h_T = (a_T q_T^2 + b_T q_T + c_T)n_T \\ h_P = (a_P q_p^2 + b_P q_p + c_P)n_P \\ (\alpha_T q_T^3 + \beta_T q_T^2 + \gamma_T q_T + \delta_T)n_T = (\alpha_P q_P^3 + \beta_P q_P^2 + \gamma_P q_P + \delta_P)n_P \end{cases} \tag{6}$$

Equation (6) expresses the relationship between the four quantities q_T, q_P, h_T and h_P. A single curve relating q_P/q_T and h_P/h_T can be obtained if the number of stages of the pump and PAT are assigned. Because $q_P/q_T = Q_P/Q_T$ and $h_P/h_T = H_P/H_T$, Equation (6) shows that the relationship between the ratio of the delivered discharge and the ratio of the delivered head is independent of the rotational speed. In Figure 5, the values of H_P/H_T are plotted versus Q_P/Q_T for the 1 to 5 pump stages with $n_T = 1$. In Figure 6, the plant efficiency η is plotted versus Q_P/Q_T. The P&P working conditions are different depending on the number of pump stages. In the presence of a single stage, the pump head is slightly greater than the available head drop at the PAT and the flow rate ratio ranges between 0 and 0.38. With two pump stages, the pump head significantly increases (about 2.5 times the head drop in the flow range 0–0.15) and also the range of flow rate ratio is larger (up to 0.36). With an increasing number of pump stages, the range of flow rate ratio decreases (up to 0.3 with five stages), while the head ratio increases (up to 4.8 with five stages). The efficiency of the P&P system also depends on the number of stages. The lowest efficiency occurs for a single stage pump, when the P&P efficiency reaches its maximum value out of the range of the flow rate ratio. The efficiency significantly increases (from less than 0.35 up to more than 0.45) for a higher number of pump stages. Considering a P&P efficiency larger than 0.4, the range of working conditions that can be obtained with the selected pump is quite variable: up to 30% of the water coming from district 1 can be pumped into district 2 with a 50% increase in the pressure head, using a double stage pump, or, at the opposite extreme, with a four stage pump, 15% of the water coming from district 1 can be pumped into district 2 with a pressure head being three times larger than the residual head of district 1. The best efficiency occurs for a three stage pump where 20% of the incoming discharge can be pumped with a 220% increase in head.

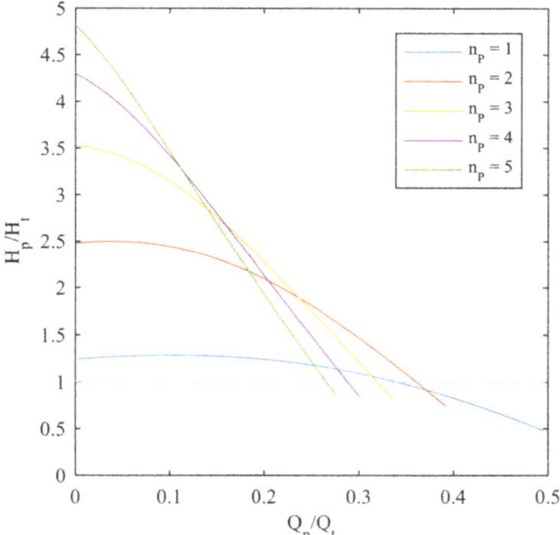

Figure 5. Head ratio of the P&P plant for different numbers of pump stages with $n_t = 1$.

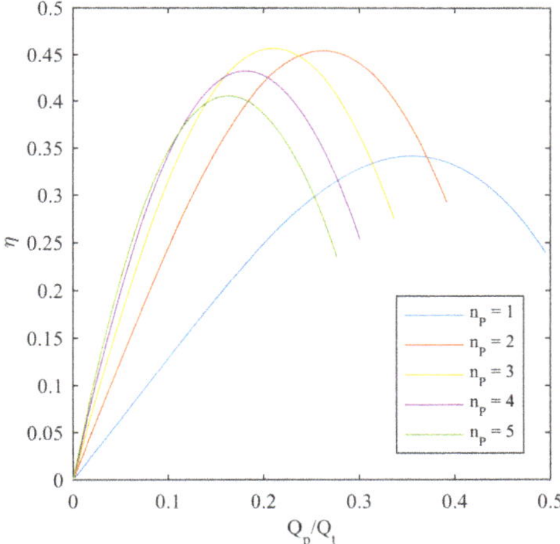

Figure 6. Efficiency of the P&P plant for different pump stages with $n_t = 1$.

5. Case Study

The operating conditions in district 1 exhibit a daily pattern determined by the variation in user demand. Therefore, the Q_T and H_T values depend on both the time and system configuration. The operating conditions in district 2 could be similar to those of district 1 if direct pumping in the network is provided. More frequently, the pumping station supplies water to an elevated tank, with fairly constant values of both Q_P and H_P. As a case study, a simplified system is analyzed. The discharge approaching the PAT is imposed and the head losses along the pipe downstream of the pumping station are completely neglected. Two different supply conditions have been considered for the feeding of district 2:

1. Elevated tank—Variable Q_T, constant H_P
2. Direct pumping—Variable Q_T, constant Q_P

Equation (6) has been solved in order to find the three unknowns (H_T, Q_P, N for scenario 1 and H_T, H_P, N for scenario 2). The daily pattern of a measuring station of the urban water distribution system of Pompei (Campania—Italy) has been scaled to obtain the approaching discharge Q_T. Figure 7 shows the non-dimensional 4 days pattern of flow rate c_q, i.e., the instantaneous discharge Q_T, scaled with the average discharge. For both supply condition, two cases have been considered:

(a) A high difference in ground elevation between district 1 and district 2 and large turbined discharge
(b) A low difference in ground elevation between district 1 and district 2 and small turbined discharge

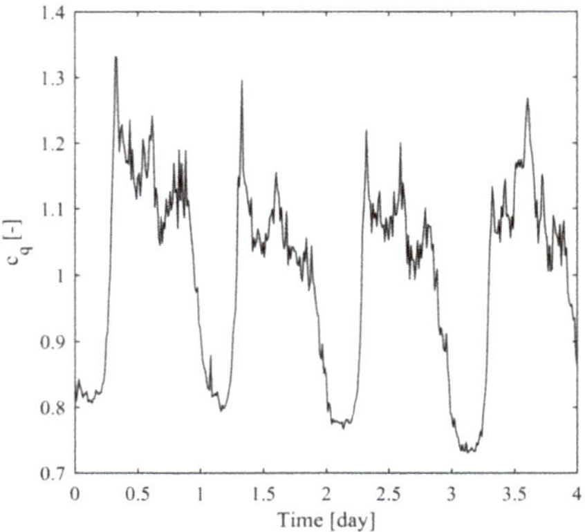

Figure 7. Discharge pattern.

The average turbined discharge, ($\overline{Q_T}$), is 40 L/s in case *a* and 25 L/s in case *b*. In the scenarios 1*a* and 1*b* the value of H_P is assigned and set to 50 m in case *a* and 15 m in case *b*. In the scenarios 2*a* and 2*b*, the pumped discharge is assigned and equal to 4.4 L/s in case *a* and 6.4 L/s in case *b*. The first case corresponds to a four stage pump, allowing a maximum flow rate in district 2 equal to about $0.3Q_T$ with the head being $4.25H_T$ maximum. In the second case, a two stage pump is considered, allowing a maximum flow rate in district 2 equal to about $0.38Q_T$ with the head being $2.5H_T$ maximum. Table 4 shows the configuration of the four different scenarios.

Table 4. Presentation of the different scenarios.

	Case *a*	Case *b*
Supply condition 1	Variable Q_T ($\overline{Q_T} = 40$ L/s) Variable H_T Variable Q_P $H_P = 50$ m $n_P = 4$	Variable Q_T ($\overline{Q_T} = 25$ L/s) Variable H_T Variable Q_P $H_P = 15$ m $n_P = 2$
Supply condition 2	Variable Q_T ($\overline{Q_T} = 40$ L/s) Variable H_T $Q_P = 4,4$ L/s Variable H_P $n_P = 4$	Variable Q_T ($\overline{Q_T} = 25$ L/s) Variable H_T $Q_P = 6,4$ L/s Variable H_P $n_P = 2$

In Figure 8, the turbined head is shown for the four scenarios, as well as the pumped head for scenarios 1*a* and 1*b* and the pumped flow rate for scenarios 2*a* and 2*b*. The results show the performance of the P&P system in the assigned conditions specified in Table 4. Figure 8I,III,V,VII shows the values of the turbined head depending on time. The calculated values of H_t should be less than the available residual pressure while an additional valve could be placed to dissipate the excess head, if necessary.

When the pumping head is fixed (scenarios 1*a* and 1*b*) the pumped discharge is variable, according to Equation (5), as shown by Figure 8II,IV. If the average values of Q_p comply with the needs of district 2, the elevated tank could compensate for the variable discharge. Otherwise, the district should be provided with a second pumping system to supply the missing discharge. Figure 8VI,VIII shows the pumped head when the pumped discharge is fixed. Such a scenario can happen, for example, when a direct pumping is performed and the district is provided with an end-line tank to receive the excess discharge or feed the network during the high demand hours. In such a case, the pumped pressure should be contained within a certain range: a minimum pressure should be guaranteed by the pump to supply the end users, and a maximum value should not be exceeded to avoid any structural or leakage problems. When the average turbined discharge is higher (scenario 2*a*), Figure 8VI shows that the pumped head is always higher than 40 m and can also exceed 130 m in certain moments. In such a case, a dissipation valve could be placed to reduce the head when the pressure exceeds the upper limit. Instead, when the turbine discharge is lower (scenario 2*b*), the pumped head is always lower than 40 m and its minimum value is lower than 10 m. In this last case, if the minimum pressure is not guaranteed, a second ordinary pump can be installed to increase the pressure when necessary.

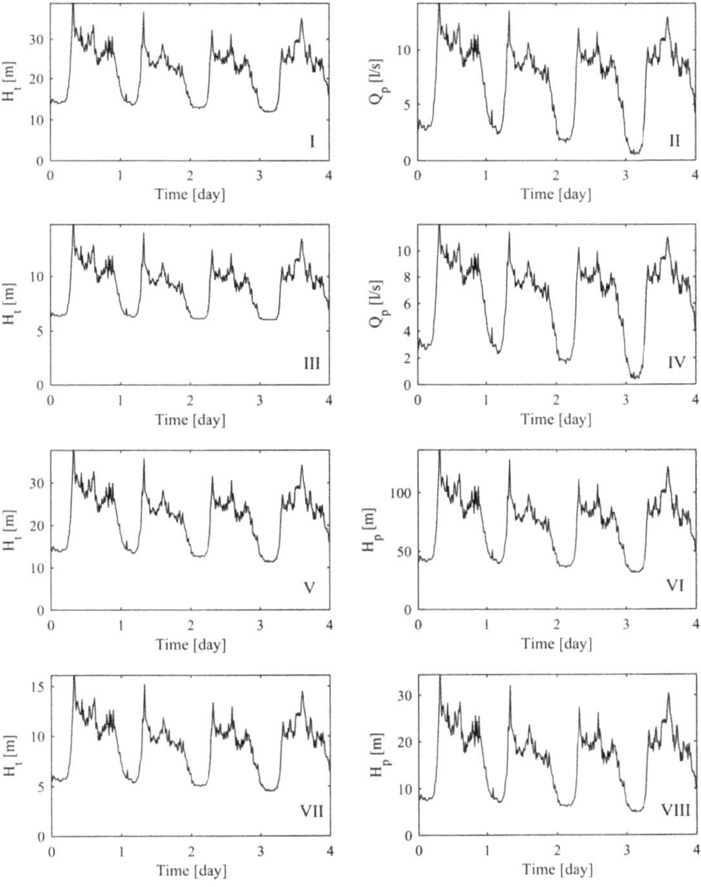

Figure 8. Behaviour of the P&P system for scenario 1*a* (**I,II**); 1*b* (**III,IV**); 2*a* (**V,VI**); and 2*b* (**VII,VIII**).

Table 5 shows the values of the hydraulic power involved in the turbining and in the pumping at the P&P station. Both the average and the maximum values demonstrate that the power values of case *a* are higher than case *b*. Scenario 1*a* corresponds to the highest power values while scenario 1*b* exhibits the lowest power. The ratio between the average pumped power and the average turbined power represents the average efficiency of the P&P station. It can be considered satisfactory in all the four cases, since it is quite close to the maximum values of Figure 6. The last column of Table 5 shows the annual energy saving range, considering an average efficiency of the pumping group ranging between 0.4 and 0.8. Such a saving is always relevant and it is obviously larger in scenarios 1*a* and 2*a*, where the pumped power is larger.

Table 5. Hydraulic power values of the different scenarios.

Scenario	Turbined Average Power (kW)	Turbined Maximum Power (kW)	Pumped Average Power (kW)	Pumped Maximum Power (kW)	Average Efficiency	Annual Energy Saving (MWh)
1*a*	8.95	20.21	3.52	6.93	0.39	48.1–77.0
1*b*	2.27	4.87	0.93	1.76	0.41	12.7–20.3
2*a*	8.8	19.70	3.10	5.88	0.35	42.5–68.0
2*b*	2.28	5.26	1.01	2.16	0.44	13.8–22.0

6. Conclusions

A new P&P technology, coupling a PAT and a pump, has been studied on the basis of the experimental data of the pump working in direct or in inverse mode. A P&P is suitable in all cases when a certain amount of the water at the end of a water supply system, presenting a residual energy, has to be pumped to a second network at a higher pressure level. In such a case, P&P is a low cost technology allowing the recovery of the residual energy of the water in the first network and the increase of the pressure level of the water in the second network. Equations for P&P design have been presented, allowing a simple analysis of the P&P working conditions. Depending on the number of stages of the pumping, different operating conditions in the ancillary supply network can be obtained, in terms of water demand and pressure level. For the pump model used in the simulation, up to 40% of the water can be transmitted by the P&P to a pressure level 20% higher than the initial level. If the water to be transmitted is limited to 20%, the pressure level can be increased by the P&P by more than three times. This high potentiality is determined by the total efficiency of the P&P, even greater than 40%. The case study shows the viability of the system in different scenarios, considering different supply conditions as well as different values of turbined discharge. The results, depending on the network configuration and the considered scenario, show that the P&P system can completely replace an ordinary pumping system or supply the network when the available hydraulic power is high. The calculated annual energy saving demonstrates the economic relevance of the P&P plant. As a future implementation of this study, an experimental investigation on a laboratory prototype could be undertaken and the results would be used for the design of a full scale plant, to investigate the interactions with a WDN in a real situation.

Author Contributions: Lauro Antipodi and Oreste Fecarotta performed the experiments on the PAT and on the pump. Armando Carravetta, Oreste Fecarotta and Umberto Golia developed the mathematical model. Umberto Golia ran the simulations. Armando Carravetta and Oreste Fecarotta wrote the paper.

References

1. Pahl-Wostl, C. Transitions towards adaptive management of water facing climate and global change. *Water Resour. Manag.* **2007**, *21*, 49–62.

2. Ramos, H.; Covas, D.; Araujo, L.; Mello, M. Available energy assessment in water supply systems. In Proceedings of the XXXI IAHR Congress, Seoul, Korea, 11–16 September 2005.

3. Gallagher, J.; Styles, D.; McNabola, A.; Williams, A.P. Life cycle environmental balance and greenhouse gas mitigation potential of micro-hydropower energy recovery in the water industry. *J. Clean. Prod.* **2015**, *99*, 152–159.

4. Almandoz, J.; Cabrera, E.; Arregui, F.; Cabrera, E.; Cobacho, R. Leakage Assessment through Water Distribution Network Simulation. *J. Water Resour. Plan. Manag.* **2005**, *131*, 458–466.

5. Xue, X.; Hawkins, T.R.; Schoen, M.E.; Garland, J.; Ashbolt, N.J. Comparing the Life Cycle Energy Consumption, Global Warming and Eutrophication Potentials of Several Water and Waste Service Options. *Water* **2016**, *8*, 154.

6. Abbott, M.; Cohen, B. Productivity and efficiency in the water industry. *Util. Policy* **2009**, *17*, 233–244.

7. Filion, Y.; MacLean, H.; Karney, B. Life Cycle Energy Analysis of a Water Distribution System. *J. Infrastruct. Syst.* **2004**, *10*, 120–130.

8. Carravetta, A.; Fecarotta, O.; Sinagra, M.; Tucciarelli, T. Cost-benefit analysis for hydropower production in water distribution networks by a Pump As Turbine. *J. Water Resour. Plan. Manag.* **2013**, *140*, 04014002.

9. Vairavamoorthy, K.; Lumbers, J. Leakage Reduction in Water Distribution Systems: Optimal Valve Control. *J. Hydraul. Eng.* **1998**, *124*, 1146–1154.

10. Tucciarelli, T.; Criminisi, A.; Termini, D. Leak Analysis in Pipeline Systems by Means of Optimal Valve Regulation. *J. Hydraul. Eng.* **1999**, *125*, 277–285.

11. Nicolini, M.; Zovatto, L. Optimal location and control of pressure reducing valves in water networks. *J. Water Resour. Plan. Manag.* **2009**, *135*, 178–187.

12. Liberatore, S.; Sechi, G. Location and Calibration of Valves in Water Distribution Networks Using a Scatter-Search Meta-heuristic Approach. *Water Resour. Manag.* **2009**, *23*, 1479–1495.

13. Ramos, H.; Mello, M.; De, P. Clean power in water supply systems as a sustainable solution: From planning to practical implementation. *Water Sci. Technol.* **2010**, *10*, 39–49.

14. Gallagher, J.; Harris, I.; Packwood, A.; McNabola, A.; Williams, A. A strategic assessment of micro-hydropower in the UK and Irish water industry: Identifying technical and economic constraints. *Renew. Energy* **2015**, *81*, 808–815.

15. Samora, I.; Manso, P.; Franca, M.J.; Schleiss, A.J.; Ramos, H.M. Energy Recovery Using Micro-Hydropower Technology in Water Supply Systems: The Case Study of the City of Fribourg. *Water* **2016**, *8*, 344.

16. Ormsbee, L.E.; Lansey, K.E. Optimal control of water supply pumping systems. *J. Water Resour. Plan. Manag.* **1994**, *120*, 237–252.

17. Zhang, H.; Xia, X.; Zhang, J. Optimal sizing and operation of pumping systems to achieve energy efficiency and load shifting. *Electr. Power Syst. Res.* **2012**, *86*, 41–50.

18. Punys, P.; Dumbrauskas, A.; Kvaraciejus, A.; Vyciene, G. Tools for Small Hydropower Plant Resource Planning and Development: A Review of Technology and Applications. *Energies* **2011**, *4*, 1258–1277.

19. Carravetta, A.; Fecarotta, O.; Ramos, H. Numerical simulation on Pump As Turbine: Mesh reliability and performance concerns. In Proceedings of the International Conference on Clean Electrical Power, Ischia, Italy, 14–16 June 2011; Franco Angeli: Milano, Italy, 2011.

20. Fontana, N.; Giugni, M.; Glielmo, L.; Marini, G. Real Time Control of a Prototype for Pressure Regulation and Energy Production in Water Distribution Networks. *J. Water Resour. Plan. Manag.* **2016**, *142*, doi:10.1061/(ASCE)WR.1943-5452.0000651.

21. Derakhshan, S.; Nourbakhsh, A. Theoretical, numerical and experimental investigation of centrifugal pumps in reverse operation. *Exp. Therm. Fluid Sci.* **2008**, *32*, 1620–1627.

22. Singh, P.; Nestmann, F. An optimization routine on a prediction and selection model for the turbine operation of centrifugal pumps. *Exp. Therm. Fluid Sci.* **2010**, *34*, 152–164.

23. Pugliese, F.; De Paola, F.; Fontana, N.; Giugni, M.; Marini, G. Experimental characterization of two Pumps As Turbines for hydropower generation. *Renew. Energy* **2016**, *99*, 180–187.

24. Carravetta, A.; Fecarotta, O.; Golia, U.M.; La Rocca, M.; Martino, R.; Padulano, R.; Tucciarelli, T. Optimization of Osmotic Desalination Plants for Water Supply Networks. *Water Resour. Manag.* **2016**, *30*, 3965–3978.

25. Carravetta, A.; Conte, M.; Fecarotta, O.; Ramos, H. Evaluation of PAT Performances by Modified Affinity Law. *Procedia Eng.* **2014**, *89*, 581–587.

26. Carravetta, A.; Fecarotta, O.; Martino, R.; Antipodi, L. PAT efficiency variation with design parameters. *Procedia Eng.* **2014**, *70*, 285–291.
27. Simpson, A.R.; Marchi, A. Evaluating the approximation of the affinity laws and improving the efficiency estimate for variable speed pumps. *J. Hydraul. Eng.* **2013**, *139*, 1314–1317.
28. Fecarotta, O.; Carravetta, A.; Ramos, H.M.; Martino, R. An improved affinity model to enhance variable operating strategy for pumps used as turbines. *J. Hydraul. Res.* **2016**, *54*, doi:10.1080/00221686.2016.1141804.
29. Afshar, A.; Jemaa, F.; Marino, M. Optimization of hydropower plant integration in water supply system. *J. Water Resour. Plan. Manag.* **1990**, *116*, 665–675.
30. Dannier, A.; Del Pizzo, A.; Giugni, M.; Fontana, N.; Marini, G.; Proto, D. Efficiency evaluation of a micro-generation system for energy recovery in water distribution networks. In Proceedings of the 2015 IEEE International Conference on Clean Electrical Power (ICCEP), Taormina, Italy, 16–18 June 2015; IEEE: Piscataway, NJ, USA, 2015; Volume 1, pp. 689–694.
31. McNabola, A.; Coughlan, P.; Corcoran, L.; Power, C.; Williams, A.P.; Harris, I.; Gallagher, J.; Styles, D. Energy recovery in the water industry using micro-hydropower: An opportunity to improve sustainability. *Water Policy* **2014**, *16*, 168–183.
32. Carravetta, A.; Del Giudice, G.; Fecarotta, O.; Ramos, H. Energy Production in Water Distribution Networks: A PAT Design Strategy. *Water Resour. Manag.* **2012**, doi:10.1007/s11269-012-0114-1.
33. Rodrigues, A.; Singh, P.; Williams, A.; Nestmann, F.; Lai, E. Hydraulic analysis of a pump as a turbine with CFD and experimental data. In Proceedings of the IMechE Seminar Computational Fluid Dynamics for Fluid Machinery, London, UK, 18 November 2003.
34. Ramos, H.M.; Borga, A.; Simão, M. New design solutions for low-power energy production in water pipe systems. *Water Sci. Eng.* **2009**, *2*, 69–84.
35. Carravetta, A.; Del Giudice, G.; Fecarotta, O.; Ramos, H. PAT Design Strategy for Energy Recovery in Water Distribution Networks by Electrical Regulation. *Energies* **2013**, *6*, 411–424.
36. Fecarotta, O.; Aricò, C.; Carravetta, A.; Martino, R.; Ramos, H.M. Hydropower potential in water distribution networks: Pressure control by PATs. *Water Resour. Manag.* **2015**, *29*, 699–714.
37. Carravetta, A.; Del Giudice, G.; Fecarotta, O.; Ramos, H.M. Pump As Turbine (PAT) design in water distribution network by system effectiveness. *Water* **2013**, *5*, 1211–1225.
38. Carravetta, A.; Fecarotta, O.; Del Giudice, G.; Ramos, H. Energy recovery in water systems by PATs: A comparisons among the different installation schemes. *Procedia Eng.* **2014**, *70*, 275–284.
39. Corcoran, L.; McNabola, A.; Coughlan, P. Optimization of water distribution networks for combined hydropower energy recovery and leakage reduction. *J. Water Resour. Plan. Manag.* **2015**, *142*, 04015045.
40. Samora, I.; Franca, M.J.; Schleiss, A.J.; Ramos, H.M. Simulated annealing in optimization of energy production in a water supply network. *Water Resour. Manag.* **2016**, *30*, 1533–1547.

Article

Energy Recovery Using Micro-Hydropower Technology in Water Supply Systems: The Case Study of the City of Fribourg

Irene Samora [1,2,*], Pedro Manso [2], Mário J. Franca [2], Anton J. Schleiss [2] and Helena M. Ramos [1]

[1] Civil Engineering Research and Innovation for Sustainability, Instituto Superior Técnico, Universidade de Lisboa, Lisboa 1049-001, Portugal; helena.ramos@tecnico.ulisboa.pt
[2] Laboratory of Hydraulic Constructions, École Polytechnique Fédérale de Lausanne, Lausanne 1015, Switzerland; pedro.manso@epfl.ch (P.M.); mario.franca@epfl.ch (M.J.F.); anton.schleiss@epfl.ch (A.J.S.)
* Correspondence: irene.a.samora@epfl.ch; Tel.: +41-21-69-32405

Academic Editor: Yung-Tse Hung
Received: 21 June 2016; Accepted: 9 August 2016; Published: 12 August 2016

Abstract: Water supply systems (WWSs) are one of the main manmade water infrastructures presenting potential for micro-hydropower. Within urban networks, local decentralized micro-hydropower plants (MHPs) may be inserted in the regional electricity grid or used for self-consumption at the local grid level. Nevertheless, such networks are complex and the quantification of the potential for micro-hydropower other than that achieved by replacing pressure reducing valves (PRVs) is difficult. In this work, a methodology to quantify the potential for hydropower based on the excess energy in a network is proposed and applied to a real case. A constructive solution is presented based on the use of a novel micro-turbine for energy conversion, the five blade tubular propeller (5BTP). The location of the MHP within the network is defined with an optimization algorithm that maximizes the net present value after 20 years of operation. These concepts are tested for the WSS in the city of Fribourg, Switzerland. The proposed solution captures 10% of the city's energy potential and represents an economic interest. The results confirm the location of PRVs as potential sites for energy recovery and stress the need for careful sensitivity analysis of the consumption. Finally, an expedited method is derived to estimate the costs and energy that one 5BTP can produce in a given network.

Keywords: micro-hydropower; water supply networks; energy potential; tubular propeller turbine; energy recovery

---

1. Introduction

A water supply system (WSS) is a set of civil infrastructures (tanks, pipes and others), hydro-mechanical facilities, electrical equipment and services that extracts, conveys and distributes water to consumers. This distribution must be compatible with the demand in both quantity and quality [1]. A sub-grid of a water supply network, also termed a "district metered area", is a part of the network between two nodes where the head pressure is controlled. This pressure is maintained constant or follows a certain schedule [2] to ensure that there is an optimal pressure distribution in all the branches downstream. This is normally ensured by pressure reducing valves (PRVs), but recent studies have proposed their replacement by turbines. Pumps operated as turbines and Kaplan turbines are the technologies proposed by [3,4] for low-head sites. The study presented in [4] has shown that there is an economical interest in micro-hydropower in WSSs. This study was performed at a regional scale, identifying existing infrastructure such as PRVs and reservoirs where there is potential for energy recovery. Moreover, there is potential for energy production within the urban network itself.

Besides the locations where PRVs are installed, there are also areas within the sub-grids that have more pressure than necessary, even if not excessive, because of their connection to other higher areas [5].

The use of PRVs or even micro-turbines in a WSS may seem illogical, since most systems need pumping. Nevertheless, the use of operational pressure control has proven to be cost-effective for reducing leakage problems [1,6]. Water loss occurs due to the structural deterioration of the pipes [7]. Local energy dissipation is seen to reduce the frequency of pipe bursts [8], ensure the quality standards of water [9] and protect roads and building foundations from underground cavities originating from water losses [10]. From the perspective of hydropower, the implementation of production units in existing infrastructure has the economic advantage of low costs [11] given the synergies with the existing facilities of the WSS, of consumption dependent discharges, which are predictable and almost guaranteed, and of benefiting several countries with respect to legal and financial incentives for self-generation and self-consumption (e.g., feed-in tariffs provided for renewable energies). However, it is worth mentioning that the feed-in-tariff regulations vary considerably between countries and, for example, in the UK, a micro-hydropower plant (MHP) downstream from any pumping station is not eligible for the subvention [12]. In addition to the abovementioned synergies with the existing WSS facilities, the energy produced may justify the installation of a grid-connected generator or local consumption [13]; the latter provides financial savings by avoiding external consumption and expensive electrical connections [14]. The installation of these systems is highly dependent on the costs associated with the turbine, construction and grid connection [4].

Examples of solutions for the installation of MHPs in WSSs can be found in the literature. For example, the installation of PATs as replacement for a PRV in the an urban water distribution system of Pompei, with discharges between 20 and 50 L/s and heads between 35 and 90 m, would produce between 20 and 94 MWh/year [3]. In Portland, Oregon, a PRV was replaced by a 10″ (approx. 250 mm) micro turbine that generates 150 MWh/year with 30 kW of installed power [15]. In Hong Kong, an eight blade spherical inline turbine is expected to produce 700 kWh/year in the city's water main pipes [16]. Finally, the installation of a turbine replacing a break pressure tank in Kildare in Ireland could generate approximately 237 MWh/year from 200 kPa to 17,910 m^3/day [17].

Although the potential for energy production within urban WSSs exists, it is difficult to quantify unless focusing on a particular location. Distribution networks are complex systems, usually composed of multiple loops and asymmetric consumption in both time and space. The optimal location of turbines within such systems is a subject of research [5,18,19]. Moreover, before enduring deep analyses and simulations of turbine operation and water demand scenarios, a measure of the excess energy available within the city network, indicating the potential for that network to be used for hydropower, would be useful for WSS managers.

In this research, a method for the evaluation of the potential for hydropower in a city is presented. This method is then applied to a case study city and compared with the actual energy production with a proposed hydropower scheme. The energy production is estimated using an algorithm for the identification of optimal location of the micro-hydropower units considering the whole network where infrastructure may or may not exist. In Section 2, the two methodologies are presented, for potential assessment and for the actual energy production simulation. The case study, the city of Fribourg in Switzerland, and the hydropower scheme are described in Section 3. The results are presented and discussed in Section 4. Section 5 presents a straightforward methodology that is proposed for the presented hydropower scheme and the main conclusions are drawn in Section 6.

2. Methodology

2.1. Availability of Excess Energy

In any pipe of a network at a certain instant with a steady flow regime, the total energy line (Figure 1) can be defined as a straight line, if there are no local head losses, between the head in each boundary nodes. The head in each node will be higher than or equal to a minimum pressure p_{min}, usually imposed by regulators to guarantee a quality service to the population. In each point of the pipe, whenever the head is higher than this minimum pressure, there is excess energy. This excess energy will vary in time, as the demand in the network is not constant, affecting both the pipe flow and the pressure in the nodes.

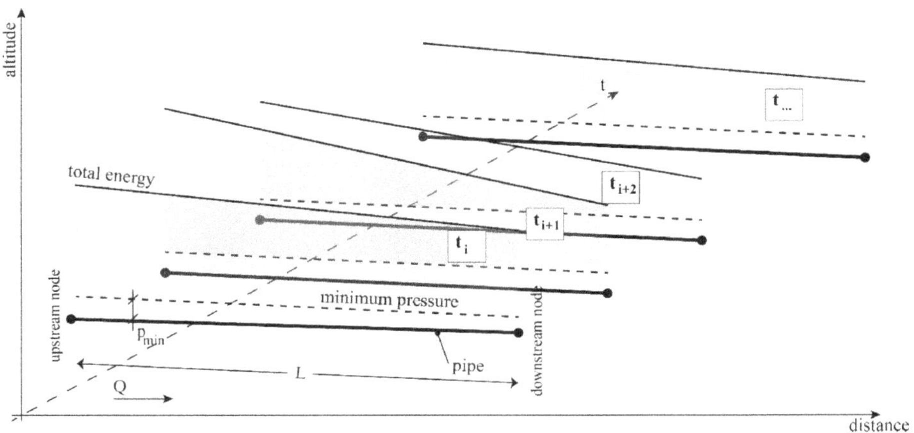

Figure 1. Energy excess at each point of a pipe, defined as the hydraulic head above minimum pressure, for several instances where a steady state is observed.

The hydraulic energy, E (Wh), at any point of the pipe is defined as

$$E = \rho g Q H \Delta t \tag{1}$$

where ρ is the water density (kg/m^3), g is the gravitational acceleration (m^2/s), Q is the flow rate (m^3/s), H is the hydraulic head (m), meaning the total energy flow subtracted by the topographic elevation of the point, and Δt (h), a time interval. This time interval can be considered as the duration of a steady state.

To estimate the excess energy in the upstream and downstream nodes of Figure 1, the minimum pressure is subtracted from the hydraulic charges:

$$\begin{aligned} E_{up} &= \rho g Q_t \left(H_{up} - p_{min} \right) \Delta t \\ E_{dn} &= \rho g Q_t \left(H_{dn} - p_{min} \right) \Delta t \end{aligned} \tag{2}$$

However, it can be argued that the application of this definition of the excess energy in an entire network cannot be done without taking into account the topography. The excess energy at a given point is often needed to ensure the minimum pressure in another and hence, is not available.

Considering the reach, defined as a sequence of pipes, sketched in Figure 2a, excess energy exists in every point of its water path, but none of this excess is available, since the most downstream point is at the minimum pressure and extra head losses would cause this pressure to decrease. Instead, if we consider the reach sketched in Figure 2b, the minimum pressure is not limited at the downstream extremity but at a high point along the path. The excess downstream from this point is partly available. During the network design, there is usually an effort to minimize this difference by using smaller diameters, as a means to control the pressure. Alternatively, the installation of turbines could be used to dissipate this excess energy.

The available energy at a point in a WSS can hence be defined as the excess energy that can be extracted from the flow without causing pressure below the minimum at any other point. To quantify how much of the excess energy is available for hydropower production, critical points must then be identified. The critical point corresponds to the position in a network where the difference between the total energy and the minimum pressure is minimum but higher than zero. This difference is the head that can be taken from the total energy line. This head and the downstream total energy line are represented by a dotted line in Figure 2b.

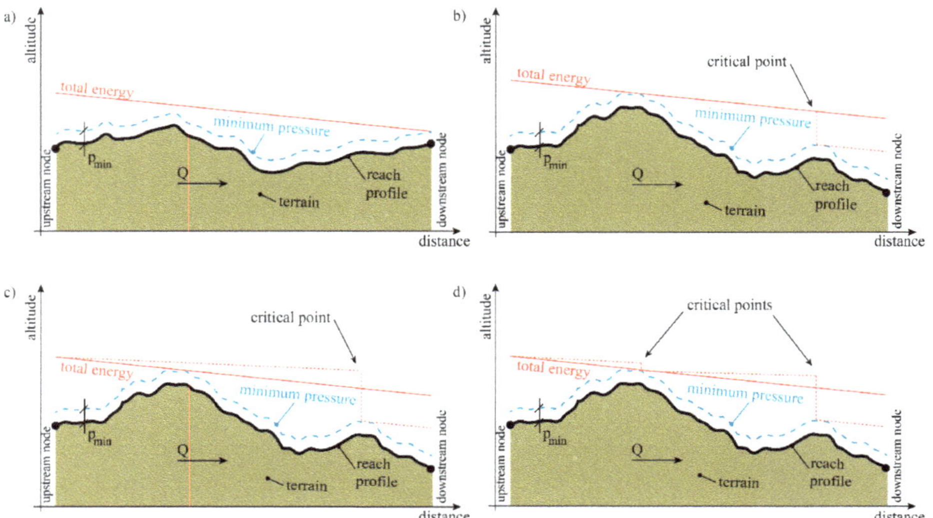

Figure 2. Examples of reaches. (**a**) Reach without available energy; (**b**) Reach with available energy; (**c**) Reach belonging to a closed network with available energy: effect of the extraction in one critical point; (**d**) Reach belonging to a closed network with available energy: effect of the extraction in two critical points.

The former examples always considered a single reach with no other connecting path between the upstream and the downstream sections either than the presented reach. The positioning of a valve or a turbine in Figure 2b would not have an impact in the discharge along the reach. However, if the reach was in a closed network, which is common in urban WSSs, the introduction of an energy converter and consequent adjustment of the energy distribution would have an impact in the discharge, as represented in Figure 2c. Moreover, the following critical point in the reach is now the point that was at minimum pressure before. If the new critical point is considered for the extraction of energy, a new readjustment occurs and the resulting total energy distribution is presented in Figure 2d.

The available energy in a network is given by the sum of the available energy at every critical point and hence its assessment requires a dynamic process. Also, as mentioned, the available head varies with the demand, which in a WSS is variable throughout the day. Nevertheless, a simpler and more immediate way for estimating available energy is to consider the steady state conditions with maximum flows, and hence the minimum heads.

An algorithm was developed to access the potential for hydropower in water supply networks taking into account the previous considerations for available energy in a network (Figure 3). Input data are composed of the network geometry, available consumption data (discharges distributed along the WSS) and minimum pressures to be assured at every node.

The consumption that leads to the average lowest pressure in the network is identified in the analysis. The nodes that do not fulfill the criteria of minimal pressure under these conditions are considered to be non-consumptive nodes, where the pressure is only required to be positive to avoid cavitation.

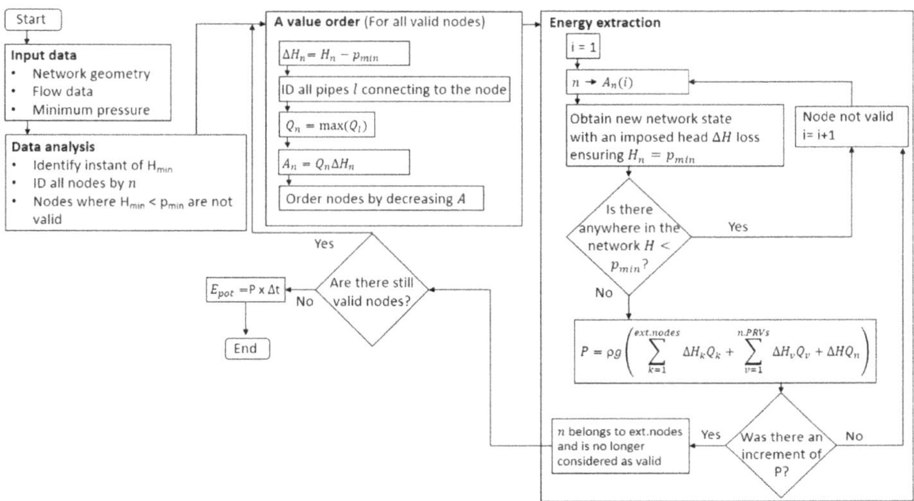

Figure 3. Algorithm to estimate the potential for hydropower of a network, given by the total available energy.

Since energy depends on both head and flow rate, the available energy cannot be evaluated solely based on available head in the critical points. For all the nodes in the network, a value A is defined, if the node is valid, as:

$$A_n = \Delta H_n \max(Q_l) \tag{3}$$

where the index l refers to the ID of all pipes connecting to the node n,

$$\Delta H_n = H_n - p_{min} \tag{4}$$

and H_n is the hydraulic head over the node n in the current hydraulic state. If the node is not valid, A is zero. The nodes are then re-ordered in decreasing order of A.

For the node with the highest A, the head from Equation (4) is extracted by applying a local head loss in the pipe connecting to the node with the highest discharge. This extraction implies a new energetic equilibrium in the network that needs to be calculated. If the minimum pressure is satisfied in all the consumptive nodes after the network recalculation, the extracted head is given by the imposed head loss, and the available power in that node is given by the head and $\max(Q_l)$. The power P of the new state is given by the sum of the available power in the new node, the available power in the all the nodes where a head loss was previously imposed and the power that is dissipated in each PRV, if they exist, multiplied by the water volumetric weight. The new node is accepted only if there is an increment in the calculation of the power, to ensure that it does not negatively affect the nodes previously gathered. If it is not accepted, the following highest A is tested and so forth until there is an acceptance or the valid nodes are exhausted.

The procedure of ordering the nodes according to the value of A is repeated with each new accepted equilibrium in the network. Moreover, when extracting the head in a new node, the flow discharge in the previous ones will be affected, and so the potential of both has to be calculated.

When all valid nodes have been evaluated, an estimation of the annual available energy E_{pot} can be given by the power P multiplied by the considered time window Δt. The potential for hydropower of a network is hence evaluated through the total available energy.

2.2. Optimum Site Locations

Even if potential for hydropower has been recognized in a certain city, the ideal location of turbines within its network is not straightforward. It depends on numerous factors such as the flow rate and respective velocity restrictions which have daily variations, the head which is dependent not only on the minimum service pressures but also on the chosen turbine, and the geometry of the network that conditions the distribution of the flow within its closed network.

To answer this problem, a search algorithm in which a simulated annealing technique was used to optimize the economic value of the installation of micro-hydropower plants (MHP) in a WSS was applied. In the recently developed search algorithm [5,18], in each iteration, a full year is simulated with an hourly time step considering the installation of a given number of turbines in several locations or combined in one location. The produced energy is obtained by coupling the EPANET hydraulic model [20] that calculates the hydraulic state of the network for each time step with the characteristics of the turbines. The position of the turbines in the WSS is changed in each iteration, in the search for the best output. Only the solutions that respect minimum pressure and maximum velocity constraints are accepted.

The best output is given by the objective function that has been defined as the maximization of the net present value of the project annual cash flows over 20 years of operation:

$$f(X) = \min\left(1/NPV_{20}\right) \tag{5}$$

where X is the solution vector, representing the placement of N_t turbines, and NPV_{20} is the net present value of the project discounted cash-flows.

In a concise way, the net present value is given by

$$NPV = Revenues - Capital\ Costs - Operational\ Costs - Maintenance\ Costs \tag{6}$$

where all components are transposed to the year 0 of operation.

In the considered economic model, the capital costs include all investment costs for the construction of the MHPs. The maintenance costs are considered negligible when compared with the maintenance costs of the entire network. The operational costs of a network are given by the electricity bought for pumping operations and the personnel costs for managing the WSS. In this case, since the focus is only on the construction and operation of the turbines, and not the entire network, only the costs for pumping that are superior to the original operational costs are considered. The operational costs are thus given by the electricity buying tariff and the difference between the electrical energy needed for pumping in the situation with the turbines installed and the energy needed for pumping in the initial situation (without turbines).

Two different types of remuneration can be considered depending on the economic model assumed: selling to the grid or self-consumption. When selling to the grid, the revenues are given by the produced energy and the considered sell-tariff. For a self-consumption scheme, it is assumed that the generated energy is consumed in operations within the network. In this case, the gain is in the savings in the electricity bill of the network. To provide an appreciation of the solution, independently of the remuneration type, another economic index, the cost price, is considered:

$$Cost\ price = \cfrac{Capital\ Costs}{\cfrac{(1+r)^{20} - 1}{r(1+r)^{20}} E_{annual}} \tag{7}$$

where r is the discount rate and E_{anual} is the annual energy production.

3. Case Study

3.1. The City of Fribourg

The proposed methodology for potential analysis and economic value assessment was applied to the case study of the city of Fribourg, Switzerland. The urban water supply network model of Fribourg, which can be seen in Figure 4, has 2972 links and 2805 nodes covering 135 km of pipes and a total elevation difference of 216 m. The network, whose model was provided by the Industrial Services of Fribourg, possesses nine PRVs, seven water tanks and four pumping stations. The population served by the WSN is about 38,000 persons, making this a small sized European city.

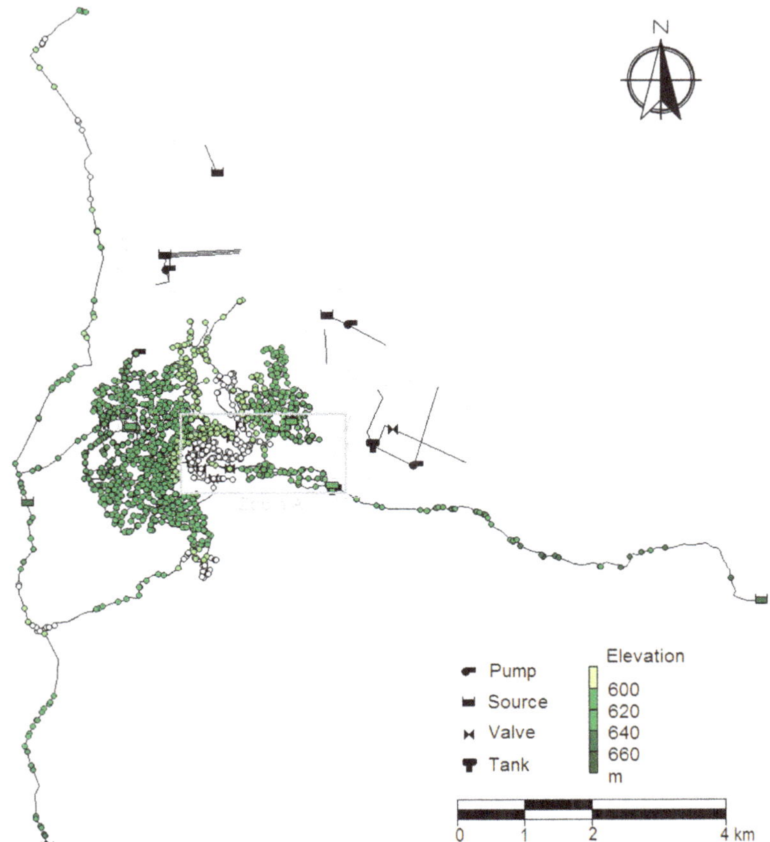

Figure 4. Water supply network of Fribourg. The nodes are represented by circles with a fill-color representing the elevation of the node.

According to the gathered data, an average daily consumption of $Q_{d,average} = 0.108$ L/s was associated with each node, and the daily pattern of consumption presented in Figure 5 was applied to obtain an hourly variation.

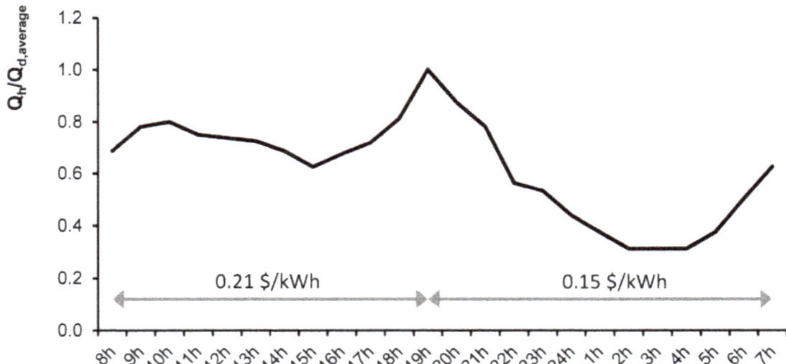

Figure 5. Pattern of hourly variation of consumption along the day, adapted from [21], with the indication of the energy buy-prices which vary within the day.

Six of the water tanks are considered as water sources, with fixed constant levels. The seventh water tank, however, has a known geometry and its level varies along the simulation. In each iteration, the initial level in the seventh water tank was defined in order to have the same initial level at the end of a 24 h cycle. The capacity and elevation of this tank also represents a restriction to the algorithm.

A minimum pressure of 30 m was assumed as a restriction in every consumptive node. According to the Swiss Directive for Water Supply [22], the minimum service pressure in Switzerland is a function of the average number of building stories. The assumed value implies that a 6-story building with 3 m per floor would not require a pump to provide 10 m of water pressure in the upper floor. A restriction of the maximum velocity in the pipes was defined as 2 m/s.

The considered electricity sell-prices is fixed at 0.33 $/kWh and the considered buy-price depends on the period as indicated in Figure 5. The sell-price is the current feed-in-tariff in Switzerland for this type of power plant [23,24].

The operations of the pumping station for the network consume 250 MWh/year, according to Table 1, assuming a constant efficiency of 60%.

Table 1. Operating conditions of the pumping stations and annual energy consumption in the network of Fribourg.

Pump Station	H_{med} (m)	Q_{med} (L/s)	P_{med} (kW)	E (MWh)
1	80	5.4	7	60
2	45	18.8	13	111
3	60	5.8	6	48
4	58	3.7	3	31
Total	244	33.7	29	250

3.2. The Micro-Hydropower Technology

The MHP considered for installation in the city was for a micro-turbine currently under development by the authors [25]. The five blade tubular propeller (5BTP), initially designed under the scope of the EU Project HYLOW (2008–2012), has been recently experimentally tested for the characterization of its performance [25].

It is a suitable turbine for energy recovery in a WSS since it can be installed directly into the pressurized pipe. Furthermore, the operation of the 5BTP is possible with variable flow rates and low heads. The machine consists of a bulb followed by a runner with five fixed blades (Figure 6). It is connected to a rotating axis that leaves the pipe through a 45° curve. The rotating axis is connected to the external generator which controls the rotational speed of the turbine. A minimum diameter of 85 mm was considered.

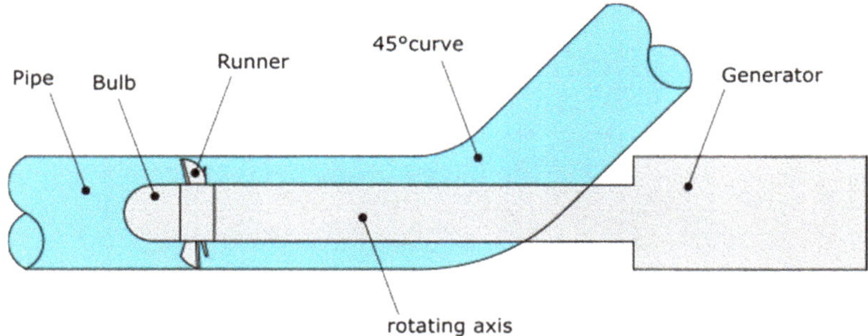

Figure 6. Detailed sketch representation of the inline installation of a 5BTP in a pipe.

The installation of the turbine within the network is proposed according to the design presented in Figure 7. It is based on a buried concrete chamber constructed in line with an existing pipe, whose dimensions depend on the diameters of both the existing pipe (D_P) and turbine runner (D_T). The by-pass is only necessary if the considered branch has no redundancy in its supply, as it has been defined in [18].

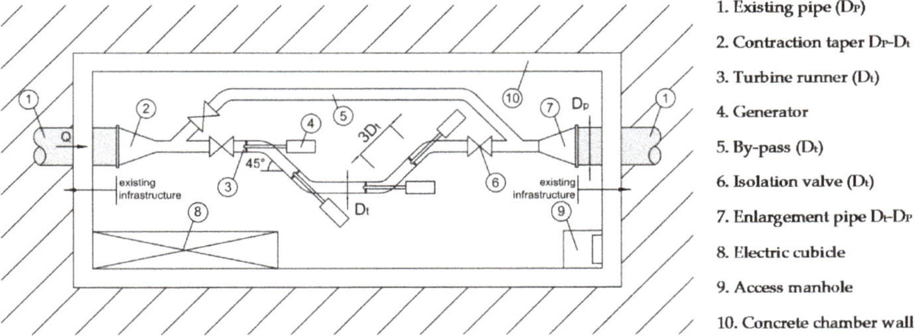

Figure 7. Sketch of the concrete chamber for the installation of a 5BTP (adapted from [18]). Up to four turbines can be installed with this design.

The choice of diameter of the runner, with a minimum of 85 mm, is a function of the maximum flow rate in the pipe and the experimental characteristic curve [25], which is scaled according to the similarity laws.

Up to four turbines can be installed in the same concrete chamber. The lay-out from Figure 7 is used to estimate the equipment and civil works costs by calculating the main quantities presented in Table 2. A surplus of 25% of this sub-total was added to account for engineering and construction supervision plus 15% for miscellaneous items not quantified at this early phase. The costs for connections to the grid and site access were considered negligible, since the MHPs are installed within urban areas. Savings of 2% were considered for the construction of more than one chamber to take into account group ordering prices. All percentages were applied independently of the number of turbines in the chamber. For the feasibility analysis, a linear cost function was adopted for the electromechanical equipment as a single item. This is justified by the fact that the 5BTP turbine is not yet commercialized, and hence a lumped price based on current technologies for this scale of installed power was considered.

Table 2. Unit prices considered for the Fribourg WSS case study.

Element	Unit Price
Stainless steel	7 $/kg
Reinforced concrete	250 $/m^3
Excavation	30 $/m^3
Earth fill	20 $/m^3
Electromechanical equipment	1 $/W
Maintenance valve w/wheel drive	190,000 $/m^2
Flowmeter	550 $/unit

In the cases of PRV sites, it was assumed that the valve ensures a constant downstream pressure unless the pressure immediately upstream from it is already lower. Since there is a defined minimum pressure throughout the network, it was considered that this constraint was enough to ensure adequate pressure levels in the network. However, in the case of a mandatory constant downstream pressure value, the hydraulic regulation strategy developed by [3] could be adopted. According to this strategy, when the pressure drop needs to be smaller than the head taken by the turbine, a by-pass is opened to divide the flow discharge.

4. Results and Discussion

4.1. Energy Recovery Potential in the WSS of Fribourg

The algorithm for the evaluation of the available energy in an urban water network (Figure 3) was applied to the case study of Fribourg. The results presented in Table 3 show that there is approximately 170 MWh/year in the network not being used. If accounting for the 430 MWh/year extracted from the WSS by PRVs, a total of approximately 600 MWh/year of available energy exists. The PRV energy contribution represents 72% of the total.

Table 3. Results from the evaluation of available energy in the city of Fribourg partitioned by the network, PRVs and the total.

	E_{pot} (MWh/year)
Network	168
Existing PRV	430
Total	598

Table 4 shows the pipes of the network where the energy is extracted and Table 5 shows the energy extracted by the PRVs. Some locations in Table 4 show available heads although the corresponding available energy is relatively low. This indicates that these locations are served with low discharges, thus hardly good position for the installation of a turbine.

Table 4. Pipes where energy has been extracted.

Pipe ID	H (m)	E_{pot} (MWh/year)
2986	12.3	101.9
2824	59.2	4.4
2914	11.1	8.9
1928	1.5	1.2
2973	39.4	29.2
786	40.8	20.6
2427	5.9	2.8
2415	0.6	0.3
Total	170.7	168.3

Table 5. Power extracted in RPV.

PRV ID	H (m)	E_{pot} (MWh/year)
2978	49.2	23.6
2979	43.2	17.3
2981	49.2	16.0
2982	49.6	24.4
2983	49.5	9.3
2984	50.0	4.4
2985	50.0	2.2
2986	40.3	331.2
2987	0.3	1.1
Total	381.3	429.5

In pipe 2986, despite the existence of a PRV (Tables 4 and 5), 12.3 m of head is still available which corresponds to an available energy of 102 MWh/year.

4.2. Capacity for Generation Using 5BTP Turbine

The search algorithm was applied to the city of Fribourg network model to obtain the optimal locations for the installation of 1, 2, 3 and 4 turbines. A discount rate of 4% was considered in the calculation of the net present value. Some interesting results were identified. The results for the selling to the grid scheme are presented in Table 6, showing also the 2nd and the 3rd best results in terms of NPV_{20} for each case. These latter options were defined by restricting one of the pipes from the previous best solution.

Table 6. Results from the search algorithm applied to the Fribourg network model. The three best solutions in terms of $NPV_{20years}$ are shown.

Best Solution	X	E (MWh/year)	D_t (mm)	P_{max} (kW)	NPV_{20} (k\$)	Cost Price (cts\$/kWh)	Payback Period (years)
Best solution	2986	60.5	165	8.1	258	2	0.7
	(2986, 2986)	120.9	(165, 165)	16.2	521	1	0.5
	(2986, 2986, 2986)	131.7	(155, 155, 155)	17.8	569	1	0.5
	(2986, 2986, 2986, 2987)	136.2	(165, 165, 165, 135)	18.5	586	1	0.6
2nd best solution	2730	60.5	165	8.1	250	3	1.1
	(2730, 2730)	120.9	(165, 165)	16.2	513	2	0.8
	(2730, 2730, 2730)	131.8	(155, 155, 155)	17.8	561	2	0.7
	(2730, 2730, 2730, 2987)	136.2	(165, 165, 165, 135)	18.5	575	2	0.8
3rd best solution	2987	1.5	85	0.2	5	7	2.9
	(2987, 2987)	1.3	(85, 85)	0.2	5	7	3.1
	(2987, 2987, 2987)	1.1	(85, 85, 85)	0.2	4	8	3.5
	(2982, 2982, 2987, 2987)	1.6	(85, 85, 95, 95)	0.3	5	12	5.6

In an initial phase, whenever the solution includes a location where a PRV concrete chamber already exists, the turbine is assumed to be installed either in line with the valve, or replacing it in the same chamber. Hence, the construction costs (concrete, excavation and earth fill) are omitted. Since these solutions tend to be cheaper than installing in new locations, they were given priority within the search algorithm.

The location in the network of the retained solutions is shown in Figure 8. In Table 6 the annual energy production, turbine runner diameters, average head, average turbinated flow, installed power, net present value after 20 years of operation and respective cost price of the best solutions are shown. No increase in pumping energy was necessary. All the energy generated was representative of excess pressure in the network. As expected from the analysis of Tables 4 and 5, the pipe 2986, with a PRV installed, was the best location to install the energy converters.

Figure 8 shows that the replacement of PRVs is often the best solution. In these, excess pressure is already recognized. Furthermore, the considered exemption of construction costs for existing chambers made these solutions more economically viable.

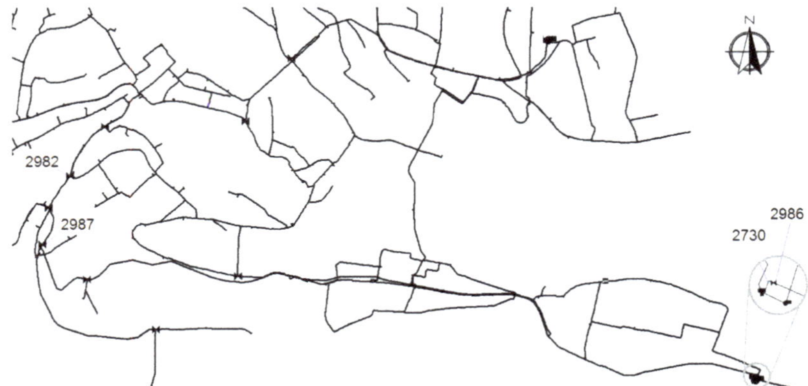

Figure 8. Localization of selected pipes from Table 6 (Zoom A from Figure 4).

The best solutions can extract a considerable quantity of the available energy of the network. One, two, three and four 5BTP turbines would recover approximately 10%, 20%, 22% and 23% of the available energy, respectively. According to these results, it can also be concluded that the installation of three turbines in one pipe represents a smaller increase of energy production from two turbines than the increase of energy production of installing two turbines when compared to one turbine. This is due to the effect of obstruction of flow discharge, in particular when the extracted head is bigger than the original head dissipated in the PRV.

The best and second best solutions, for all the number of turbines, are identical in terms of energy production and of NPV_{20}. Pipes 2896 and 2730 (Figure 8) are presented in the best and second best solutions, respectively. Since they share one node and are inline, the expected production is similar. The differences between both NPV_{20} values are mainly due to the construction costs.

The third best solution has a very small energy production when compared with others, since the discharges and heads are lower. It can be concluded that the pipes where the best and second best solutions are located, upstream from one of the main water tanks, is the most interesting area for hydropower production.

Comparing with the examples of other studies on MHPs in WSSs presented in the introduction, the obtained production in the city of Fribourg is within the same order of magnitude.

Not considering construction costs for the sites where PRVs are located may be optimistic. Hence, a second batch of simulations was carried out considering that additional construction works will be necessary to enlarge and adapt the existing chamber. Site conditions being very varied and coupling old and new chambers being sometimes cumbersome, it was assumed that the construction costs would be equivalent to that of a new chamber. Under these conditions, the best solution from Table 6 became equivalent in terms of NPV_{20} to the 2nd best solution, since the difference between the two was the construction costs. For the 3rd best solution, no locations were found were it would be feasible to install turbines. The construction costs have hence a considerable weight with respect to the feasibility of these chambers.

4.3. Response to Changes in Water Consumption

Based on the three best solutions previously identified, a 20% decrease in water consumption was imposed on the network. A new energetic equilibrium was computed for these conditions, leading to a new energy production for the network. The results of this sensitivity analysis are presented in Table 7, which can be directly compared with Table 6.

Table 7. Previous solutions with 80% of the consumption.

Best Solution	E (MWh/year)	NPV_{20} (k$)	Payback Period (years)
Best solution	60.5	258	0.7
	120.9	521	0.5
	130.3	563	0.5
	124.9	538	0.6
2nd best solution	60.5	250	1.1
	120.9	513	0.8
	130.4	555	0.7
	124.9	527	0.8
3rd best solution	2.0	8	2.1
	1.6	6	2.6
	1.5	6	2.7
	2.1	7	4.4

Considering that consumption decreased and that the energy production is highly dependent on the flow discharge, a decrease in energy production was expected. However, the 1st and 2nd best solutions present negligible changes and in the 3rd best solution, there is an increase of energy production. For the best and 2nd best solutions, the MHPs are installed immediately upstream from a regulation water tank. The 20% reduction in the consumption does not strongly influence the flow discharges in this area, which are highly dependent on the levels of the water tank and water source. For the 3rd best solution, the flow discharge increased due to the new network equilibrium. The majority of pipes in the network suffered a decrease in flow discharge with the smaller consumption. However, pipe 2987 was one of the exceptions. These results illustrate the complexity of installing MHPs in urban networks and evoke the need for careful sensitivity analysis with respect to the consumption. Considering the small differences in the results, a sensitivity analysis for a reduction of 10% in the consumption is not shown. Under the actual conditions, the network does not support an increase in the consumption (negative pressures appear in the network even for a slight increase). Thus, an analysis of an increase in the consumption is not presented. However, a complete sensibility analysis should always be envisaged. Also, carrying out long-term simulations is recommended, in order to achieve a robust estimation of the produced energy and economic value of the installation [11].

5. Methodology for Expedited Assessment of Energy Recovery

The obtained results were achieved through an optimization process where a considerable amount of data and also time are needed. Based on the experience gained in the course of this work, an expedited method to preliminarily evaluate the interest of placing one turbine in a given location of a network is provided here. The topography of the network, the maximum discharge in the pipes and the temporal variation of the consumption are assumed to be known.

Since the choice of diameter of the turbine is dependent on the maximum discharge in the pipe, the corresponding head is given by the characteristic curve of the turbine according to the similarity law. Figure 9 presents the variation of head with the maximum pipe discharge, which is obtained considering different runner diameters of the 5BTP. Figure 10 plots investment costs and the installed power as a function of the maximum pipe discharge and, consequently, the diameter of the runner. This figure was obtained considering the unit prices from Table 2 and the existing pipe has a diameter which allows a design velocity of 1 m/s.

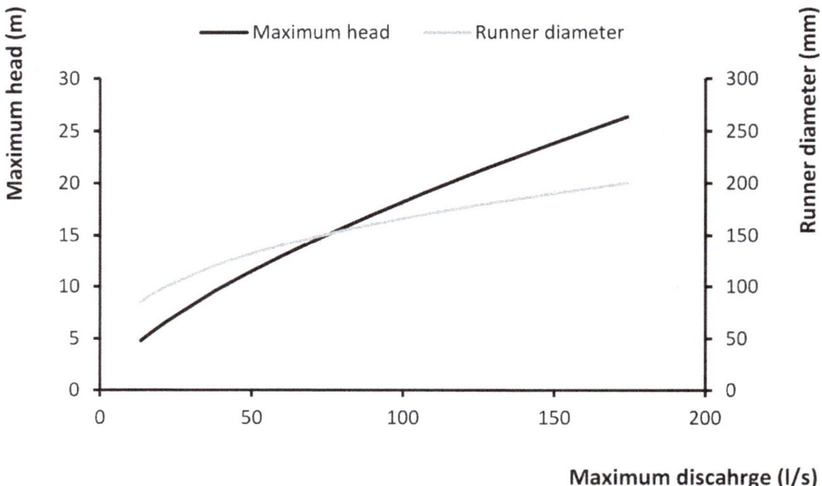

Figure 9. Characteristic curve of maximum discharges, for different diameters, of the 5BTP.

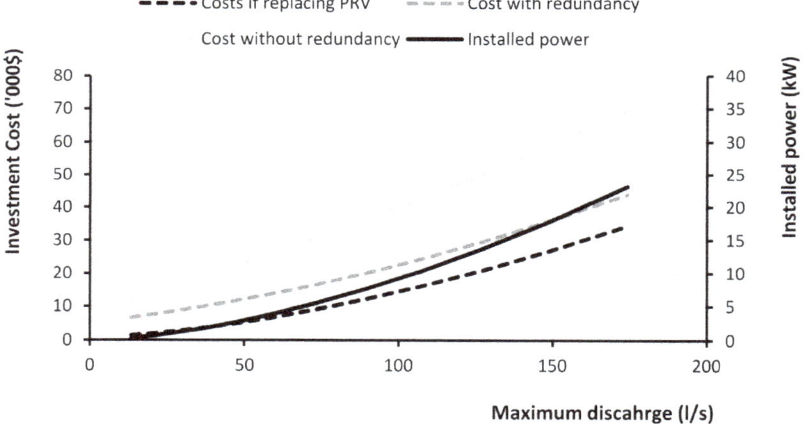

Figure 10. Variation of average investment cost and installed power of the 5BTP with maximum discharge in the pipe (assuming design head from Figure 9).

The expedite method follows the following steps:

1. Identify all PRVs in the network and obtain the respective maximum discharge.
2. Order all pipes in the network by decreasing discharge and select a feasible number of pipes with highest discharges to analyze (n = 20 for example).
3. In the n selected pipes, and considering the value of the maximum discharge, verify if the difference between the lowest pressure in the downstream node of each selected pipe and the limit minimum pressure is larger than the maximum head according to Figure 9.
4. Estimate for each PRV site and selected pipe the cost of the MHP and generation potential according to Figure 10.

5. Estimate an energy production based on the characteristic curves of the 5BTP [25] and the temporal variation of the consumption. The estimation can be obtained considering the flow data or, if not available, a consumption pattern such as Figure 5.

These steps allow for the preliminary identification of potential locations in the network to install a MHP with one turbine. However, a more detailed simulation, as proposed in the methodology, is required to ensure that the minimum pressure in all nodes is maintained and account for possible discharge variation due to the redundancy of the network, estimating with higher precision both energy production and costs. Combinations with more than one turbine also require detailed simulations.

6. Conclusions

A methodology has been presented in this paper to estimate the potential for hydropower in urban water supply networks based on an available energy concept. It is applied to the city of Fribourg, Switzerland and, for comparison, an optimization algorithm was used to estimate the actual energy production in the best locations of the network with a proposed micro-turbine, the 5BTP. The optimization maximizes the economic value of the installation, and combinations with one to four turbines were tested. It is concluded that one turbine can produce 60.5 MWh/year in this network, representing 10% of the available energy, and that four 5BTP turbines would extract 136.2 MWh/year, which is less than four times the production with one turbine. Locations where PRVs are already in place are attractive, especially if a civil infrastructure that can be adapted is already in place. The costs for civil construction were seen to have an important weight on the feasibility results. Sensitivity analysis of the demand should be considered to verify the impact of this in the energy production and in the behavior of the network. Finally, a method for an expedited assessment of the location and energy production with one turbine is proposed based on discharge data in the network and on typical design assumptions.

Acknowledgments: This research is supported by Ph.D. grant ref. SFRH/BD/51931/2012 issued by FCT under the IST-EPFL Joint Ph.D. initiative. Sponsoring by the Swiss Competence Center for Energy and Research—Supply of Electricity (SCCER-SoE) under contract CTI.2013.0288 is also acknowledged. The authors thank the Industrial Services of Fribourg, Switzerland, for the data provided to support this work and Ribi SA for sharing their knowledgeable insight on the Fribourg water supply network.

Author Contributions: Irene Samora is a Ph.D. Student supervised by Helena M. Ramos and Anton J. Schleiss; Mário J. Franca and Pedro Manso contributed to the developed concepts; all authors contributed to the writing of the paper.

Conflicts of Interest: The authors declare no conflict of interest. The funding sponsors had no role in the design of the study; in the collection, analyses and interpretation of data; in the writing of the manuscript; and in the decision to publish the results.

Abbreviations

The following abbreviations are used in this manuscript:

WSS	Water supply system
MHP	Micro-hydropower plant
PRV	Pressure reducing valve
NPV	Net present value
5BTP	Five blade tubular propeller

References

1. Vilanova, M.R.N.; Balestieri, J.A.P.; Filho, P.M. Energy and hydraulic efficiency in conventional water supply systems. *Energy Policy* **2013**, *30*, 701–714.
2. Ulanicki, B.; Bounds, P.L.M.; Rance, J.P.; Reynolds, L. Open and closed loop pressure control for leakage reduction. *Urban Water* **2000**, *2*, 105–114. [CrossRef]
3. Carravetta, A.; Giuseppe, G.; Fecarotta, O.; Ramos, H. Energy production in water distribution networks: A PAT design strategy. *Water Resour. Manag.* **2012**, *26*, 3947–3959. [CrossRef]

4. Gallagher, J.; Harris, I.M.; Packwood, A.J.; McNabola, A.; Williams, A.P. Strategic assessment of energy recovery sites in the water industry for UK and Ireland: Setting technical and economic constraints through spatial mapping. *Renew. Energy* **2015**, *81*, 808–815. [CrossRef]

5. Samora, I.; Franca, M.J.; Schleiss, A.J.; Ramos, H.M. Simulated annealing in optimization of energy production in a water supply network. *Water Resour. Manag.* **2016**, *30*, 1533–1547. [CrossRef]

6. Vicente, D.J.; Garrote, L.; Sánchez, R.; Santillán, D. Pressure management in water distribution systems: Current status, proposals, and future trends. *J. Water Resour. Plan. Manag.* **2016**. [CrossRef]

7. Xu, Q.; Chen, Q.; Ma, J.; Blanckaert, K.; Wan, Z. Water saving and energy reduction through pressure management in urban water distribution networks. *Water Resour. Manag.* **2014**, *28*, 3715–3726. [CrossRef]

8. Fantozzi, M.; Calza, F.; Kingdom, A. Experience and results achieved in introducing District Metered Areas (DMA) and Pressure Management Areas (PMA) at Enia utility (Italy). In Proceedings of the IWA International Specialized Conference Water Loss, Cape Town, South Africa, 26–29 April 2009.

9. Carravetta, A.; Giugni, M. Functionality factors in the management and rehabilitation of water networks. In *Management of Water Networks*, Proceedings of the Conference Efficient Management of Water Networks, Ferrara, Italy, 17–19 May 2006.

10. Fecarotta, O.; Aricò, C.; Carravetta, A.; Martino, R.; Ramos, H.M. Hydropower potential in water distribution networks: Pressure control by PATs. *J. Water Resour. Plan. Manag.* **2014**, *29*, 699–714. [CrossRef]

11. Sitzenfrei, R.; von Leon, J. Long-time simulation of water distribution systems for the design of small hydropower systems. *Renew. Energy* **2014**, *72*, 182–187. [CrossRef]

12. OFGEM. *Feed-in Tariff: Guidance for Renewable Installations*, 10.2th ed.; Gas and Electricity Markets Authority: London, UK, 2016.

13. Ramos, H.; Kenov, K.; Vieira, F. Environmentally friendly hybrid solutions to improve the energy and hydraulic efficiency in water supply systems. *Energy Sustain. Dev.* **2011**, *15*, 436–442. [CrossRef]

14. Williams, A.; Smith, N.P.A.; Bird, C.; Howard, M. Pumps as turbines and the inductions motors as generators for energy recovery in water supply systems. *Water Environ. J.* **1998**, *12*, 175–178. [CrossRef]

15. Lisk, B.; Greenberg, E.; Bloetscher, F. *Implementing Renewable Energy at Water Utilities*; Case Studies; Water Research Foundation: Denver, CO, USA, 2012.

16. Hong Kong Polytechnic University, Novel Inline Hydropower System for Power Generation from Water Pipelines. Available online: http://phys.org/news/2012-12-inline-hydropower-power-pipelines.html (accessed on 24 May 2016).

17. McNabola, A.; Coughlan, P.; Williams, A.P. Energy recovery in the water industry: An assessment of the potential of micro hydropower. *Water Environ. J.* **2014**, *28*, 294–304. [CrossRef]

18. Samora, I.; Manso, P.; Franca, M.J.; Schleiss, A.J.; Ramos, H.M. Opportunity and economic feasibility of inline micro-hydropower units in water supply networks. *J. Water Resour. Plan. Manag.* **2016**. [CrossRef]

19. Corcoran, L.; McNabola, A.; Coughlan, P. Optimization of water distribution networks for combined hydropower energy recovery and leakage reduction. *J. Water Resour. Plan. Manag.* **2015**. [CrossRef]

20. Rossman, L.A. *EPANET 2 Users' Manual. Water Supply and Water Resources Division, National Risk Management Research Laboratory, Cincinnati*; United States Environmental Protection Agency: Cincinnati, OH, USA, 2000.

21. Hickey, H. *Water Supply Systems and Evaluation Methods Vol II*; U.S. Fire Administration: Emmitsburg, MD, USA, 2008.

22. Swiss Gas and Water Industry Association (SVGW). *W4—Directive for Water Supply*; SVGW: Zurich, Switzerland, 2013. (In French)

23. Swiss Federal Council (SFC). *Directive for Energy*; SFC: Bern, Switzerland, 1998.

24. Swiss Federal Office of Energy (SFOE). *Directive for Feed-in-Tariff of Injected Current*; SFOE: Bern, Switzerland, 2015.

25. Samora, I.; Hasmatuchi, V.; Münch-Alligné, C.; Franca, M.J.; Schleiss, A.J.; Ramos, H.M. Experimental characterization of a five blade tubular propeller turbine for pipe inline installation. *Renew. Energy* **2016**, *95*, 356–366. [CrossRef]

Article

Experimental and Numerical Analysis of a Water Emptying Pipeline Using Different Air Valves

Oscar E. Coronado-Hernández [1,2,*], Vicente S. Fuertes-Miquel [2], Mohsen Besharat [3,4] and Helena M. Ramos [4]

1   Facultad de Ingeniería, Universidad Tecnológica de Bolívar, Cartagena 131001, Colombia
2   Departamento de Ingeniería Hidráulica y Medio Ambiente, Universitat Politècnica de València, Valencia 46022, Spain; vfuertes@upv.es
3   Department of Civil Engineering, Saghez Branch, Islamic Azad University, Saghez 66819-73477, Iran; mohsen.besharat@iausaghez.ac.ir
4   Department of Civil Engineering and Architecture, CERIS, Instituto Superior Técnico, University of Lisbon, Lisbon 1049-001, Portugal; mohsen.besharat@tecnico.ulisboa.pt (M.B.); hr@civil.ist.utl.pt (H.M.R.)
*   Correspondence: ocoronado@unitecnologica.edu.co; Tel.: +57-301-371-5398

Academic Editor: Arjen Y. Hoekstra
Received: 5 December 2016; Accepted: 3 February 2017; Published: 8 February 2017

Abstract: The emptying procedure is a common operation that engineers have to face in pipelines. This generates subatmospheric pressure caused by the expansion of air pockets, which can produce the collapse of the system depending on the conditions of the installation. To avoid this problem, engineers have to install air valves in pipelines. However, if air valves are not adequately designed, then the risk in pipelines continues. In this research, a mathematical model is developed to simulate an emptying process in pipelines that can be used for planning this type of operation. The one-dimensional proposed model analyzes the water phase propagation by a new rigid model and the air pockets effect using thermodynamic formulations. The proposed model is validated through measurements of the air pocket absolute pressure, the water velocity and the length of the emptying columns in an experimental facility. Results show that the proposed model can accurately predict the hydraulic characteristic variables.

Keywords: air-water; air pocket; air valve; hydraulic model; pipeline; emptying; water supply; water hammer

1. Introduction

The emptying procedure in a pipeline generates hydraulic events that can cause problems if air valves are not well sized. In practical applications, engineers follow typical recommendations from the American Water Works Association (AWWA) [1] or manufacturers about sizing and location of air valves along a pipeline in order to avoid subatmospheric pressure that can cause the collapse of the system. It is recommended in a pipeline that the air volume admitted by air valves should be the same as the water volume drained [2]. Consequently, air valves should work in subsonic flow conditions. Air valves should be located at high points, in long horizontal pipe branches, long descents, long ascents, at the decrease in an up-slope and increase in a down-slope of a pipeline, and on the discharge side of deep well pumps and at vertical turbines/pumps [1]. An inappropriate selection of the air valve size and location produces not only subatmospheric pressure but also a slow drainage of the system.

Presently, there are only few studies related to the emptying process in the literature, but they are not focused on practical applications because they do not consider a pipeline with an irregular profile [3] and air valves [4,5].

The analysis of an emptying process is not trivial because it requires the study of a two-phase flow [6–9]. This problem can be studied using one (1D) [10,11], two (2D) [12] or three-dimensional (3D) [13] models. The water phase in the 1D model can be analyzed considering two types of models [14]: (i) elastic models [15,16], which consider the elasticity of the pipe and the water; or (ii) rigid models [17], which ignore the elasticity of them. Normally, elastic models are solved by using the method of characteristics [18,19] and rigid models by using the numerical solutions of ordinary differential equations [3,11,17,20]. In pressurized systems, the air effect can be analyzed as a single-phase flow, where the absolute pressure of the air pocket is computed between two water columns [10,11,21]. Regarding the analysis of 2D and 3D CFD modeling of air-water interface in closed pipes, they are still unyielding in the application of pipeline draining because length and time scales are not appropriate.

This research develops a general 1D mathematical model that can be used for analyzing the behaviour of the main hydraulic variables during the emptying process in a pipeline with an irregular profile and with several air valves installed along it based on formulations of previous works [3,10,11,17,20,22,23]. The proposed model can give important information in real systems about: (i) the risk of collapse of pipeline installations, by checking in a pipe manufacturer characteristics if the stiffness class is appropriate to support the subatmospheric pressure reached during the hydraulic event depending on the type of soil in natural conditions, the type of backfill and the cover depth; (ii) the appropriate selection of the air valve during the emptying process; (iii) the size and the maneuver time of the drain valves; and (iv) the estimation of the drainage time of a pipeline.

2. Mathematical Model

Figure 1 shows a typical configuration of an irregular profile in a pipeline which consists of n entrapped air pockets, d air valves, p pipes, and o drain valves located at low points for draining the system, D being the internal pipe diameter (m); A being the cross section area of the pipe (m^2); f being the Darcy–Weisbach friction factor; and g being the gravity acceleration (m/s^2). L_j ($j = 1, 2, ..., p$) is the total length of each pipe, AV_m ($m = 1, 2, ..., d$) are the air valves, and K_s ($s = 1, 2, ..., o$) represents the flow factor for each drain valve. The emptying process starts when the drain valves are opened, thus air valves start to admit air into the pipeline. After that, the air pocket i will begin to expand generating subatmospheric pressure. At the same time, the drainage of the water column starts until the entire pipeline is completely emptied. Figure 1 gives the evolution at time t, the length of the emptying column $L_{e,j}$ ($j = 1, 2, ..., p$), the absolute pressure of the air pocket p_i^* ($i = 1, 2, ..., n$), and the water velocity $v_{e,j}$ ($j = 1, 2, ..., p$). The expansion of the air pocket i can be computed as $x_i = L_j - L_{e,j} + L_{j+1} - L_{e,j+1}$.

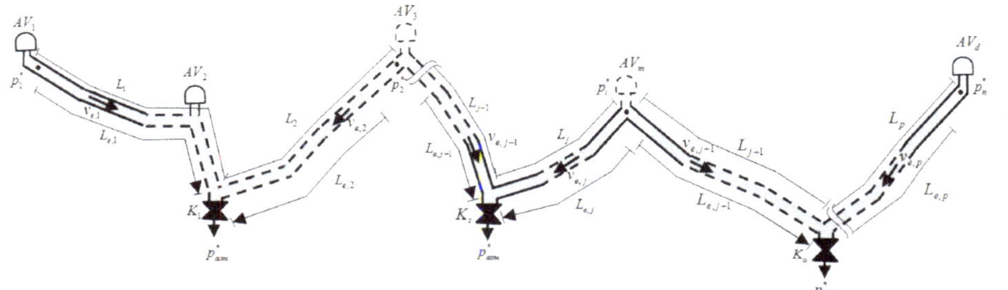

Figure 1. Scheme of entrapped air pockets in a pipeline with irregular profile while water empties.

The one-dimensional (1D) proposed model has the following assumptions: (1) water column has been modeled by the rigid model; (2) the Darcy–Weisbach equation was considered to evaluate friction losses; (3) the thermodynamic behaviour of the air pocket is analyzed using a polytropic model;

and (4) the air–water interface is perpendicular to the pipe direction. The proposed model can be used for pipelines with small diameters and with hydraulic slopes enough to prevent downstream air intrusion where open-channel flow does not occur [20,24].

Under these hypotheses, the problem is modeled by the following set of equations.

2.1. Equations for the Water Phase

- Mass oscillation equation:

The rigid model was used in order to compute the evolution of the water column [11,25], considering that the elasticity of the air is much higher than the elasticity of the pipe and the water [12]. Applying the rigid model to the emptying column j and considering that the drain valve s joins pipes L_j and L_{j-1}, which is a common point in a pipeline, then:

$$\frac{dv_{e,j}}{dt} = \frac{p_i^* - p_{atm}^*}{\rho_w L_{e,j}} + g\frac{\Delta z_{e,j}}{L_{e,j}} - f_j\frac{v_{e,j}|v_{e,j}|}{2D} - \frac{gA^2(v_{e,j} + v_{e,j-1})|v_{e,j} + v_{e,j-1}|}{L_{e,j}K_s^2}, \tag{1}$$

where $\Delta z_{e,j}$ is the elevation difference (m), ρ_w is the water density (kg/m^3), p_{atm}^* is the atmospheric pressure (101,325 Pa) and g is the gravity acceleration (m/s^2).

The expression $h_{m,s} = Q_{w,s}^2/K_s^2$ was used to estimate the local loss of the drain valve s in Equation (1). K_s is the flow factor and $Q_{w,s}$ is the water discharge.

If the drain valve only connects pipe L_j, thus $v_{e,j-1} = 0$.

- Air–water interface position:

The position of the air–water interface is considered perpendicular to the pipe direction [11,18,26]. The continuity equation for the moving air-water interface j is:

$$\frac{dL_{e,j}}{dt} = -v_{e,j} \quad \left(L_{e,j} = L_{e,j,0} - \int_0^t v_{e,j}dt\right), \tag{2}$$

where subscript 0 refers to the initial condition.

2.2. Equations for Air Pockets

- Continuity equation [3,17]:

$$\frac{dm_{a,i}}{dt} = \rho_{a,nc}v_{a,nc,m}A_{adm,m} \tag{3}$$

and by deriving:

$$\frac{dm_{a,i}}{dt} = \frac{d(\rho_{a,i}V_{a,i})}{dt} = \frac{d\rho_{a,i}}{dt}V_{a,i} + \frac{dV_{a,i}}{dt}\rho_{a,i}, \tag{4}$$

where $m_{a,i}$ is the air mass and $V_{a,i}$ is the air volume of the air pocket i.

Due to the air pocket i located between pipes L_j and L_{j+1}, then $V_{a,i} = (L_j - L_{e,j})A + (L_{j+1} - L_{e,j+1})A$ and $dV_{a,i}/dt = -(dL_{e,j}/dt)A - (dL_{e,j+1}/dt)A = A(v_{e,j} + v_{e,j+1})$, thus:

$$\frac{d\rho_{a,i}}{dt} = \frac{\rho_{a,nc}v_{a,nc,m}A_{adm,m} - (v_{e,j+1} + v_{e,j})A\rho_{a,i}}{A(L_j - L_{e,j} + L_{j+1} - L_{e,j+1})}, \tag{5}$$

where $\rho_{a,i}$ is the air density of the air pocket i, $\rho_{a,nc}$ is the air density in normal conditions (1.205 kg/m^3), $A_{adm,m}$ is the cross section area (m^2) of the air valve m and $v_{a,nc,m}$ is the air velocity in normal conditions admitted by the air valve m.

- Expansion equation for the air pocket i:

the thermodynamic behaviour of the air pocket [27] is treated by using a polytropic model [22,28]

$$\frac{dp_i^*}{dt} = -k\frac{p_i^*}{V_{a,i}}\frac{dV_{a,i}}{dt} + \frac{p_i^*}{V_{a,i}}\frac{k}{\rho_{a,i}}\frac{dm_{a,i}}{dt}. \tag{6}$$

Considering the equations presented before, then:

$$\frac{dp_i^*}{dt} = \frac{kp_i^*}{A(L_j - L_{e,j} + L_{j+1} - L_{e,j+1})}\left(\frac{\rho_{a,nc}v_{a,nc,i}A_{adm,m}}{\rho_{a,i}} - A(v_{e,j+1} + v_{e,j})\right), \tag{7}$$

where k is the polytropic coefficient: if $k = 1.0$, then the process is isothermal, but, if $k = 1.4$, the process is adiabatic.

In Equations (5) and (7), if the air pocket i is only located in pipe L_j, then $v_{e,j+1} = 0$ and $L_{j+1} = L_{e,j+1} = 0$.

- Air valve characterization:

The formulation proposed by Wylie and Streeter [23,29] was used to represent the air admission into the system for air valves. Ideally, the air valve m should be working in subsonic flow ($p_{atm}^* > p_i^* > 0.528p_{atm}^*$), then:

$$Q_{a,nc,m} = C_{adm,m}A_{adm,m}\sqrt{7p_{atm}^*\rho_{a,nc}\left[\left(\frac{p_i^*}{p_{atm}^*}\right)^{1.4286} - \left(\frac{p_i^*}{p_{atm}^*}\right)^{1.714}\right]}, \tag{8}$$

where $Q_{a,nc,m}$ is the air discharge in normal conditions admitted by the air valve m and $Q_{a,nc,m} = v_{a,nc,m}A_{adm,m}$.

When air valves are not located in high points of the system, then the position of them should be specified. When the air–water interface reaches an air valve, then it starts to admit air inside the system.

In summary, a set of $2p + 2n + d$ equations describes the whole system. Together with the corresponding boundary conditions, it can be solved for the $2p + 2n + d$ unknowns $v_{e,j}, L_{e,j}, p_i^*, \rho_{a,i}, v_{a,nc,m} (j = 1, 2, ..., p; i = 1, 2, ..., n; m = 1, 2, ..., d)$.

2.3. Initial and Boundary Conditions

When the pipeline is at rest, the initial conditions are described as follows: $v_{e,j}(0) = 0(j = 1, 2, .., p)$, $L_{e,j}(0) = L_{e,j,0}(j = 1, 2, .., p)$, $p_i^*(0) = p_{i,0}^*(i = 1, 2, .., n)$, $\rho_{a,i}(0) = \rho_{a,i,0}(i = 1, 2, .., n)$, and $v_{a,nc,m}(0) = 0(i = 1, 2, .., d)$.

The upstream boundary condition is given by $p_i^* = p_{i,0}^*$, the downstream is given by the flow factor K_s of the drain valve s, and the atmospheric pressure p_{atm}^* due to the free discharge.

2.4. Gravity Term

Figure 2 describes the evolution of the gravity term (see Equation (1)) along the emptying column j. $L_{j,r}$ $(r = 1, 2, ..., h)$ is the length of the pipe reach r, and L_j is the total pipe length j $(L_j = \sum_{r=1}^{r=h} L_{j,r})$. Subscript u is used to identify the pipe reaches (1 to h) where the air–water interface is located.

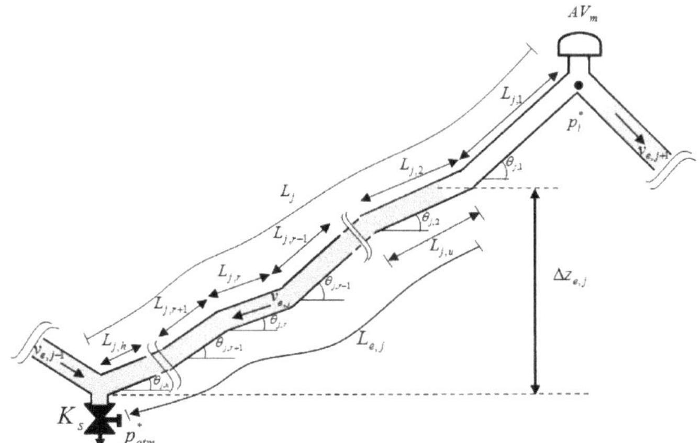

Figure 2. Evolution of the gravity term of the emptying column *j*.

The gravity term of the emptying column *j* is computed by:

- When the air–water interface has not arrived the last reach ($r \neq h$):

$$\frac{\Delta z_{e,j}}{L_{e,j}} = \frac{\sum_{r=u+1}^{r=h} L_{j,r}\sin(\theta_{j,r})}{L_{e,j}} + \left(1 - \frac{\sum_{r=u+1}^{r=h} L_{j,r}}{L_{e,j}}\right)\sin(\theta_{j,u}). \tag{9}$$

- When the air–water interface is located on the last reach ($r = h$):

$$\frac{\Delta z_{e,j}}{L_{e,j}} = \sin(\theta_{j,h}). \tag{10}$$

3. Application to Two Emptying Columns

Figure 3 presents a case of two emptying columns. More complex systems can be treated in the same way based on the proposed model.

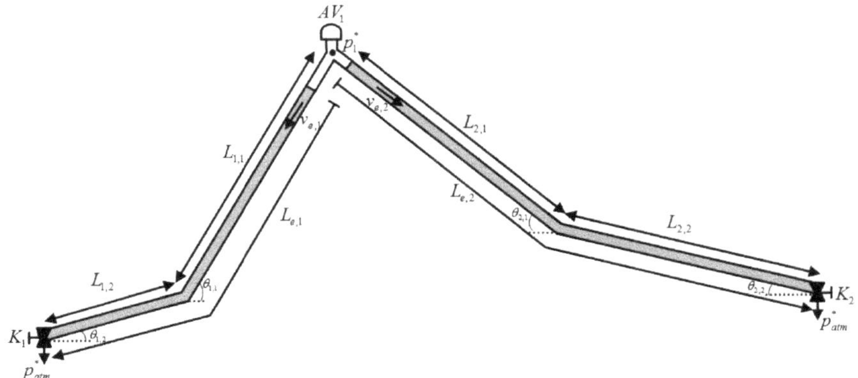

Figure 3. Two emptying columns in a pipeline.

The corresponding equations of the pipeline are:

1. Mass oscillation equation applied to the emptying column 1

$$\frac{dv_{e,1}}{dt} = \frac{p_1^* - p_{atm}^*}{\rho_w L_{e,1}} + g\frac{\Delta z_{e,1}}{L_{e,1}} - f_1\frac{v_{e,1}|v_{e,1}|}{2D} - \frac{gA_1^2 v_{e,1}|v_{e,1}|}{K_1^2 L_{e,1}}. \tag{11}$$

2. Emptying column 1 position

$$\frac{dL_{e,1}}{dt} = -v_{e,1} \quad \left(L_{e,1} = L_{e,1,0} - \int_0^t v_1 dt \right). \tag{12}$$

3. Mass oscillation equation applied to the emptying column 2

$$\frac{dv_{e,2}}{dt} = \frac{p_1^* - p_{atm}^*}{\rho_w L_{e,2}} + g\frac{\Delta z_{e,2}}{L_{e,2}} - f_2\frac{v_{e,2}|v_{e,2}|}{2D} - \frac{gA_2^2 v_{e,2}|v_{e,2}|}{K_2^2 L_{e,2}}. \tag{13}$$

4. Emptying column 2 position

$$\frac{dL_{e,2}}{dt} = -v_{e,2} \quad \left(L_{e,2} = L_{e,2,0} - \int_0^t v_{e,2} dt \right). \tag{14}$$

5. Evolution of the air pocket

$$\frac{dp_1^*}{dt} = -k\frac{p_1^*(v_{e,1}A_1 + v_{e,2}A_2)}{A_2(L_2 - L_{e,2}) + A_1(L_1 - L_{e,1})} + \frac{p_1^*\rho_{a,nc}v_{a,nc,1}A_{adm,1}}{A_2(L_2 - L_{e,2}) + A_1(L_1 - L_{e,1})}\frac{k}{\rho_{a,1}}. \tag{15}$$

6. Continuity equation of the air pocket

$$\frac{d\rho_{a,1}}{dt} = \frac{\rho_{a,nc}v_{a,nc,1}A_{adm,1} - (v_{e,1}A_1 + v_{e,2}A_2)\rho_{a,1}}{A_2(L_2 - L_{e,2}) + A_1(L_1 - L_{e,1})}. \tag{16}$$

7. Air valve characterization

$$Q_{a,nc,1} = C_{adm,1}A_{adm,1}\sqrt{7p_{atm}^*\rho_{a,nc}\left[\left(\frac{p_1^*}{p_{atm}^*}\right)^{1.4286} - \left(\frac{p_1^*}{p_{atm}^*}\right)^{1.714}\right]}. \tag{17}$$

This set of seven differential-algebraic equations (Equations (11) to (17)), together with the initial condition given by $v_{e,1}(0) = 0$, $L_{e,1}(0) = L_{e,1,0}$, $v_{e,2}(0) = 0$, $L_{e,2}(0) = L_{e,2,0}$, $p_{1,0}^* = p_{atm}^* = 101{,}325$ Pa, $\rho_{a,1,0} = \rho_{a,nc} = 1.205$ kg/m^3 and $Q_{a,nc,1}(0) = 0$, allow for describing the whole problem. The Simulink Library in Matlab (The MathWorks, Inc., Natick, MA, USA) was used in order to compute the seven unknown functions: $v_{e,1}$, $L_{e,1}$, $v_{e,2}$, $L_{e,2}$, p_1^*, $\rho_{a,1}$ and $Q_{a,nc,1}$.

3.1. Experimental Model

In order to study the emptying process in a pipeline, an experimental facility was developed (see Figure 4) at the Civil Engineering, Research and Innovation for Sustainability (CEris) Center, in the Hydraulic Lab of Instituto Superior Técnico (IST), University of Lisbon, Portugal. The experimental facility consisted of a set of transparent PVC pipes with 7.3 m length and nominal diameter of 63 mm (DN63). An air valve (AV_1) was installed at the highest point of the pipeline with a pressure transducer (PT_1) to measure the absolute pressure. The air valves S050 and D040 (manufacturer A.R.I.) and different air pocket sizes were tested for showing the effect on the hydraulic behaviour. There were four ball valves (BV_s). BV_1, BV_2 and BV_4 were opened, consequently permitting the movement of the water column. BV_3 and manual valve (MV_1) were closed and represented the system configuration extremities. The two manual ball valves (MBV_s) identified as MBV_1 and MBV_2 with nominal diameter of 25 mm (*DN*25) at the downstream ends were used to control the outflow conditions. These valves have the same level as the horizontal pipes $L_{1,2}$ and $L_{2,2}$. Two free-surface small tanks were used to collect the drainage water. The PicoScope system was used for absolute pressure data recording. The frequency of the pressure data collection was 0.0062 s. The length of the emptying columns were measured by using a Sony Camera DSC-HX200V (Sony Corporation, Minato, Tokyo, Japan) for decomposing frames for each second. The water velocities were measured with an Ultrasonic Doppler Velocimetry (UDV) device with a transducer for 4 MHz frequency (MET FLOW, Lausanne, Switzerland). The transducer was located on the horizontal pipe with an angle of 20°. To measure the water velocities, all other facilities were turned off in the Hydraulic Lab to avoid the noise, and seeding was used inside the water in order to get appropriate measurements.

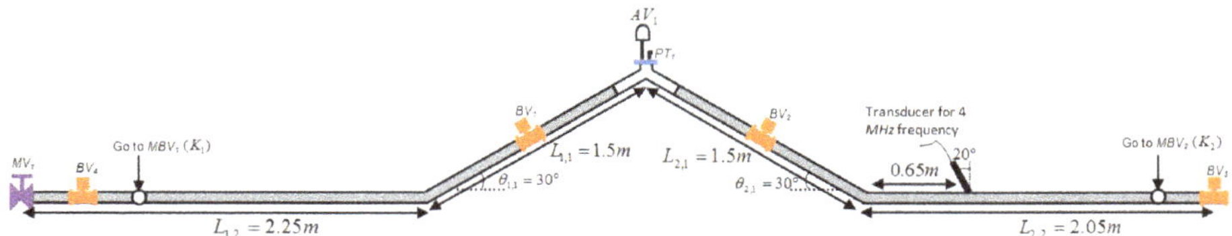

Figure 4. The pipe system and its components.

The emptying process of the experimental facility was started by an opening maneuver at the same time as valves (MBV_1) and (MBV_2). Consequently, the two emptying columns started the emptying procedure until reaching the horizontal pipes when the drainage is practically stopped, and part of the two water columns remain inside the installation because the gravity term is zero in both pipe reaches.

Equations (8)–(14) were used to simulate the emptying process of this experimental facility. The gravity terms were computed for the two emptying columns. For the emptying column 1, it depends on:

- If the air-water interface located on the sloped pipe reaches $\theta = 30°$ ($L_{e,1} \leq L_{1,1} + L_{1,2}$ and $L_{e,1} > L_{1,2}$), then:

$$\frac{\Delta z_{e,1}}{L_{e,1}} = \left(1 - \frac{L_{1,2}}{L_{e,1}}\right) \sin(\theta_{1,1}). \tag{18}$$

- If the air–water interface located on the horizontal pipe reaches $\theta = 0°$ ($L_{e,1} > 0$ and $L_{e,1} \leq L_{1,2}$), then:

$$\frac{\Delta z_{e,1}}{L_{e,1}} = 0. \tag{19}$$

The gravity term for emptying column 2 is similar in emptying column 1.

3.2. Experimental Results and Model Verification

Ten experimental tests are selected as shown in Table 1, where two different air valves and five air pocket sizes were defined. The air valve D040 admits large quantities of air during the emptying process, and it has a diameter of 9.375 mm and C_{adm} of 0.375 according to the vacuum curve presented by the manufacturer. The air valve S050 is not recommended for vacuum protection because it has a smaller orifice of 3.175 mm. The manufacturer does not present a vacuum curve because it is used specially for relief in pressurized systems. Consequently, the C_{adm} was calibrated during the tests with a value of 0.303. The selection of the appropriate air valve size is of utmost importance. The initial air pocket lengths were 0.001, 0.540, 0.920, 1.320 and 2.120 m. To avoid a numerical problem in the proposed model, a minimum air pocket size around of 1 mm is imposed instead of 0 mm.

Table 1. Characteristics of tests.

Test No.	Air Valve Model	Air Pocket Length (m)
1	S050	0.001
2	S050	0.540
3	S050	0.920
4	S050	1.320
5	S050	2.120
6	D040	0.001
7	D040	0.540
8	D040	0.920
9	D040	1.320
10	D040	2.120

In the model, a constant friction factor of $f = 0.018$ was used. Valves MBV_1 and MBV_2 were modeled by using a synthetic maneuver, with a flow factor of $K_1 = K_2 = 1.4 \times 10^{-3}$ m^3/s and a valve maneuvering time (T_m) of 1.6 s. The flow factor represents the local losses due to the opening of the valve and the reduction from $DN63$ to $DN25$. The expansion of the air pocket was modeled by using a polytropic model in adiabatic conditions ($k = 1.4$) because the event occurs very quickly.

According to the results, there are two types of behaviours that depend on the air valve: (1) air valve $S050$ (see Figure 5) and (2) air valve $D040$ (see Figure 6). In all tests, the proposed model can predict the subatmospheric pressure pattern. Test No. 1 and Test No. 6 were selected in order to compare results.

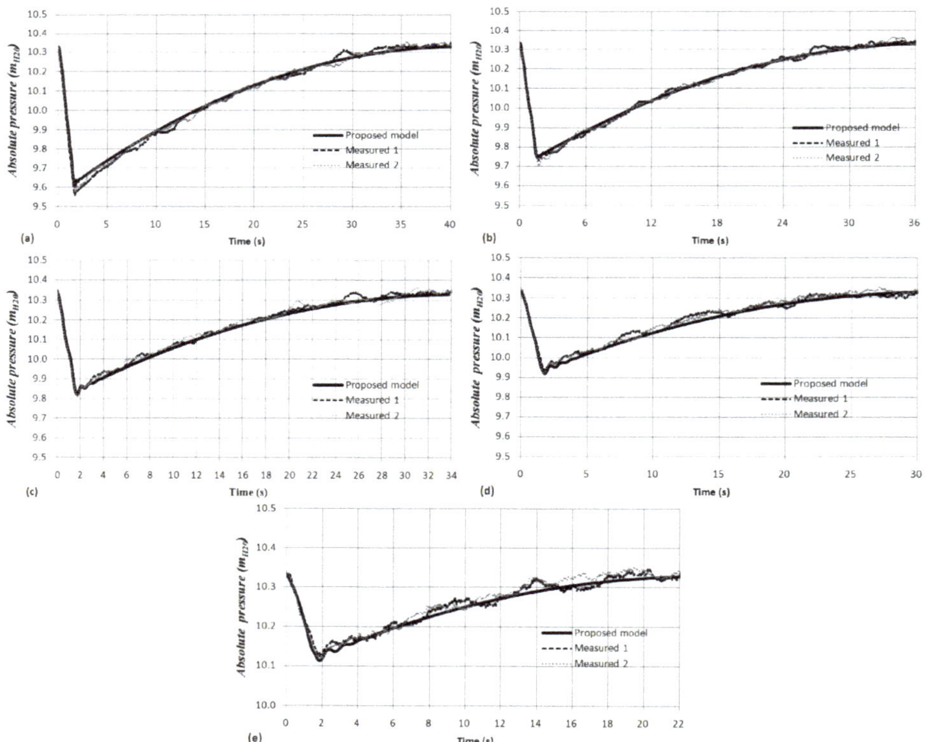

Figure 5. Comparison between computed and measured absolute pressure oscillation patterns (air valve S050): (**a**) Test No. 1; (**b**) Test No. 2; (**c**) Test No. 3; (**d**) Test No. 4; (**e**) Test No. 5.

Figure 5a shows Test No. 1 (air pocket size of 0.001 m) where the absolute pressure quickly reaches the minimum subatmospheric pressure of 9.61 m_{H20} at 1.69 s. Then, the absolute pressure pattern starts to increase slowly until it reaches the atmospheric condition. The duration of the hydraulic event is 40.3 s. In contrast, when the air valve $D040$ is used, small troughs of subatmospheric pressure occur and the hydraulic event is very short. Figure 6a shows the results for Test No. 6 (air pocket size of 0.001 m) where the absolute pressure decreases quickly until it reaches the minimum subatmospheric pressure of 10.16 m_{H20} at 1.82 s and then increases again until it reaches the atmospheric condition (10.33 m_{H20}). The duration of the hydraulic event is 8.13 s. Figures 5 and 6 show that the pressure drop is linear due to the opening of the valves MBV_1 and MBV_2. Then, subatmospheric pressure is presented and the water flow starts to decrease since the air valve can admit a better ratio of the air flow. Consequently, the pressure pattern rises.

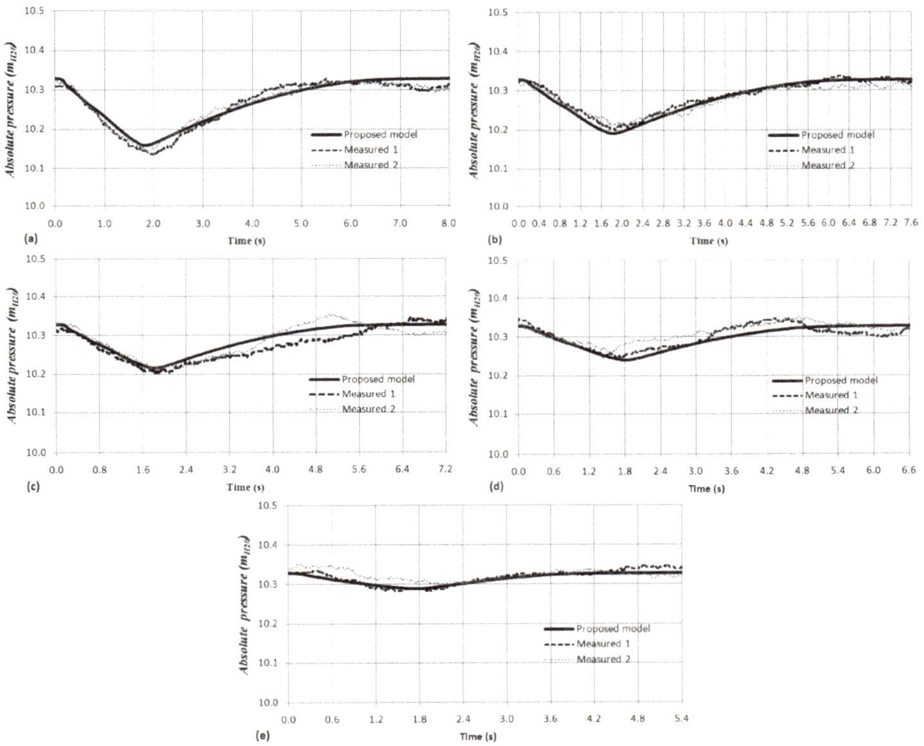

Figure 6. Comparison between computed and measured absolute pressure oscillation patterns (air valve D040): (**a**) Test No. 6; (**b**) Test No. 7; (**c**) Test No. 8; (**d**) Test No. 9; (**e**) Test No. 10.

In more complex and large systems, an air valve similar to $S050$ cannot be recommended as a protection device during the emptying process because, depending on the conditions of the installation, the subatmospheric pressure can reach excessively low values. Engineers should select an air valve similar to $D040$ for minimizing problems associated with the pressure drop to subatmospheric value.

The density of the air pocket is validated with the measurements of the absolute pressure, since these variables are related, because the temperature of the air pocket remains practically constant. Therefore, the results are similar considering an isothermal process ($k = 1.0$).

Figure 7 presents the evolution of the length of emptying columns 1 and 2 for Test No. 2 and Test No. 7. Figure 7a shows the results for Test No. 2 (air valve $S050$) where the emptying column 1 reached the horizontal pipe at 28 s, while the emptying column 2 reached it at 29 s. This difference of 1 s was caused because the valves MBV_1 and MBV_2 were not opened exactly at the same time. In contrast, Figure 7 shows the results for Test No. 7 (air valve D040), where practically the two emptying column reached the horizontal pipe in 5 s because the hydraulic event in this case is faster. In both tests, the proposed model predicted the length of emptying columns. It is important to note that when the emptying column reaches the horizontal pipe $\theta = 0°$, the proposed model cannot be applied because air–water interface is parallel to the horizontal pipe direction (as a stratified flow).

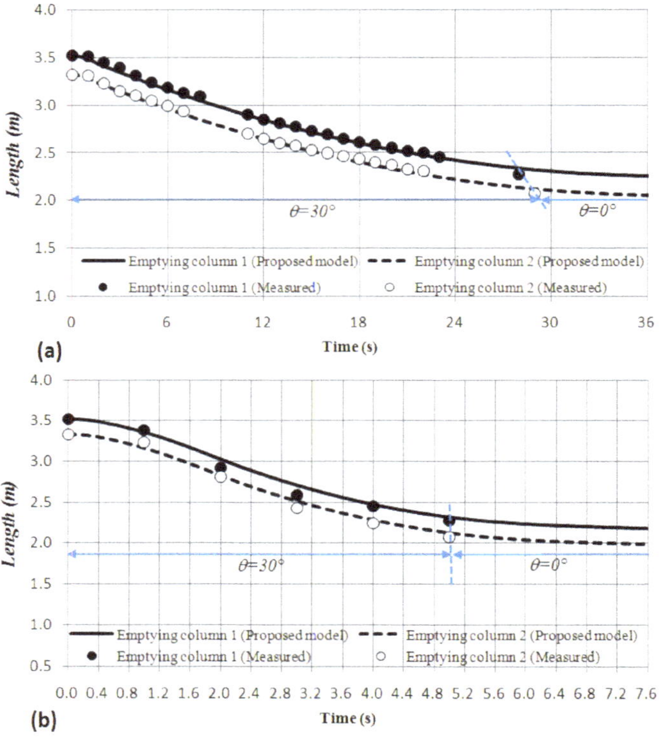

Figure 7. Comparison between computed and measured length of emptying columns: (**a**) Test No. 2 (air valve S050); (**b**) Test No. 7 (air valve D040).

Figure 8 shows the comparison between computed and measured water velocity for Test No. 3 and Test No. 8. In both tests, the water velocity in emptying column 1 is practically the same as in emptying column 2 ($v_{e,1} \approx v_{e,2}$). In addition, in all tests from 0 s to 1.6 s, the water velocity is induced by the opening of the valves MBV_1 and MBV_2. In this range of values, the measurements are not adequate because the system starts resting and the UDV cannot detect the suspended small seeding particles because of no reflection. However, after 1.6 s, the proposed model can predict adequately and give information about the system behaviour. Figure 8a presents the results for Test No. 3 (air valve S050) where the maximum water velocity is rapidly reached at 1.39 s, with a value of 0.076 m/s. According to the measurements, the maximum value is 0.0775 m/s at 1.30 s, which is very close to the proposed model. After the maximum value is attained, the water velocity decreases linearly until it reaches a value of 0 m/s at 34 s. The oscillations around 2 s are caused after the complete opening of valves MBV_1 and MBV_2. The water velocity range is very low during the emptying procedure with the air valve S050. Consequently, the UDV device with a transducer of 4 MHz frequency cannot detect appropriately the evolution of the water velocity. It records water velocity with intervals of 0.015 m/s. Figure 8b shows the comparison between computed and measured water velocities for the air valve $D040$ where the water velocity reaches its maximum value of 0.324 m/s at 1.77 s. According to the measurements, the maximum value is 0.32 m/s at 1.78 s, which is quite similar to the proposed model. After this maximum, the water velocity starts to decrease until the end of the event. During this range, the UDV device can measure appropriately the evolution of the water velocity. The volume of admitted air by $D040$ is almost the same as the water volume drained by valves MBV_1 and MBV_2 since the minimum subatmospheric pressure is 10.22 m_{H20}, practically the atmospheric pressure.

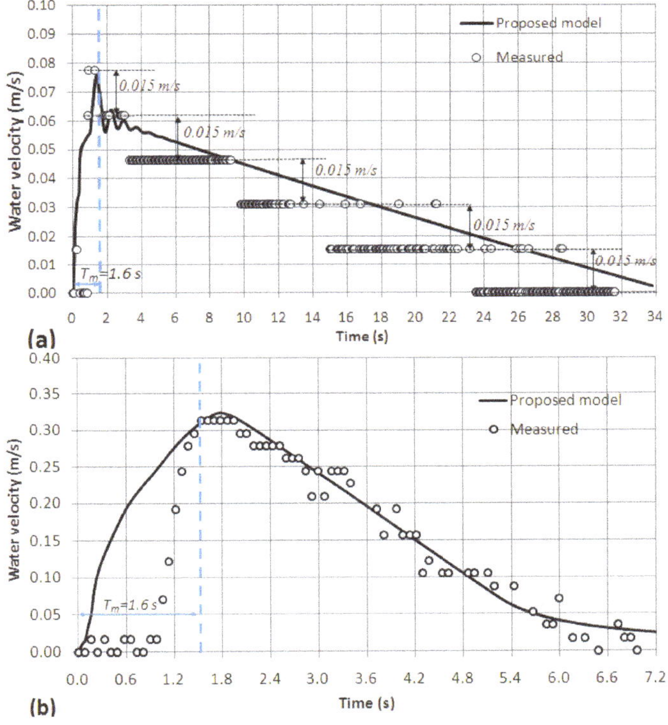

Figure 8. Comparison of water velocity between computed and measured values: (**a**) Test No. 3 (air valve $S050$); (**b**) Test No. 8 (air valve $D040$).

3.3. Sensitivity Analysis

3.3.1. Effect of Air Pocket Sizes

It is important to identify the great influence of the size of the entrapped air pocket on the minimum of the subatmospheric pressure. Figure 9 shows the results taking different air pocket lengths (0.001, 0.540, 0.920, 1.320 and 2.120 m). The smaller the air pocket size, the lower subatmospheric pressure is obtained. Equation (15) shows this situation with the comparison between the gradient of the absolute pressure (dp_i^*/dt) and the air pocket volume $(A_2(L_2 - L_{e,2}) + A_1(L_1 - L_{e,1}))$. For the air valve $S050$, subatmospheric pressures are found in the range of 9.61 m_{H20} and 10.11 m_{H20}, while, for the air valve $D040$, small variations are found between 10.16 m_{H20} and 10.32 m_{H20}, showing the adequacy of this air valve for the emptying process. It also shows the importance of the air valve size since it can induce critical conditions associated with the subatmospheric pressure occurrence.

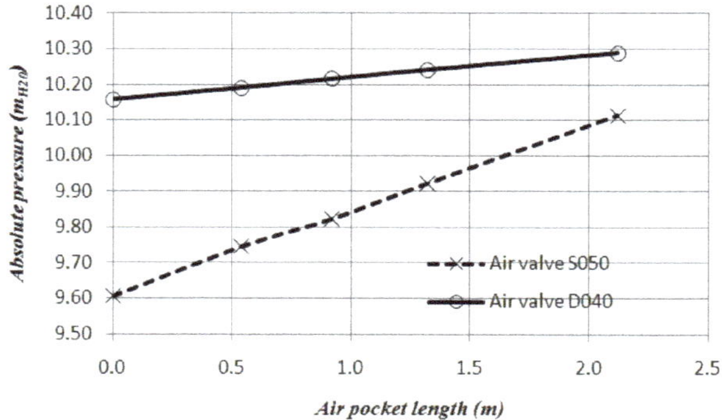

Figure 9. Effect of the air pocket sizes on the minimum pressure attained.

3.3.2. Maximum Water Velocity

Figure 10 shows a comparison between computed and measured maximum water velocities. The proposed model predicts the maximum values of the water velocity for all tests. For the air valve $S050$, maximum values of water velocity are found in the range of 0.049 to 0.079 m/s, while, for the air valve $D040$, water velocities are found in the range of 0.193 to 0.397 m/s.

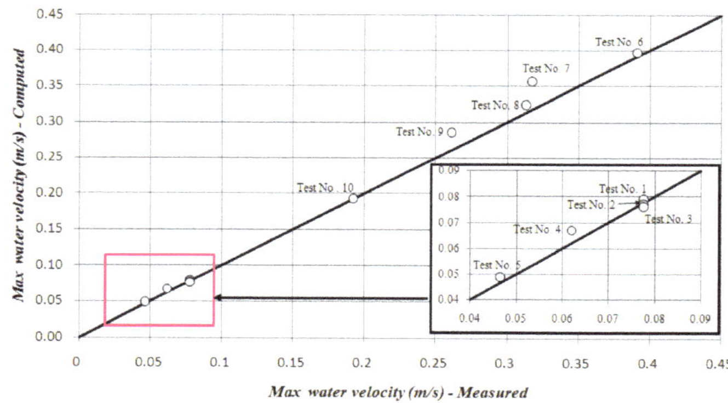

Figure 10. Comparison between computed and measured maximum water velocities.

4. Conclusions

Subatmospheric pressures occur during the emptying process in pipelines with undulating profiles, which is a typical and common operation that engineers have to face. A rigid two-phase flow model was developed for analyzing it, considering n possible air pockets, d air valves, p pipes, and o drain valves. The proposed model is validated using an experimental facility of an irregular profile of 7.3 m long and nominal diameter 63 mm ($DN63$), with an air valve located at the high point and several drain valves along the pipeline. The main hydraulic variables could be measured such as flow (both phases), pressure and air-water front position. Comparisons between computed and measured values of the absolute pressure, water velocity and the length of the emptying columns show that the proposed model can predict accurately not only the extreme values but also their patterns. According to the results, the following conclusions can be drawn:

1. The proposed model can be used for planning the emptying operations in pipelines with undulating profiles, considering their limitations.

2. The expansion of the air pocket produces minimum subatmospheric pressure. In order to control these effects, air valves should be installed in pipelines.

3. It is very important to have a correct selection of air valves for vacuum protection for emptying pipelines. An inadequate selection produces both lower values of subatmospheric pressure and a slower drainage of the system. On the other hand, a larger air valve orifice size reduces the lowest values of subatmospheric pressures.

4. Engineers should consider the initial condition in which the pipeline is completely filled, which is the most critical condition, since the smallest air pocket sizes produce the lowest subatmospheric pressures.

In real hydraulic pressurized systems, the proposed model can be used for: (i) checking the risk of a system collapse considering factors such as stiffness class, the soil in natural conditions, the type of backfill and the cover depth; and (ii) selecting the air valve sizes depending on the characteristics of each hydraulic system.

Supplementary Materials: The following are available online at www.mdpi.com/2073-4441/9/2/98/s1: The following videos show the emptying process for emptying column 1. Videos: Test No. 1, Test No. 2, Test No. 3, Test No. 4, Test No. 5, Test No. 6, Test No. 7, Test No. 8, Test No. 9, and Test No. 10. The behaviour for emptying column 2 is similar to emptying column 1.

Acknowledgments: The authors acknowledge: (1) the financial support for the first author covered by Fundación Centro de Estudios Interdisciplinarios Básicos y Aplicados (CEIBA)—Gobernación de Bolívar (Colombia); and (2) the experimental facilities for conducting the tests provided by IST (Instituto Superior Técnico), Civil Engineering Research and Innovation for Sustainability (CERIS), of the University of Lisbon (Portugal).

Author Contributions: Helena M. Ramos and Vicente S. Fuertes-Miquel conceived and designed the experimental facility and type of tests; Oscar E. Coronado-Hernández and Mohsen Besharat performed the experimental tests; Oscar E. Coronado-Hernández wrote the first draft paper; and Helena M. Ramos and Vicente S. Fuertes-Miquel revised the paper and completed the submission.

Conflicts of Interest: The authors declare no conflict of interest.

Abbreviations

The following abbreviations are used in this manuscript:

A	Cross sectional area of the pipe (m^2)
$A_{adm,m}$	Cross section area of the air valve m (m^2)
$C_{adm,m}$	Inflow discharge coefficient of the air valve m (–)
D	Internal pipe diameter (m)
f	Darcy–Weisbach friction factor (–)
g	Gravity acceleration (m/s^2)
$L_{e,j}$	Length of the emptying column j (m)
L_j	Total length of the pipe j (m)
k	Polytropic coefficient (–)
K_s	Flow factor of the drain valve s (m^3/s) with a pressure drop of 1 m_{H20}
$h_{m,s}$	Local loss of the drain valve s (m)
$m_{a,i}$	Air mass of the air pocket i (kg)
p_i^*	Absolute pressure of the air pocket i (Pa)
p_{atm}^*	Atmospheric pressure (Pa)
t	Time (s)
T_m	Valve maneuvering time (s)
$Q_{a,nc,m}$	Air discharge in normal conditions admitted by the air valve m (m^3/s)
$Q_{w,s}$	Water discharge by the drain valve s (m^3/s)
$V_{a,i}$	Air volume of the air pocket i (m^3)
$v_{e,j}$	Water velocity of the emptying column j (m/s)
$v_{a,nc,m}$	Air velocity in normal conditions admitted by the air valve m (m/s)
x_i	Length of the air pocket i (m)

$\Delta z_{e,j}$	Elevation difference (m)
$\rho_{a,i}$	Air density of the air pocket i (kg/m^3)
$\rho_{a,nc}$	Air density in normal conditions (kg/m^3)
ρ_a	Water density (kg/m^3)
BV	Ball valve
MBV	Manual ball valve
PT_1	Absolute pressure transducer
Superscripts	
*	Absolute values
Subscripts	
a	Refers to air
i	Refers to air pocket
j	Refers to pipe
m	Refers to air valve
nc	Normal conditions
w	Refers to water
0	Initial condition

References

1. American Water Works Association (AWWA). *Manual of Water Supply Practices—M51: Air-Release, Air-Vacuum, and Combination Air Valves*, 1st ed.; American Water Works Association: Denver, CO, USA, 2001.

2. Ramezani, L.; Karney, B.; Malekpour, A. The Challenge of Air Valves: A Selective Critical Literature Review. *J. Water Resour. Plan. Manag.* **2016**, *141*, doi:10.1061/(ASCE)WR.1943-5452.0000530.

3. Fuertes-Miquel, V.S.; Coronado-Hernández, O.E.; Iglesias-Rey, P.L.; Mora-Melia, D. Transient phenomenon during the emptying process of a single pipe with water-air interaction. *J. Hydraul. Res.* **2016**, submitted.

4. Tijsseling, A.; Hou, Q.; Bozkuş, Z.; Laanearu, J. Improved One-Dimensional Models for Rapid Emptying and Filling of Pipelines. *J. Press. Vessel Technol.* **2016**, *138*, 031301.

5. Laanearu, J.; Annus, I.; Koppel, T.; Bergant, A.; Vučković, S.; Hou, Q.; Tijsseling, A.; Anderson, A.; Van't Westende, J. Emptying of large-scale pipeline by pressurized air. *J. Hydraul. Eng.* **2012**, *138*, 1090–1100.

6. Fuertes, V.S. Hydraulic Transients with Entrapped Air Pockets. Ph.D. Thesis, Department of Hydraulic Engineering, Polytechnic University of Valencia, Valencia, Spain, 2001.

7. Besharat, M.; Tarinejad, R.; Ramos, H.M. The effect of water hammer on a confined air pocket towards flow energy storage system. *J. Water Supply Res. Technol. AQUA* **2016**, *65*, 116–126.

8. Apollonio, C.; Balacco, G.; Fontana, N.; Giugni, M.; Marini, G.; Piccinni, A.F. Hydraulic Transients Caused by Air Expulsion during Rapid Filling of Undulating Pipelines. *Water* **2016**, *8*, 25.

9. Balacco, G.; Apollonio, C.; Piccinni, A.F. Experimental Analysis of Air Valve Behaviour During Hydraulic Transients. *J. Appl. Water Eng. Res.* **2015**, *3*, 3–11.

10. Zhou, L.; Liu, D.; Karney, B. Investigation of hydraulic transients of two entrapped air pockets in a water pipeline. *J. Hydraul. Eng.* **2013**, *139*, 949–959.

11. Izquierdo, J.; Fuertes, V.S.; Cabrera, E.; Iglesias, P.; García-Serra, J. Pipeline start-up with entrapped air. *J. Hydraul. Res.* **1999**, *37*, 579–590.

12. Zhou, L.; Liu, D.; Ou, C. Simulation of flow transients in a water filling pipe containing entrapped air pocket with VOF model. *Eng. Appl. Comput. Fluid Mech.* **2011**, *5*, 127–140.

13. Martins, N.M.C.; Soares, A.K.; Ramos, H.M.; Covas, D.I.C. CFD modeling of transient flow in pressurized pipes. *Comput. Fluids* **2016**, *126*, 129–140.

14. Abreu, J.; Cabrera, E.; Izquierdo, J.; García-Serra, J. Flow Modeling in Pressurized Systmes Revisited. *J. Hydraul. Eng.* **1999**, *125*, 1154–1169.

15. Martins, S.C.; Ramos, H.M.; Almeida, A.B. Mathematical Modeling of Pressurized System Behaviour with Entrapped Air. In *Environmental Hydraulics: Theoretical, Experimental and Computational Solutions*; CRC Press: Boca Raton, FL, USA, 2010; pp. 61–64.

16. Martins, S.C.; Ramos, H.M.; Almeida, A.B. *Computational Evaluation of Hydraulic System Behaviour with Entrapped Air under Rapid Pressurization*; Integrating Water Systems; CRC Press: Boca Raton, FL, USA, 2010; pp. 241–247.

17. Fuertes-Miquel, V.S.; López-Jiménez, P.A.; Martínez-Solano, F.J.; López-Patiño, G. Numerical modelling of pipelines with air pockets and air valves. *Can. J. Civ. Eng.* **2016**, *43*, 1052–1061.

18. Zhou, L.; Liu, D. Experimental investigation of entrapped air pocket in a partially full water pipe. *J. Hydraul. Res.* **2013**, *51*, 469–474.

19. Covas, D.; Stoianov, I.; Ramos, H.M.; Graham, N.; Maksimoviċ, C.; Butler, D. Water hammer in pressurized polyethylene pipes:conceptual model and experimental analysis. *Urban Water J.* **2010**, *1*, 177–197.

20. Liou, C.; Hunt, W.A. Filling of pipelines with undulating elevation profiles. *J. Hydraul. Eng.* **1996**, *122*, 534–539.

21. Bousso, S.; Daynou, M.; Fuamba, M. Numerical Modeling of Mixed Flows in Storm Water Systems: Critical Review of Literature. *J. Hydraul. Eng.* **2013**, *139*, 385–396.

22. Leon, A.; Ghidaoui, M.; Schmidt, A.; Garcia, M. A robust two-equation model for transient-mixed flows. *J. Hydraul. Res.* **2010**, *48*, 44–56.

23. Wylie, E.; Streeter, V. *Fluid Transients in Systems*; Prentice Hall: Englewood Cliffs, NJ, USA, 1993.

24. Vasconcelos, J.G.; Wright, S.J. Rapid Flow Startup in Filled Horizontal Pipelines. *J. Hydraul. Eng.* **2008**, *134*, 984–992.

25. Cabrera, E.; Abreu, J.; Pérez, R.; Vela, A. Influence of Liquid Length Variation in Hydraulic Transients. *J. Hydraul. Res.* **1992**, *118*, 1639–1650.

26. Zhou, L.; Liu, D.; Karney, B. Phenomenon of white mist in pipelines rapidly filling with water with entrapped air pocket. *J. Hydraul. Eng.* **2013**, *139*, 1041–1051.

27. Martins, S.C.; Ramos, H.M.; Almeida, A.B. Conceptual analogy for modelling entrapped air action in hydraulic systems. *J. Hydraul. Res.* **2015**, *53*, 678–686.

28. Martin, C.S. Entrapped Air in Pipelines. In Proceedings of the Second International Conference on Pressure Surges, London, UK, 22–24 September 1976.

29. Iglesias-Rey, P.L.; Fuertes-Miquel, V.S.; García-Mares, F.J.; Martínez-Solano, F.J. Comparative Study of Intake and Exhaust Air Flows of Different Commercial Air Valves. In Proceedings of the 16th Conference on Water Distribution System Analysis, WDSA 2014, Bari, Italy, 14–17 July 2014; pp. 1412–1419.

Article

Experimental Study of Air Vessel Behavior for Energy Storage or System Protection in Water Hammer Events

Mohsen Besharat [1,2,*], Maria Teresa Viseu [3] and Helena M. Ramos [2]

[1]   Department of Civil Engineering, Saghez Branch, Islamic Azad University, Saghez 66819-73477, Iran
[2]   Department of Civil Engineering and Architecture, Instituto Superior Técnico, University of Lisbon, Lisbon 1049-001, Portugal; helena.ramos@tecnico.ulisboa.pt
[3]   Laboratório Nacional de Engenharia Civil (LNEC), Lisbon 1700-066, Portugal; tviseu@lnec.pt
*   Correspondence: mohsen.besharat@tecnico.ulisboa.pt or mohsen.besharat@iausaghez.ac.ir; Tel.: +351-934-667-406

Academic Editor: Marco Franchini
Received: 8 November 2016; Accepted: 14 January 2017; Published: 20 January 2017

Abstract: An experimental assessment of an air pocket (AP), confined in a compressed air vessel (CAV), has been investigated under several different water hammer (WH) events to better define the use of protection devices or compressed air energy storage (CAES) systems. This research focuses on the size of an AP within an air vessel and tries to describe how it affects important parameters of the system, i.e., the pressure in the pipe, stored pressure, flow velocity, displaced volume of water and water level in the CAV. Results present a specific range of air pockets based on a dimensionless parameter extractable for other real systems.

Keywords: water hammer; air vessel sizing; energy storage; dynamic behavior; CAES

1. Introduction

A sudden valve closure has always created complex conditions capable of causing major problems and damages to the water conveyance systems. So far, many methods have been proposed and used to overcome these problems, e.g., surge tanks, air vessels and different types of valves. Of those methods listed, air vessels have been used more frequently with better application because of their economic and the stability advantages. An air vessel is capable of controlling both negative and positive pressures and usually, the outflow action from an air vessel to the pipe has less head loss than the inflow into the air vessel to control the negative pressure [1]. Most of the previous and recent studies in air vessel sizing have investigated the entrance connection situation effect. Plenty of charts and graphs are presented by these studies for the air vessel sizing [2–6]. However, most of them are rarely used for sizing purpose because of the simplification assumptions made for their preparation. Instead, a trial and error approach for checking the minimum and maximum pressure in the pipe is used. By means of throttling the entrance in the air vessel, smaller air vessel sizes can be achieved. Based on this concept, Martino and Fontana [5] proposed a method for sizing air vessels based on optimizing the throttling. Nevertheless, the confined air in a hydraulic system can amplify the pressure peaks from a transient action and make the transient situation more complex due to the compressibility of the air. Hence, there is actually a great necessity for studies about air pocket (AP) behavior during a fast-transient occurrence that can provide great knowledge in the primary sizing of air vessels and more reliable final decisions in water pipe systems' behavior and safety. Hatcher and Vasconcelos [7] studied the behavior of an air pocket inside a pipe using an experimental approach, giving useful information to fill the gap between numerical models and real situations. They declared that few experimental studies focused on the effect of parameters related to the geometry of the system as well as air and flow conditions.

Several researchers have worked on the behavior of air in the rapid filling subject or entrapped air bubbles along a pipeline [8–10]. There are some studies that investigate the thermodynamic prediction and behavior of an AP under slow-transient which consider the heat transfer process that is negligible in fast-transient cases [1]. Usually, it is assumed that fast-transient thermodynamic behavior obeys the polytropic equation, while in experimental works, it was shown that this equation is not obeyed for a special range of AP sizes and Reynolds number (Re) [11]. For that reason, more analysis of AP behavior, specifically for small APs, seems to be required.

This present work studies the behavior of an AP in a sophisticated experimental apparatus for different major governing variables. In this study, the outflow way at the air vessel entrance is closed using a check valve (CV) to store the surge pressure from a fast-transient action induced by a water hammer (WH) event. It helps to study the compression phase of an AP as in real conditions. Furthermore, the expansion phase investigation is also provided by means of a ball valve (BV) at the lower level of the air vessel. Placing a CV in the air vessel entrance allows the examination of a micro-CAES (compressed air energy storage) system for storing the surge pressure in small-scale (SS) size. A traditional CAES system stores energy from a cheap and sustainable source inside underground caverns which is a major limitation of this technique. Energy is stored as pressure in compressed air in low-peak hours and the stored energy is recovered during high-peak hours. SS-CAES systems are defined in the net power range of 10 kW to 10 MW [12]. SS-CAES systems may present good alternatives for common generators without using electricity [13]. A water-compensated micro-CAES or SS-CAES system is the focus of this study to examine the storage ability of confined AP within a compressed air vessel (CAV). The size of the air vessel needs exact investigation, in this case affecting the applicability and efficiency of the system. Mostly, the cylindrical shape is more economic and practical for air vessels. To achieve the most effective air vessel size, the most important parameters are the maximum pressure and the AP volume fraction within the air vessel. In a previous study, Kiam and Favrat [14] proposed a SS-CAES system with constant air pressure using a column of water. In a review work, Budt et al. [15] presented a wide review of CAES systems from past to future, addressing the main challenges for their development. One challenge that they indicate appears in the lack of appropriate tools for the detailed simulation of CAES systems [15]. Experimental works can provide valuable background for similar challenges by examining various parameters, relating them together and presenting dimensionless variables.

Understanding the behavior of a confined air pocket (AP) under different events of WH seems to give substantial knowledge for future experimental studies or real field designs. In this study, a well-equipped experimental apparatus was used to test and measure AP reaction under three WH events' conditions, namely the occurrence of one WH (1WH), five WHs (5WH) and nine WHs (9WH). In this study, the stored amount of pressure inside the AP for different flow conditions has been assessed. Also, the ability of WH to relocate water to another place with a similar action to the old ram pumps was also examined. To find the behavior of the system as a WH protection vessel, the effect of each parameter on the pressure and the velocity changes along the pipe system has also been investigated. The major considered parameters in this research are constituted by flow velocity, which in the presented graphs is related with Reynolds number (Re); volume fraction ratio (VFR) of air; the storage amount of pressure within each AP; and the relocated volume of water. The VFR is defined as the ratio of air volume to water volume inside the CAV, in percentage.

2. Measurements

The experimental apparatus, as shown in Figure 1, is composed of PVC pipes and a transparent PVC compressed air vessel (CAV) with 0.10 m diameter and 0.60 m height. The CAV includes an air pocket (AP) at the top and a water column (WC) at the lower part. A pump creates a circulation flow with different flow velocities. A fast closure electro-pneumatic ball valve (EBV) has been used to create one or several consecutive WH phenomena in the system. Three check valves (CV) are installed to block the returning way of the flow. A pressure transducer (PT) and an ultrasound velocity profiler

(UVP) were used for recording the pressure and flow velocity during the tests. The UVP represents both a method and a device for measuring an instantaneous velocity profile in the water flow along the ultrasonic beam axis by detecting the doppler shift frequency of echoed ultrasound as a function of time. The doppler was installed on a horizontal pipe before the EBV location. There are two PT in the system; one located at the top of the CAV and the other upstream of the EBV. Tests are fulfilled in two different conditions of the BV located at the lower level of the CAV for discharging the water; (i) closed BV for measuring the stored pressure within the AP; (ii) open BV for measuring the relocated water volume. An open water tank is placed downstream of the outflow pipe from the CAV, which provides a tool to collect water and measure the relocated water volume after each test.

Different AP sizes (and related VFRs) and flow velocities are considered during the tests. To cover a wide range of AP sizes, seven AP lengths, i.e., 2, 3, 4, 5, 10, 20 and 40 cm were tested with corresponding VFRs of 3.17%, 4.75%, 6.33%, 7.92%, 15.84%, 31.67% and 63.35% respectively. Also, seven flow velocity values of 1.00, 1.50, 2.00, 2.50, 3.00, 3.50, 4.00 and 4.50 m/s and Re of 36,000, 56,000, 75,000, 93,000, 115,000, 132,000 and 155,000 are used for testing each VFR value with different combinations of VFR and Re.

Several WH tests, i.e., 1WH, 5WH and 9WH have been fulfilled, induced by EBV for different Re and VFR magnitudes. The numerous part of 1WH, 5WH and 9WH symbols shows the number of WH repetitions during each test. During these series of tests, the surge pressure obtained in WH action is conducted toward the CAV to store the pressure in the AP. Measurements for different tests were carried out in similar initial conditions of velocity and pressure in the pipe system and AP. The initial AP pressure was adjusted to a constant value equal to the atmospheric pressure throughout all the tests to achieve the maximum pressure during the transient action [16].

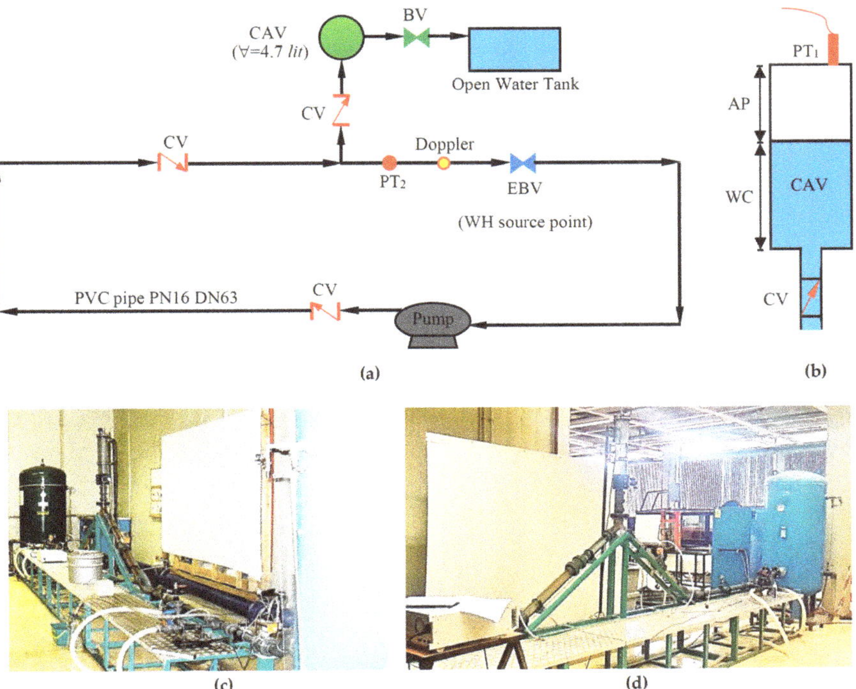

Figure 1. Scheme of experimental apparatus and components; (**a**) system; (**b**) compressed air vessel (CAV); (**c**) upstream side of the apparatus; (**d**) downstream side of the apparatus.

3. Results

3.1. Tests with Closed BV

When the BV is closed, the WH allows the observation of the storage ability of each AP. This type of transient action is important in protection vessels and also in SS-CAES applications. The measured pressure shows that the AP is able to retain an amount of pressure as stored energy after finishing the tests due to the compressibility and confinement effects. Figure 2 shows the pressure change for different velocities of an AP = 3 cm (VFR = 4.75%). The pressure related to the first peak is called "peak pressure" and is referred by P_{1p}. As in Figure 2, an amount of pressure has been stored in the air pocket after the test which is called "stored pressure" and is shown by P_s. Graphs show a large growth in the peak pressure with increasing Re, while similar growth does not occur in the stored pressure. As it can be seen in Figure 2, from Re of 36,000 to 155,000, the peak pressure is changed from 1.33 bar to 6.25 bar. However, the stored pressure in the AP varied from 2.33 bar to 4.51 bar. Previous work [17] showed that either decreasing the VFR or increasing the Re will amplify the peak pressure considerably. Similar behavior exists for the stored pressure change, i.e., stored pressure with direct correlation to the flow velocity, but this is in contrast to the VFR. Results show that decreasing the VFR will continuously increase the stored amount of energy. For the VFR = 4.75% shown in Figure 2, the pressure change during 5WH and 9WH tests starts by increasing steeply and eventually ends up at the final stored pressure with smaller increases. The final magnitude of the stored pressure is also directly related with Re.

When comparing the stored pressure and peak pressure, in most of the tests the magnitude of the stored pressure is higher than the peak pressure, as expected. However, in some tests related to small AP and high Re, the amount of peak pressure is extremely high and even after the pressure storage cycles, the stored pressure remains smaller than the peak pressure. Some general rules can be stated based on results from these tests for all the Re and VFR as shown in Figure 3:

- For Re of 36,000, 56,000 and 75,000, the stored pressure is always higher than the peak pressure ($P_s > P_{1p}$);
- For large values of an AP, i.e., VFR > 11%, the stored pressure is always higher than the peak pressure ($P_s > P_{1p}$).

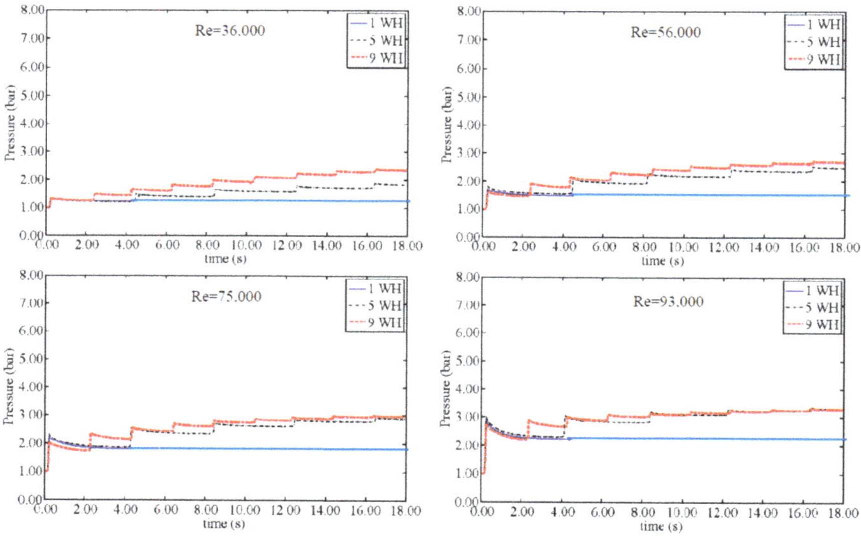

Figure 2. *Cont.*

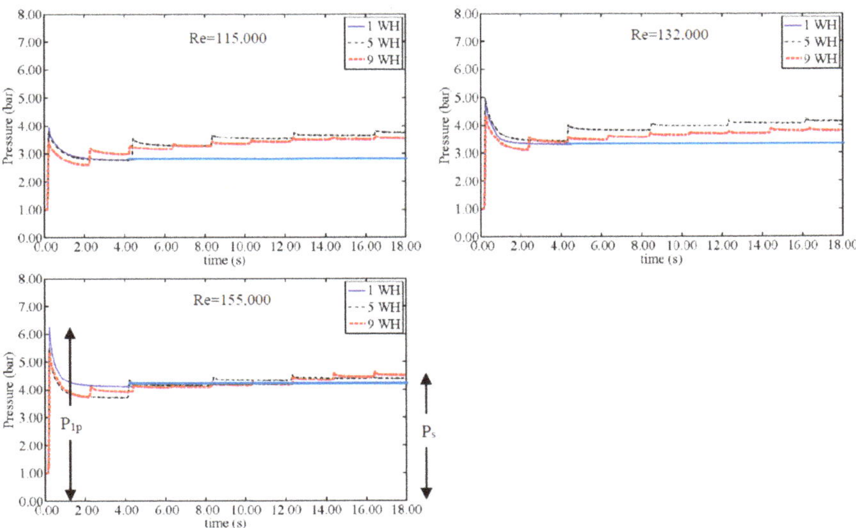

Figure 2. Pressure change in the air pocket (AP) for different water hammer (WH) events and Reynolds number (Re) for the volume fraction ratio VFR = 4.75%.

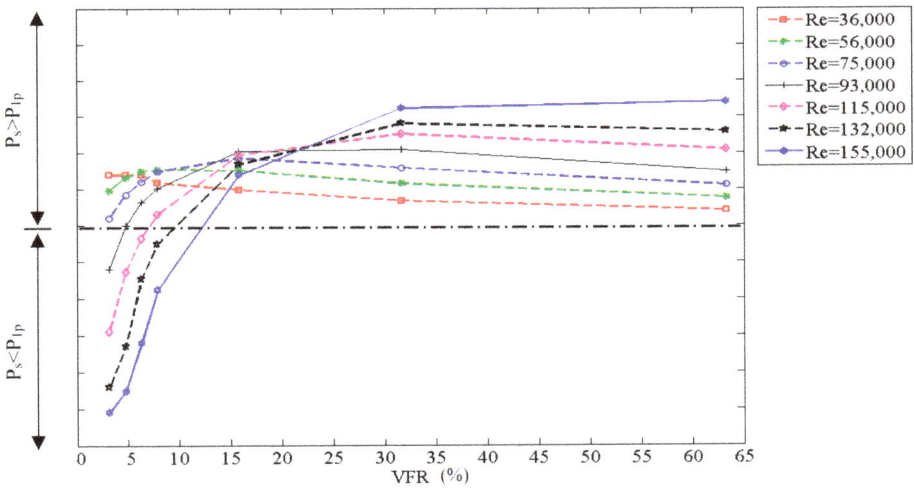

Figure 3. Comparison of the final stored pressure (P_s) with the first peak (P_{1p}).

As a logical expectation, the final stored pressure should be higher in the 9WH test in comparison to the 5WH test as a consequence of more WH events. Nevertheless, graphs show higher pressure storage magnitudes in 9WH tests when Re is low while for higher Re, the recorded pressure magnitudes of 9WH tests become less than or equal to 5WH pressure magnitudes. Large AP lengths of 20 cm and 40 cm (VFR > 30%) do not obey this rule; in other words, for these two APs, the recorded pressure of 9WH is always higher than for 5WH values. A comparison of the pressure magnitudes of 9WH and 5WH is presented in Figure 4, demonstrating the non-conforming behavior of a small VFR and high Re values.

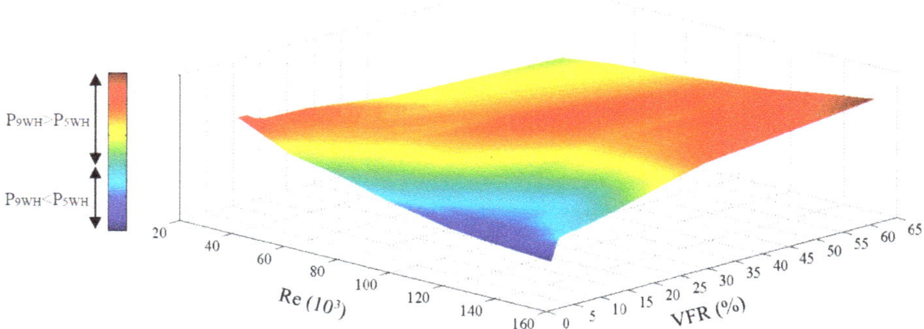

Figure 4. Comparison between the first peak for 5WH and 9WH as a function of Re and VFR.

Figures 5 and 6 show the route of change in pressure recorded from 1WH, 5WH and 9WH tests for AP lengths of 10 cm and 40 cm (VFR of 15.84% and 63.35% respectively) in order to demonstrate the behavior of an AP in more detail. For small values of VFR, as in Figures 4 and 5, pressure variation shows irregular steps during the WH events. In the case of low Re values, small and almost equal steps have been recorded for all VFR. The larger values of Re show big increments at first WH steps and fast dissipation steps at the final events. In addition to irregular steps, considerable higher stored pressure was not recorded for the 9WH in comparison with the 5WH test, representing behavior against the usual expectation for obtaining higher stored pressure as a consequence of more WH events. On the other hand, the magnitude of pressure steps remains nearly constant for every Re value related to a high VFR in measurements similar to Figure 6. Furthermore, in this range, the stored pressure of 9WH tests always stays larger than the 5WH stored pressure, as explained before. Accordingly, 1WH tests are shown to be regular and constantly increasing in the stored pressure when increasing the Re number.

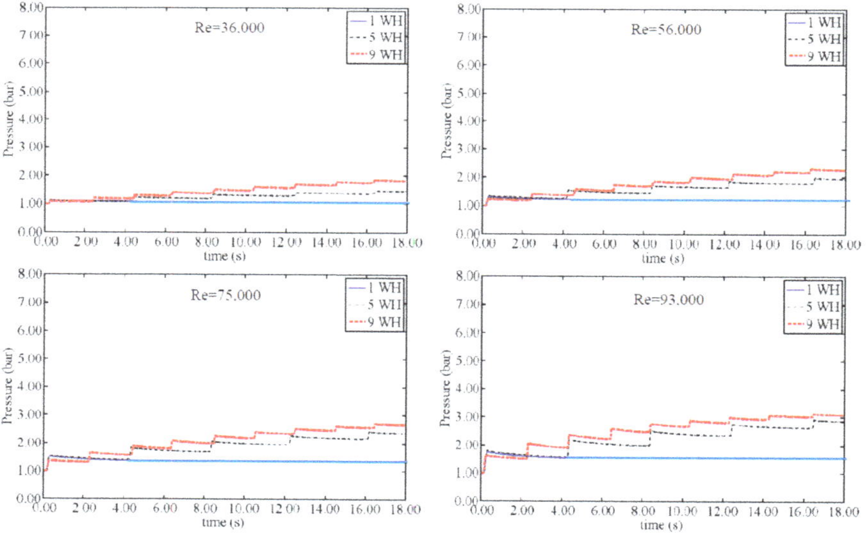

Figure 5. *Cont.*

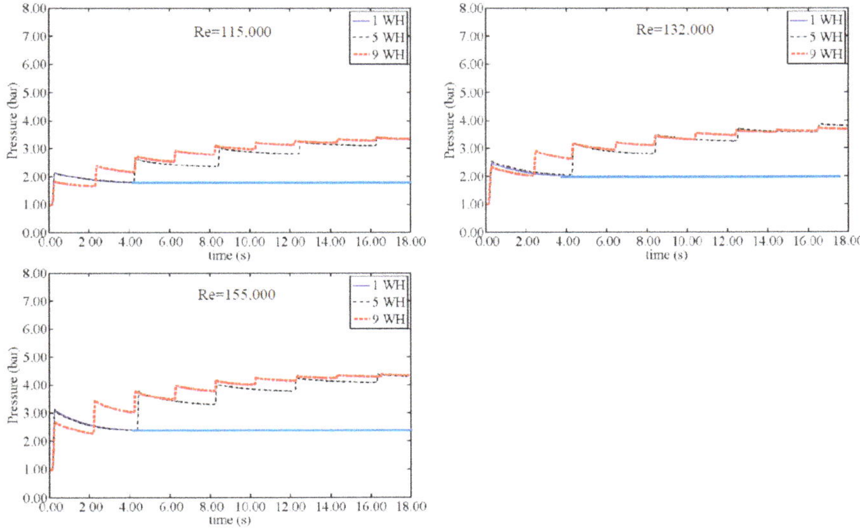

Figure 5. Pressure change in the AP for various WH events and Re for the VFR = 15.84%.

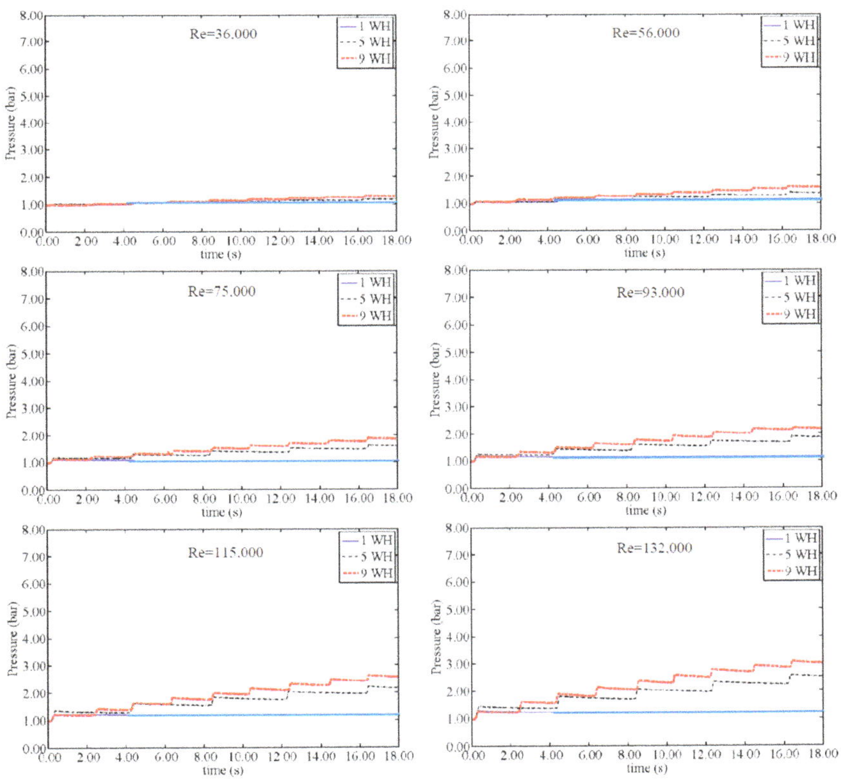

Figure 6. *Cont.*

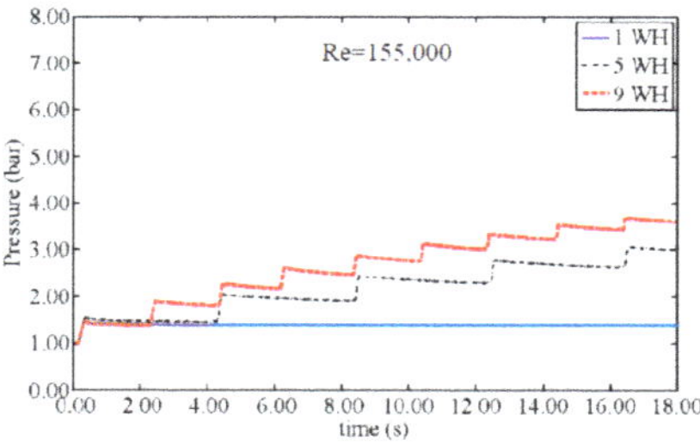

Figure 6. Pressure change in the AP for various WH events and Re for the VFR = 63.35%.

In fact, the rules for changing the stored pressure with VFR and Re are the same in all the tests, as stated below and shown in Figure 7:

- Increasing the AP size or the VFR continuously decreases the stored pressure;
- Increasing the Re number continuously increases the stored pressure;
- For each Re, the magnitude of stored pressure changes less for various VFR by increasing the number of WH events. Also, the same tendency exists for each VFR when changing the Re. In Figure 7, the VFR gradient is lowest in 9WH tests and highest in 1WH tests.

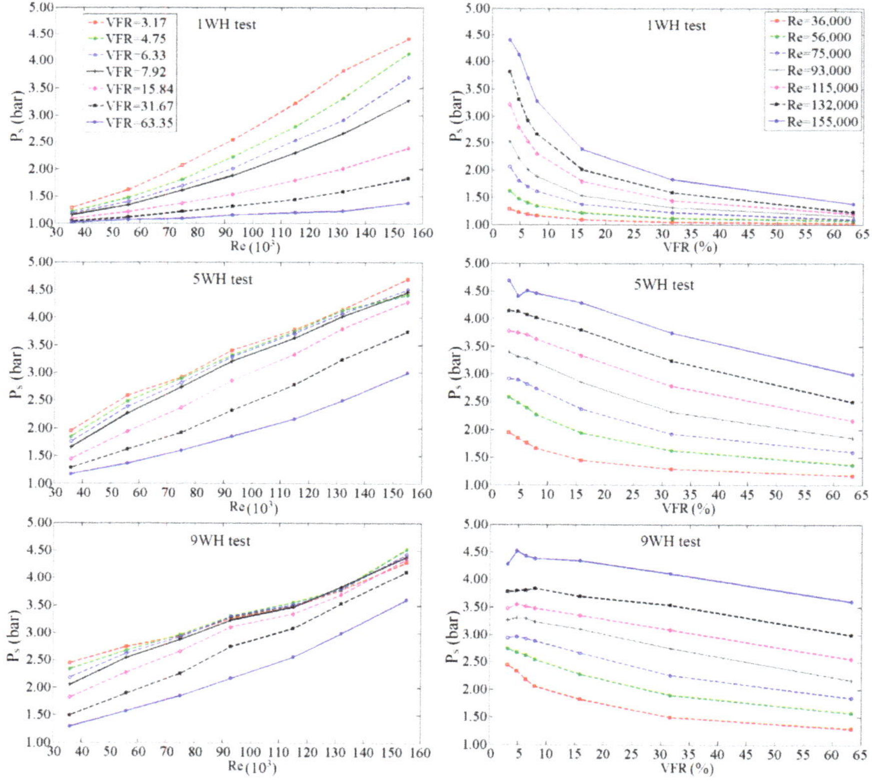

Figure 7. Stored pressure change against Re and VFR for various WH tests.

Finally, the curves in Figure 7 show that for the VFR < 15%, the stored pressure of all three groups of tests, i.e., 1WH, 5WH and 9WH do not show a substantial difference in magnitude and the WH events do not appear to play an important role in pressure storage. However, for VFRs > 15%, the stored pressure is significantly higher for 5WH and 9WH than for 1WH tests, showing the importance of number of WH events.

To keep a pressurized system in a safe and stable condition during transient events, controlling the pressure along a pipe from the transient source position plays a major role. For that purpose, the pressure inside the pipe at the PT_2 point is measured and its maximum value is named P_p. Figure 8 shows the change of maximum pressure in the pipe system for different Re values. Although the CAV does not have any discharge to the main pipe of the system, results in Figure 8 show that increasing the AP size will significantly decrease the pressure peak inside the pipe (P_p), specifically for higher Re. All WH events show the same behavior. Since the safety of pressurized systems under transient conditions matters most for high Re, a high VFR seems to be a better option to control pressure surge. However, smaller CAVs are more economic and easier to apply. Regarding this concern, Figure 8 shows the constant magnitude of pressure for the VFR > 30%.

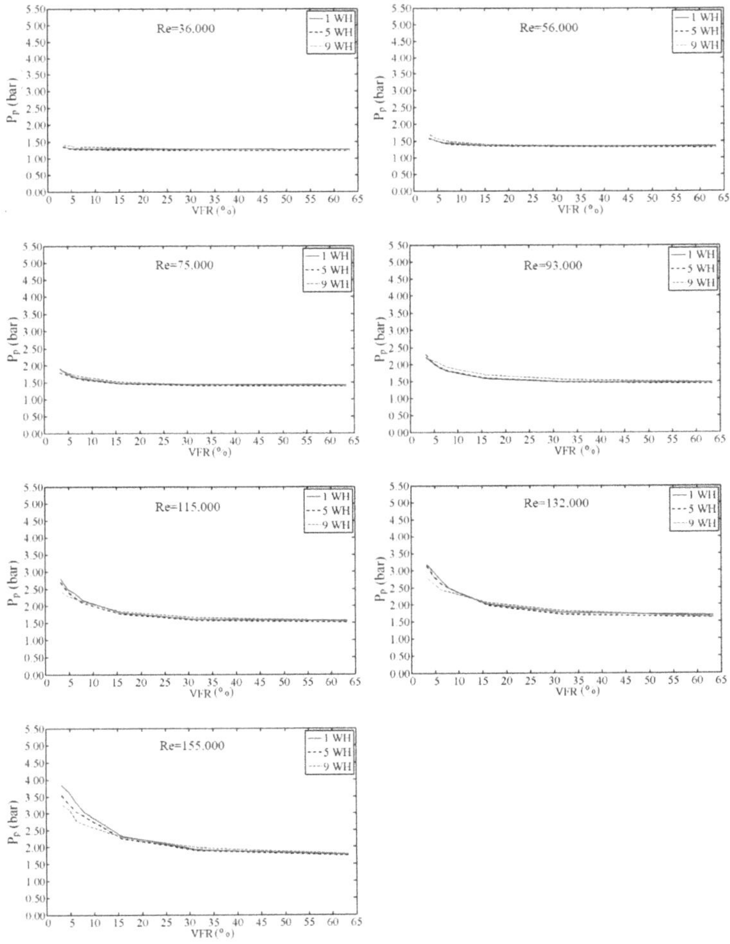

Figure 8. Pressure change inside the pipe at PT_2 against the VFR for different Re and WH events.

The water level in the CAV will rise during the WH tests as a consequence of pressure increasing by each WH action. The rising height of the water level is called Δh, a parameter that is calculated with Re and the VFR. The Δh shows a wide range of change from lower to higher values of Re and VFR (Figure 9).

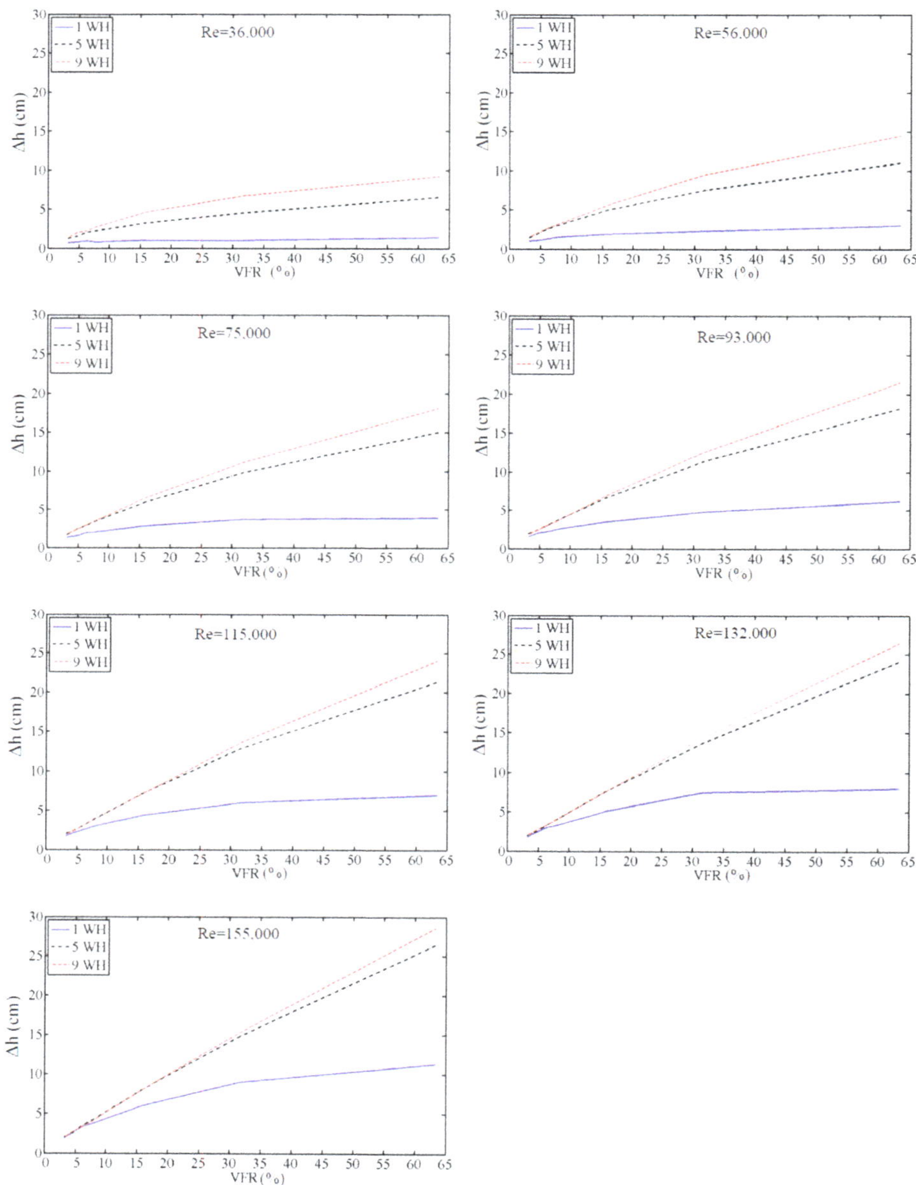

Figure 9. Water level change (Δh) in the CAV against the VFR for different Re.

The Δh magnitude starts at around 1.2 cm for a small VFR and goes up to around 28 cm for the biggest VFR and Re in the 9WH case. The VFR > 30% showed good performance as presented in Figure 8 for controlling the pressure surge in the pipe system. Since, for the purpose of controlling the pressure surge by a CAV, only one WH can occur, the Δh remains in an acceptable range for the worst case, i.e., remains below 20% of the CAV length.

Figures 10 and 11 present average velocities (V_{avg}) for different Re for the 1WH and 9WH tests, where V_{avg} around zero indicates a WH occurrence. Since the V_{avg} does not show less than zero values, it can be concluded that the velocity vectors in reverse direction are not significant compared to the remaining flow. When repeating the WH to store energy in an AP, it is important to preserve the required flow rate in the pipe. During the "closed BV tests", the flow rate downstream of the system is not altered.

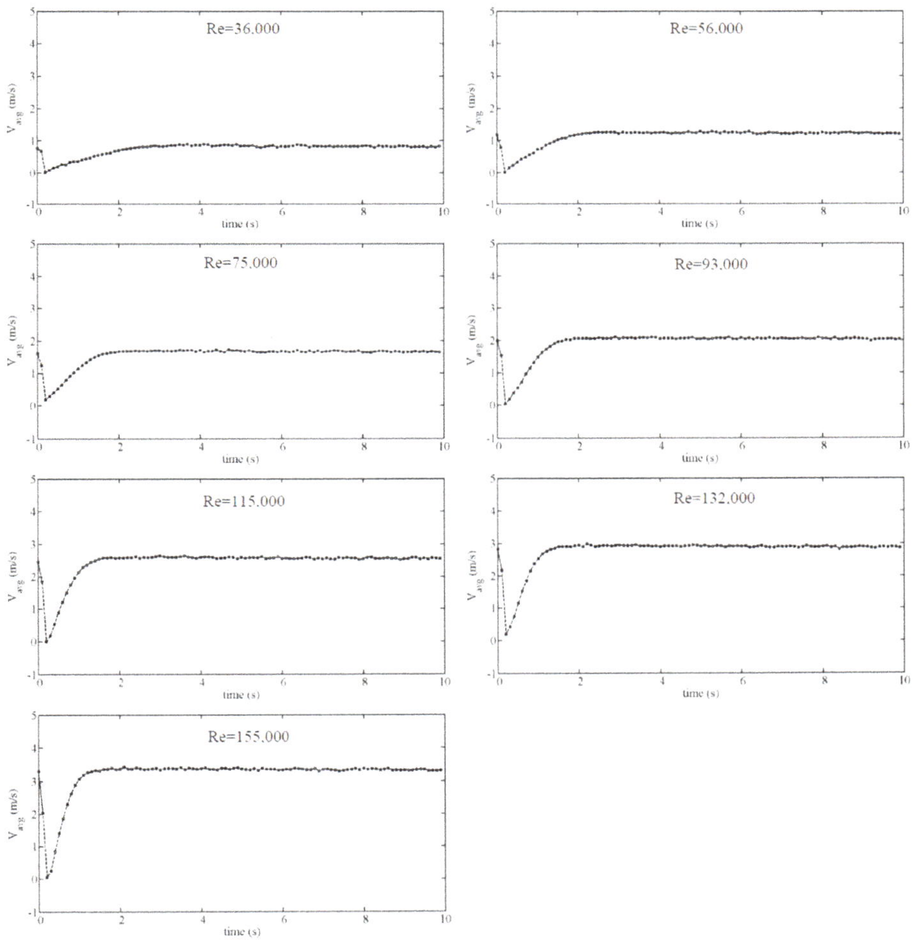

Figure 10. Average velocity for different Re during 1WH tests.

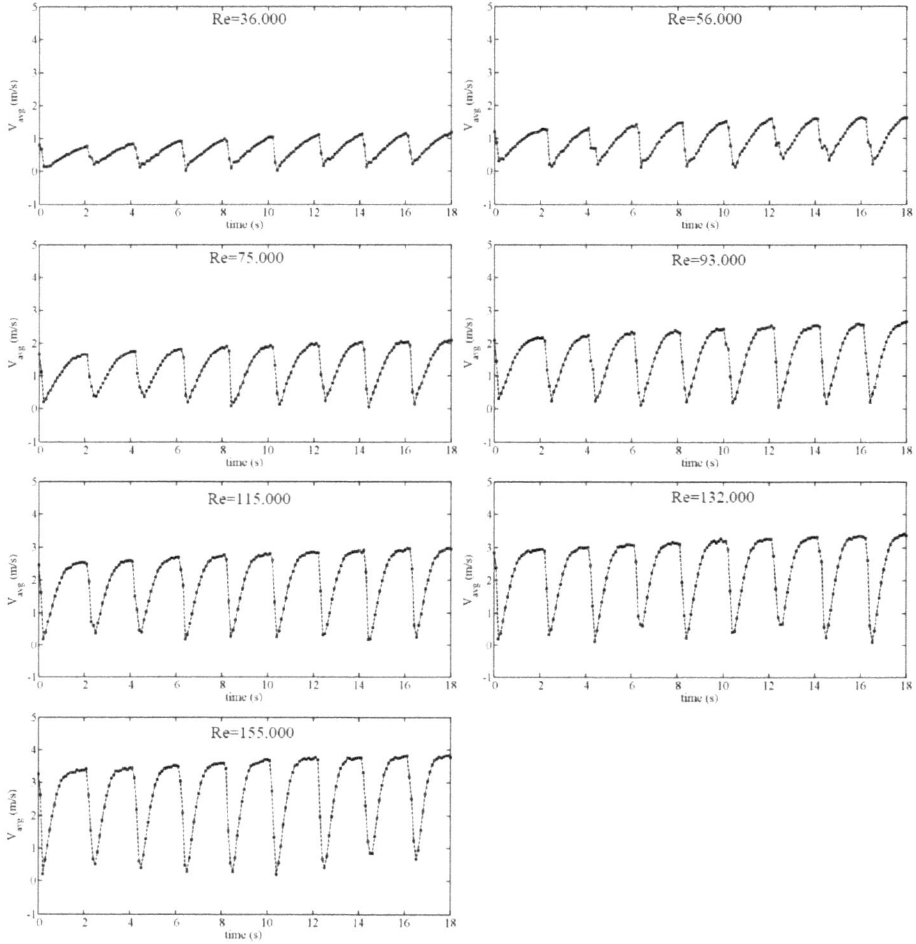

Figure 11. Average velocity for different Re during 9WH tests.

3.2. Tests with Open BV

In these series of tests, the BV located in the lower level of the CAV is left open to discharge water into an open water tank (Figure 1). The water level should be adjusted before each test to have the predefined AP length. Since the CAV is completely confined, opening the BV before starting a test does not affect the water level in the CAV because of an emerging vacuum state in the AP. After each upsurge in the AP induced by the WH, a volume of water was dislodged. After finishing the test, the water level in the CAV returns to the initial level. In fact, no change takes place in the water level when comparing the start and end of each test. As a result, the pressure of the AP returned to initial pressure at the end of every WH. Figure 12 demonstrates the pressure changing way for 5WH and 9WH tests for AP length of 3 cm (VFR = 4.75%). The results from 1WH tests are not presented in Figure 12 since they do not make a sensible change in the measured parameters. The discharged volume of water at the end of the test is shown in each graph too. The peak pressure is almost constant during the repetition of WH and the magnitude of peak pressure is approximately equal to the tests with closed

BV presented in Figure 3. As expected, by increasing the Re value, the magnitude of discharged water from the CAV rises.

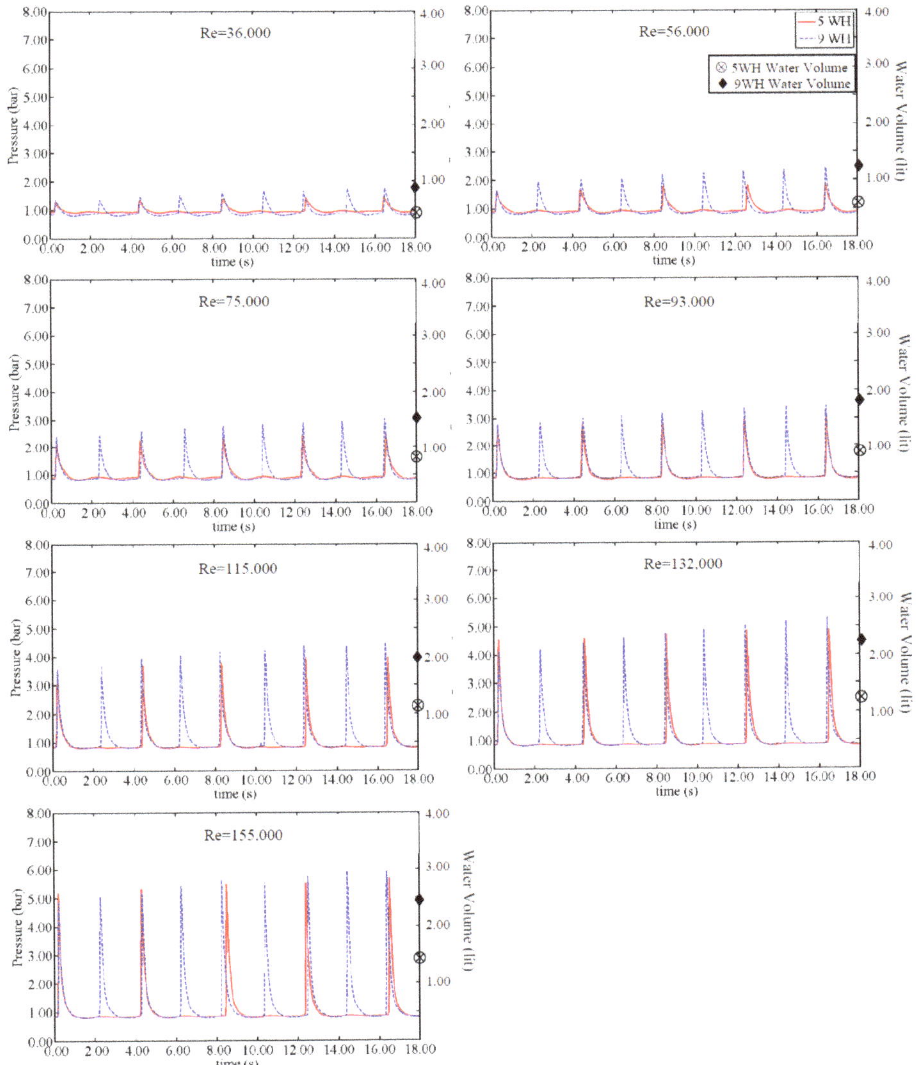

Figure 12. Pressure change in the AP and discharged water volume for 5WH and 9WH events, different Re and VFR = 4.75%.

Figure 13 shows the amount of discharged water volume after each test for different VFR, Re and WH events. There is a fast increase in discharging water volume for small VFR while it rises slowly for bigger VFR. Working with a small VFR needs more attention due to its high sensitivity to the pressure. Figure 13 shows that the discharged water volume for 9WH is considerably higher than for 5WH. In "open BV tests" the 9WH event creates a constant discharge from the CAV, while for the 5WH the discharge is not constant. In fact, the AP acts as a means of pressure storage and recovery, as a regulating flow system that adjusts the outflow from the CAV to create a constant discharge (see Supplementary Materials).

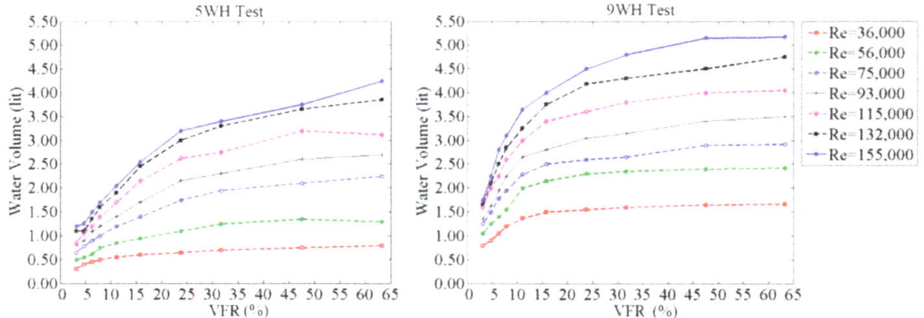

Figure 13. Discharged water volume against the VFR for different Re and WH events.

After finishing the "open BV tests", the discharge from the CAV stops and the water level returns to the initial level. For the VFR > 32% (AP length > 20 cm), the 9WH test leads to a constant discharge from the CAV which is less evident for the 5WH test. A high VFR shows very useful capability. This range of VFR seems to be very effective in SS-CAES systems when connected to the main water conveyance system.

4. Discussion

An extensive experimental work for AP behavior cognition has been carried out in Instituto Superior Técnico (IST) experimental laboratory. Tests are done with three WH events: 1WH, 5WH and 9WH. Also, two situations of the BV in the CAV, i.e., closed BV and open BV were conducted. In the closed BV tests, both peak pressure and stored pressure were raised by increasing the Re number and decreasing the VFR. In addition, stored pressure is much lower than peak pressure for small VFR, i.e., VFR < 11%. The pressure storage process is very regular and predictable for the VFR > 11%, but for smaller VFR values, the pressure storage starts with a high peak at the first WH occurrence. The stored pressure is always higher than the peak pressure ($P_s > P_{1p}$) for Re ≤ 75,000. Also, the stored pressure is always higher than the peak pressure ($P_s > P_{1p}$) if VFR > 11%. Despite being more frequent in 9WH tests, the stored pressure is not always higher than for other events. In fact, for the VFR < 30% and Re > 93,000, the stored pressure from 9WH is smaller than for 5WH. For reasoning, it can be stated that in one level of WH repetitions, the air reaches maximum compression for this pressure range. After that, more WH actions do not compress the air but lead to pressure loss by solving some portion of air into the water. Regular and equal pressure steps can be seen in the rising pressure of several WH events for Re < 75,000. For higher Re numbers, the pressure change route starts with a peak and dissipates fast, continuing with smaller steps. The stored pressure is in correlation with Re but in contrast to the VFR.

The pressure in the pipe system shows that increasing the VFR or AP size will diminish the pressure rising within the pipe. The final pressure in the pipe is more or less the same for the VFR > 20%. So, it seems that the VFR in the range of around 20% to 30% presents good ability to control the pressure without expanding the air vessel size.

It is very important for the optimal size of a CAV that the rising water level (Δh) stays in an acceptable range. The VFR from 20% showed good performance in controlling the pressure surge of WH events. For this range of VFR, the Δh remains below 20% of the CAV length.

In the "open BV test", the condition of using the pressure energy of the WH to supply water as a ram pump was examined. By increasing the Re number, the outflow water volume will be increased too. The water level inside the CAV keeps the same elevation due to confinement and vacuum conditions. For the VFR > 32%, the WH events create a constant outflow from the CAV with regular compression and expansion in the AP in the 9WH tests.

This work presents an extensive experimental study of different WH events that can be used as a benchmark for future calculations using WH equations.

Finally, authors believe that the main effective parameter in pressure change inside the AP is the volume of the AP and flow velocity which have been referred by the dimensionless parameters VFR and Re respectively in the graphs. However, for different CAV sizes, namely diameter or height, the results may change. As a suggestion, researchers may examine the effect of different air vessel sizes on pressure change and other variables of an AP.

Supplementary Materials: The following videos from "open BV test" and "closed BV test" for 5WH and 9WH events are available showing the continuity of the flow discharge for APs of 20 and 40 cm in case of 9WH events. Videos AP = 5_5WH, AP = 5_9WH, AP = 10_5WH, AP = 10_9WH, AP = 20_5WH, AP = 20_9WH, AP = 40_5WH, AP = 40_9WH. The videos are available online at www.mdpi.com/2073-4441/9/1/63/s1.

Acknowledgments: Authors appreciate the financial support from Ph.D. grant ref. PD/BD/114459/2016 issued by FCT under the IST-LNEC Joint Doctoral Initiative supported by the Civil Engineering, Research, and Innovation for Sustainability (CEris) Center, Department of Civil Engineering, Architecture and Georesources, Instituto Superior Técnico (IST), University of Lisbon (ULisbon), Portugal and providing the experimental facilities for conducting the tests.

Author Contributions: Mohsen Besharat has fulfilled the experimental activities. Results are analyzed and the paper is written with the contribution of Helena M. Ramos and Maria Teresa Viseu.

Conflicts of Interest: The authors declare no conflict of interest. The funding sponsors had no role in the design of the study; in the collection, analyses and interpretation of data; in the writing of the manuscript; and in the decision to publish the results.

Nomenclature

AP	Air Pocket
WC	Water Column
CAV	Compressed Air Vessel
CAES	Compressed Air Energy Storage
WH	Water Hammer
VFR	Volume Fraction Ratio
Re	Reynolds Number
PT	Pressure Transducer
CV	Check Valve
EBV	Electro-Pneumatic Ball Valve
BV	Ball Valve
P_{1p}	Maximum pressure related to the first peak of pressure oscillation
P_s	Stored pressure in the AP after finishing the tests
P_p	Maximum pressure in the pipe at PT_2 point during WH test

References

1. Chaudhry, M.H. *Applied Hydraulic Transients*, 3rd ed.; Springer: New York, NY, USA, 2014; pp. 351–352.
2. Ruus, E.; Karney, B. *Applied Hydraulic Transients*; Friesens Corporation: Altona, MB, Canada, 1997.
3. Stephenson, D. Simple guide for design of air vessels for water hammer protection on pumping lines. *J. Hydraul. Eng.* **2002**, *128*, 792–797. [CrossRef]
4. Izquierdo, J.; Lopez, P.A.; Lopez, G.; Martinez, F.J.; Perez, R. Encapsulation of air vessel design in a neural network. *Appl. Math. Model.* **2006**, *30*, 395–405. [CrossRef]
5. De Martino, G.; Fontana, N. Simplified approach for the optimal sizing of throttled air chambers. *J. Hydraul. Eng.* **2012**, *138*, 1101–1109. [CrossRef]
6. Ramalingam, D.; Lingireddy, S. Neural network-derived heuristic framework for sizing surge vessels. *J. Water Resour. Plan. Manag.* **2014**, *140*, 678–692. [CrossRef]
7. Hatcher, T.M.; Vasconcelos, J.G. Peak pressure surges and pressure damping following sudden air pocket compression. *J. Hydraul. Eng.* **2016**. [CrossRef]

8. Pozos-Estrada, O.; Sánchez-Huerta, A.; Breña-Naranjo, J.A.; Pedrozo-Acuña, A. Failure analysis of a water supply pumping pipeline system. *Water* **2016**, *8*, 395. [CrossRef]

9. Malekpour, A.; Karney, B.; Nault, J. Physical understanding of sudden pressurization of pipe systems with entrapped air: Energy auditing approach. *J. Hydraul. Eng.* **2015**, *142*. [CrossRef]

10. Apollonio, C.; Balacco, G.; Fontana, N.; Giugni, M.; Marini, G.; Piccinni, A.F. Hydraulic transients caused by air expulsion during rapid filling of undulating pipelines. *Water* **2016**, *8*, 25. [CrossRef]

11. Besharat, M.; Ramos, H.M. Theoretical and experimental analysis of pressure surge: In a two-phase compressed air vessel. In Proceedings of the 12th International Conference on Pressure Surges, Dublin, Ireland, 18–20 November 2015; Code 119785. Available online: http://www.scopus.com/inward/record.url?eid=2-s2.0-84966351254&partnerID=MN8TOARS (accessed on 17 January 2017).

12. Proczka, J.J.; Muralidharan, K.; Villela, D.; Simmons, J.H.; Frantziskonis, G. Guidelines for the pressure and efficient sizing of pressure vessels for compressed air energy storage. *Energy Convers. Manag.* **2013**, *65*, 597–605. [CrossRef]

13. Ibrahim, H.; Younès, R.; Basbous, T.; Ilinca, A.; Dimitrova, M. Optimization of diesel engine performances for a hybrid wind-diesel system with compressed air energy storage. *Energy* **2011**, *36*, 3079–3091. [CrossRef]

14. Kim, Y.M.; Favrat, D. Energy and exergy analysis of a micro-compressed air energy storage and air cycle heating and cooling system. *Energy* **2010**, *35*, 213–220. [CrossRef]

15. Budt, M.; Wolf, D.; Span, R.; Yan, J. A review on compressed air energy storage: Basic principles, past milestones and recent developments. *Appl. Energy* **2016**, *170*, 250–268. [CrossRef]

16. Martins, S.C. Dynamic of Hydraulic System Pressurization with Entrapped Air. Ph.D. Thesis, Instituto Superior Técnico, University of Lisbon, Lisbon, Portugal, 2013. (In Portuguese)

17. Besharat, M.; Tarinejad, R.; Ramos, H.M. The effect of water hammer on a confined air pocket towards flow energy storage system. *J. Water Suppl. Res. Technol.-AQUA* **2015**, *65*, 116–126. [CrossRef]

Article

Design Criteria for Suspended Pipelines Based on Structural Analysis [†]

Mariana Simão [1,*], Jesus Mora-Rodriguez [2,‡] and Helena M. Ramos [1,‡]

[1] Civil Engineering Research and Innovation for Sustainability (CEris) of the Department of Civil Engineering, Architecture and Georesources, Instituto Superior Técnico (IST), Universidade de Lisboa (ULisboa), Lisbon 1049-001, Portugal; helena.ramos@tecnico.ulisboa.pt

[2] Geomatics and Hydraulic Engineering Department, University of Guanajuato, Guanajuato 36094, Mexico; jesusmora@ugto.mx

[*] Correspondence: m.c.madeira.simao@tecnico.ulisboa.pt; Tel.: +351-218-418-151

[†] Based on Orynyak, I.; Vasylyuk, V.; Radchenko, S.; Batur, A. The modelling of the dynamical behavior of the aerial pipeline bridge at the rupture of its underground part during the water pressure testing. *Ukr. J. Oil Gas. Ind.* **2006**, *4*, 27–32. (In Ukrainian)

[‡] These authors contributed equally to this work.

Academic Editor: Arjen Y. Hoekstra

Received: 4 May 2016; Accepted: 12 June 2016; Published: 17 June 2016

Abstract: Mathematical models have become the target of numerous attempts to obtain results that can be extrapolated to the study of hydraulic pressure infrastructures associated with different engineering requests. Simulation analysis based on finite element method (FEM) models are used to determine the vulnerability of hydraulic systems under different types of actions (e.g., natural events and pressure variation). As part of the numerical simulation of a suspended pipeline, the adequacy of existing supports to sustain the pressure loads is verified. With a certain value of load application, the pipeline is forced to sway sideways, possibly lifting up off its deadweight supports. Thus, identifying the frequency, consequences and predictability of accidental events is of extreme importance. This study focuses on the stability of vertical supports associated with extreme transient loads and how a pipeline design can be improved using FEM simulations, in the design stage, to avoid accidents. Distribution of bending moments, axial forces, displacements and deformations along the pipeline and supports are studied for a set of important parametric variations. A good representation of the pipeline displacements is obtained using FEM.

Keywords: design criteria; structure analysis; suspended pipelines; finite element method (FEM)

---

1. Introduction

Accidents associated with natural events or human actions are a common cause of hydraulic system failure. As a result, unwelcome accidents lead to enormous damage, the destruction of networks and the abruptly stoppage of supply. These events can also be classified as external or internal. External ones may result from extreme events (*i.e.*, such as storms, floods, landslides, earthquakes, sudden releases or ruptures) and can be a consequence of structural characteristics, maintenance state, or hydraulic operation devices [1]. Pipe systems, especially those installed above the ground, are under relevant dynamic forces during the occurrence of transients (water-hammer). When these forces are associated with system movement, a fluid structure interaction is generated, which means that the liquid and the pipe must be analysed as a whole [2]. Transient pressures and dynamic vibrations generated by a water-hammer or by the closing of a valve (internal causes) lead to new loads on the system such as internal and external pressures created by the soil and/or through the pipe supports [3].

The vulnerability and susceptibility of a pipeline system can be better understood after the occurrence of accidents. Such hazards may be associated with several factors causing numerous problems, particularly when they are neglected during different design stages: project concept, sizing, implementation and operation [4–7].

Extreme transient pressure variations may cause deformations and displacements in pipes which, in turn, will interact with the hydraulic system, modifying its own active loading and, consequently, its reaction [8,9], resulting in vibration or resonance phenomena that may cause the system to fail. As a consequence, the vibrations can cause the structure to rupture due to fatigue or through the increase of structural deformations. Hence, the structural response under different load combinations is provided by using an FEM representing the characteristics involved in hydraulic systems exposed to external and internal load vibrations [10]. The behaviour of the fluid and the structure can be studied as a whole, with a structural model, if appropriate load combinations are adopted. This approach allows the possible consequences to be shown in terms of displacements and the vulnerability of the pipeline and its supports.

In the classic theory of finite elements it is assumed that movements and deformations are small and the material has a linear elastic behaviour. In some cases this condition cannot be satisfied, leading to the inclusion of non-linearity action/reaction to the model [11]. In this work, the model Robot Autodesk (Autodesk, Inc., San Rafael, CA, USA) is used to study the behaviour of a suspended pipeline under different loads, using nonlinear theory. The results obtained by the computational simulation allow us to verify the adequacy of the supports' design in a presence of transient phenomena. The importance of detailed studies of identical system behaviour for different loads in the infrastructure design is also emphasized.

2. Background

The design of a suspended pipeline requires two alternative structure solutions: pipelines supported by auxiliary structures or self-supporting pipelines. In this research, a real case of auxiliary structures supporting a pipeline (*i.e.*, pipeline bridge) is studied. The preferred conceptual design of a pipe bridge consists of a straight configuration that can use a restrained mechanical joint ductile iron pipe (DIP) or a butt welded steel pipe with a pipe expansion joint, on a roller system, allowing the pipe to act independently of the superstructure [12]. This scenario is considered the ideal case because it is free of fittings and minimizes pipe joint deflections, which create thrust forces [13].

A stress analysis of the pipelines needs to be conducted to verify the integrity of the suspended structure, as the expected pressure extremes which ensure the applied pre-tension guaranteed that the columns remain in tension [14].

Many pipe failures in natural events have resulted from the sliding, rocking or overturning of large equipment to which the pipe is attached. During these events, the tanks may twist and slide in the concrete saddles, resulting in ruptures. It is therefore important to verify the adequacy of the anchorage equipment and tie-downs as part of design or retrofit of pipeline systems [15]. The same type of failure can take place when a pipe is connected to two separate structures (e.g., vertical pipe supports and lateral bracing).

The vibration caused by external or internal forces is transmitted to the pipe, and if the pipe is too stiff, this motion may cause the pipe to fail. Some building codes advocate the use of "flexible assemblies" to absorb differential building motion, such as placing a flexible assembly in the pipe where it crosses building joints [16]. However, most pipe spans are sufficient flexibles to absorb this differential movement. Otherwise, unless required by a building code, it is prudent to avoid placing these flexible assemblies at the preliminary design stage and only use them if there is no other alternative solution confirmed by detailed stress analysis [17].

In a suspended pipeline, this system can be idealized as continuous beams spanning lateral braces [18]. Commonly, vertical supports for gravity and operational loads are adequate to resist the vertical seismic forces, since the vertical component of seismic force is often lower than other vertical loads [19]. However, where a support resists to the vertical component of a lateral or longitudinal brace force, it should be designed explicitly to resist all applied forces, such as transient events.

3. Formulation of the Problem

3.1. Description of the Accident

Knowledge on the effect of a flowing mass on the dynamic response of suspended pipelines simultaneously under moving loads is still lacking from the point of view of structural design. The dynamic stability of suspension structures has been the focus of several studies [20–22]. However, few studies has been focused on the effect of interaction between fluid and structure, especially under different actions of the pipeline's suspension with pressurized flows. The purpose of this research is to investigate the dynamic effects involved in a suspended pipeline bridge by combining the maximum overloading and water-hammer phenomena.

The case study is based on an accident that happened in Ukraine 2005, where due to water-hammer phenomena the pipeline, used to transport oil to refineries, was set in motion, bent at the opposite ground entrance section (*i.e.*, displacements in upper and transverse directions) and thrown from its supports. The measured pressures registered values higher than 10 MPa. This overpressure was responsible for the failure at an underground section, *i.e.*, 50 m from the surface section of the pipeline to the source (Figure 1).

Figure 1. Pipeline accident (adapted from [23]).

The material of the pipeline is steel and has a nominal diameter (ND) of 500 mm. The steel supports, fixed in concrete footings, have U-configuration with maximum spans of 15 m. After the water hammer phenomena, the pipeline moved about 1.50 m in the pipe direction (axis-x) and 0.75 m in the transverse direction (axis-y).

This accident highlights the importance of assessment of factors which cause instability and lead to eventual rupture, with analyses of the safety condition of strategic pipeline systems under pressure, mainly due to effects of water-hammer and earthquakes responsible for the rupture and fatigue of the structure. In the case of sudden changes (*i.e.*, transient events, earthquakes), a pipeline's design rarely remains in good performance conditions. Thus, the vulnerability of the pipeline system to transient actions is analysed, through an advanced FEM structural analysis (Robot Autodesk), to simulate the consequences of a water-hammer in terms of displacements and loads in the pipes and support structures. Subsequently, comparisons are made between the results obtained at the accident and in simulations to better understand the causes and effects herein reported.

3.2. Definition of Physical Characteristics

The system, based on the real case, is a pipeline with a total length of 74 m, composed of 6 steel supports with a 500 mm internal diameter (Figure 2). Three of the 6 supports are 3 m in height with a 15 m span, followed by 2 supports 1.20 m in height in the downstream section (Figure 2). The remaining upstream section has spans of 5 (*i.e.*, in the bend) and 7.4 m (*i.e.*, in the initial section). The supports consist of 3 steel tubes with wall thickness of 50 mm. The pipeline is attached to two separate structures forming the supports. The pipeline has a thickness of 26.2 mm.

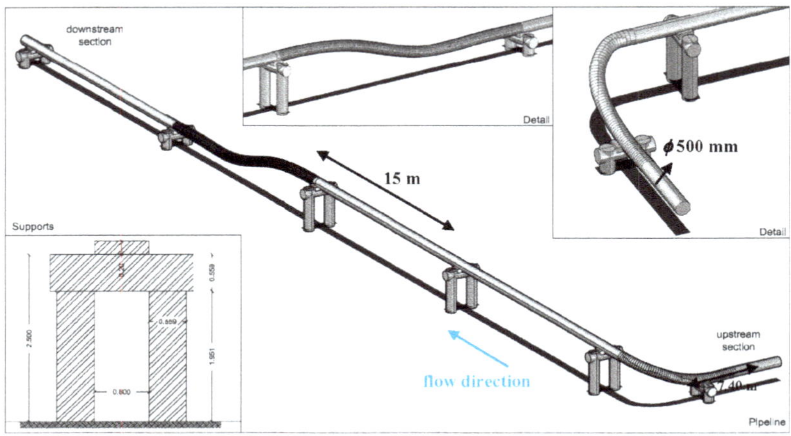

Figure 2. Pipeline system: part under analysis.

The chapters below present finite element results of the stresses and displacements in the pipeline and in supports due to internal pressure under the combination of the water-hammer event. Two combinations and nonlinear effects associated with each load are also studied separately. The material used for the pipeline, according to specifications ANSI/ASME B.31.4, is carbon steel API-5L Gr. B, with the characteristics presented in Table 1.

Table 1. Material specifications of the pipeline according to ANSI/ASME B.31.4.

Material	API-5L Gr B
Schedule/Serie	60
ND (mm)	500
thickness (mm)	20.0
Max. allowable operating pressure (MPa)	10.1
Max. allowable stress (MPa)	173.8 *
Moment of inertia (cm^4)	91,428
Elastic section modulus (cm^3)	3600
Plastic section modulus (cm^3)	4766

Note: * The maximum allowable stress accounts for the welding efficiency factor.

The selected material was estimated based on the maximum pressure achieved during the accident. Thus, Sch/serie 60 was chosen since the maximum allowable operating pressures is higher than that registered *in situ* and because the overpressure induced by the water-hammer did not cause the rupture of the pipeline.

4. Structural Model

4.1. Introduction

In this research, the Autodesk Robot Structural Analysis Professional 2015 program was used to determine the static and dynamic performance of structural systems and analyse the linearity and non-linearity of the system behaviour. This is a program for calculation by finite element structures that includes a wide range of design codes of all types of metal and concrete structures, with the possibility of contemplating other structural materials [24].

For different structures, the beam element is a slender structural member that offers resistance to forces and bending under applied loads [25]. It is represented by line elements used to create a 1D idealization of a 3D structure, and is computationally more efficient than solids or shells [18].

The use of a beam element is highly applicable to most beam structures, e.g., bridges, roadways, building constructions, and people movers (railcars, trams and buses), since they support linear and nonlinear analyses, including plasticity, large deformations, and nonlinear collapse [26]. It is easy to use especially when:

- the length of the element is much greater than the width or depth;
- the element has consistent cross-sectional properties;
- the element must be able to transfer moments;
- the element must be able to handle a load distributed across its length.

Thus, to reproduce the accident, the structure was modelled using beam elements. All combinations were performed using the European section database (EURO code). The geometry was defined in Autocad and then exported to Robot, where the boundary conditions were set. Herein, fixed supports are assigned to the two pillars of the 6 support structures (Figure 3). The transfer of forces, induced by the pipeline and the fluid to the support, are made by intermediate connectors welded to the steel beam tube.

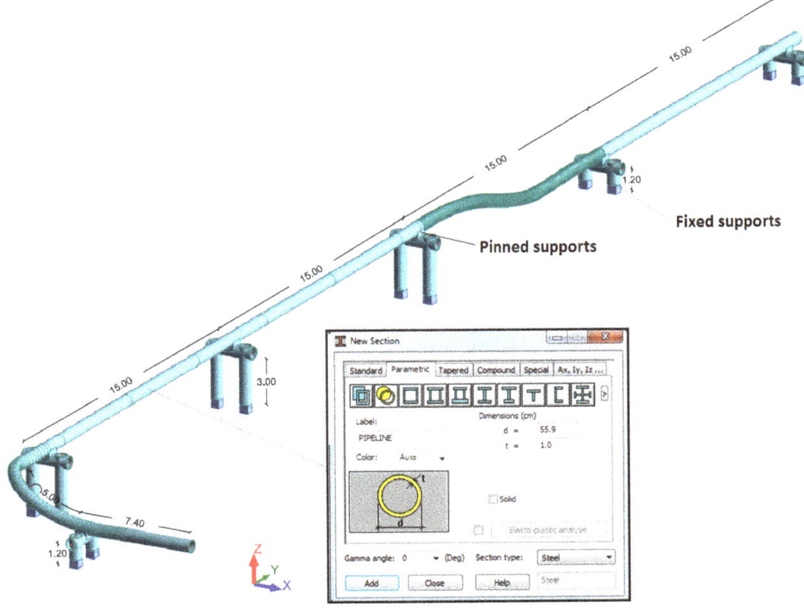

Figure 3. Pipeline geometry defined in Robot structural analysis.

4.2. Design Considerations

4.2.1. Load Combinations

The primary pipe loading on the bridge structure is the pipe dead load which includes the weight of the pipe, the water inside the pipeline, pipe accessories (*i.e.*, expansion joints, couplings), and the pipe support system. Other loads that should be considered are wind, earthquake and impact. In the present study, only an internal load is evaluated: the water-hammer phenomenon.

In limit states, the loads are multiplied by safety factors and grouped into load combinations. For different codes (e.g., BSI 8010 UK; ASME B31.4 US; ISO 13623), different combinations of loads are established. The general requirement is that the defined load combination should be appropriate to the ultimate limit state. The Canadian code CSA-Z662 (1996) uses the following general equation for calculating the load combinations ($i = 1, 2, 3, ...$):

$$Lci = \alpha \left(\gamma_G G + \gamma_Q Q + \gamma_E E + \gamma_A A \right) \tag{1}$$

where, α = safety factor class; $\gamma_G, \gamma_Q, \gamma_E, \gamma_A$ = loading factors (Table 2) for G = permanent loads (*i.e.*, weight of the water inside the pipeline); Q = operational loads (overloading, *i.e.*, internal pressure during normal conditions); E = environmental loads (*i.e.*, seismic load); A = live loads (*i.e.*, transient events).

Table 2. Load factors according to the Canadian code (adapted from CSA-Z662, 1996).

Load Combination	Load Factor (γ)				
SLS—Serviceability Limit States ULS—Ultimate Limit States	γ_G	γ_Q		γ_E	γ_A
		Internal Pressure	Other		
ULS1: max. overloading	1.35	1.50	1.25	0.70	0.00
ULS2: max. environmental	1.05	1.05	1.05	1.35	0.00
ULS3: live	1.00	1.00	1.00	0.00	1.00
ULS4: fatigue	1.00	1.00	1.00	1.00	0.00
ULS5: faulted *	1.00	1.00	1.00	1.00	1.00
SLS	1.00	1.00	1.00	0.00	0.00

Note: * For a faulted combination, the condition is associated to fault fluid transients resulting from an earthquake [27].

For each load combination, its effects are calculated as the level of stress and deformation. The resulting value of these effects is compared with the threshold value, to determine if the limit state is respected.

The safety factor class depends on the risk to which the structure is subjected. In the case of pipelines the safety factor is 2.0. According to the case study, two types of combinations were selected and applied as presented in Table 3.

Table 3. Selected combinations.

Load Combination *	Load Factor (γ)			
ULS—Ultimate Limit States	γ_G	γ_Q	γ_E	γ_A
ULS1: max. overloading	1.35	1.50	0.00 *	0.00
ULS3: live	1.00	1.00	0.00	1.00

Note: * The performed simulations do not take into account seismic events.

4.2.2. Internal Loads

The first simulation is performed for combination ULS1 (without transient and natural events), where the overloading is the internal pressure, during normal conditions, plus 2000 N applied in the vertical direction, for pipelines located 3 m from the ground according to [28]. In order to verify whether the pipeline bridge is designed appropriately, it is necessary to satisfy the following conditions:

i   the maximum pipeline bending stress, due to pressure, weight and other sustained mechanical loads, is below the ultimate material load limits, given by [29]:

$$S_v = \frac{L}{10Z}\left[qL + 2\left(C + W\right)\right] < 35\,\text{MPa} \tag{2}$$

where L = length between supports (m); Z = moment of resistance (cm^4); q = total distributed loads (N/m); C = total concentrated loads (N), in this case $C = 0$; W = overload imposed at mid-span;

ii   the maximum deflection at mid-span is below 25 mm for the limit value of pipelines with diameters higher than 100 mm [29], calculated as:

$$\delta = \frac{2400L^3}{EI}\left[\frac{C + W}{3} + \frac{qL}{4}\right] \tag{3}$$

where E = elastic modulus (MPa); I = moment of inertia of the pipeline cross section (cm^4);

iii   the maximum stress due to internal pressure ($S_{p\,max}$) is below the acceptable stress of the pipe (*i.e.*, $S_h = 173.8\ MPa$ (Table 1)):

$$S_{p\,max} \leqslant S_h \tag{4}$$

and the total longitudinal load stresses (S_l) (*i.e.*, pressure, weight, overloading) is below S_h:

$$S_l \leqslant S_h \Leftrightarrow \frac{PD}{4e} + S_v \leqslant S_h \tag{5}$$

iv   the vertical or lateral motion is below to the limit motion given by [29]:

$$\delta_{max} = \frac{10^7 qL^4}{24EI} \tag{6}$$

where δ_{max} = maximum displacement (mm); E = elastic modulus (kgf/cm^2).

For combination ULS1, all the conditions are verified. The maximum deflection Equation (3) obtained at mid-span is ~11.5 mm (Figure 4), which is below the allowable limit value.

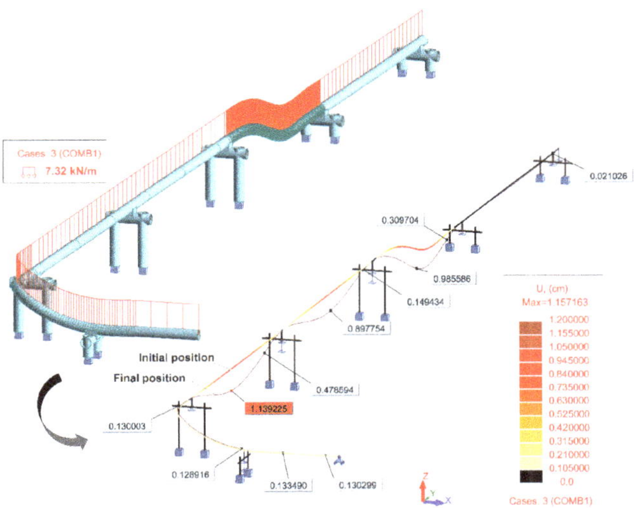

Figure 4. Maximum displacement achieved for permanent loads (ULS1) during normal conditions.

4.2.3. Transient Loads

For combination ULS3, the effects of pressure, weight, other sustained loads and occasional loads including earthquakes shall meet the requirements of Equation (7):

$$S_l \leqslant S_h \Leftrightarrow \frac{PD}{4e} + S_v + S_o \leqslant kS_h \tag{7}$$

where k = 1.2 for occasional loads acting less than 1% of the operating period; S_o = stress due to occasional loads, such as thrusts from pressure, flow transients and earthquakes.

The other assumption is related to the critical pressure (*i.e.*, burst pressure) [29]:

$$p_c = \frac{2E}{1 - \lambda^2} \left(\frac{t}{D} \right)^3 \tag{8}$$

where E the elastic modulus of steel, λ the poisson modulus, D and t the outer diameter and the thickness of the pipe, respectively. By Equation (8) with the specified parameters of the pipeline (Table 1), the critical pressure is 24 MPa, which is higher than the maximum pressure of the transient event. This means that the characteristics of the pipeline (*i.e.*, diameter and material) guarantee extreme conditions between 10 MPa and 24 MPa.

Considering combination ULS3, the water-hammer event is defined as a concentrated force applied at the upstream point of the pipeline. To simulate the transient event, a path was set to the concentrated force (*i.e.*, F = 3456.4 kN) obtained through the maximum pressure registered. This generates a moving load starting along the pipeline (Figure 5).

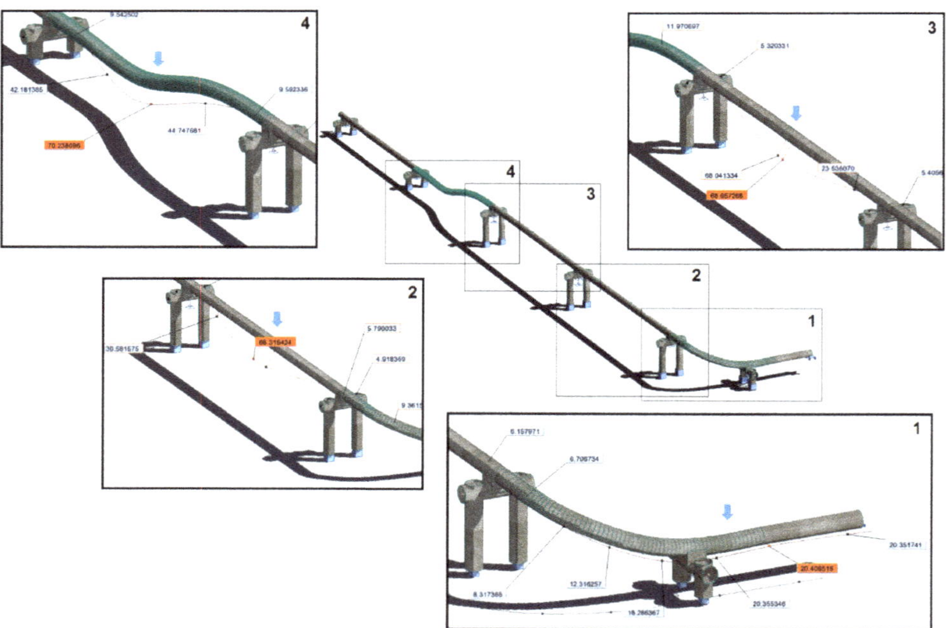

Figure 5. Combination ULS3: water-hammer defined as a moving load starting in 1. Maximum displacements achieved under the transient event (in cm).

Calculating the design assumptions described in Section 4.2.2, it is verified that this combination fails in the ULS3 conditions presented in Table 4.

Table 4. Results obtained for the two selected load conditions.

Combination	$S_v < 35\ Mpa$	$\delta < 25\ mm$	$S_{p\ max},\ S_l < 173.8\ Mpa$	$\delta_{struct} < \delta_{max} = 65.74\ mm$
ULS1	31.41	9.46	56.04	–
ULS3	2943.18	588.44	2961.99	588.44

The vibration of the suspended pipeline under the action of a transient phenomenon described by a moving load gives results similar to the ones presented *in situ*. Moreover, the numerical results indicate that the maximum deflection at mid-span is 702.40 mm, which is 1.19 times the theoretical value (588.44 mm) and 0.94 less than the real value (750 mm). The difference between FEM and *in situ* measurements is not significant, confirming the dynamic behaviour of the structure under an impulsive load can be simulated using this model.

Figure 6 illustrates the maximum moving effect in each point of the structure under the moving load (*i.e.*, a water-hammer event).

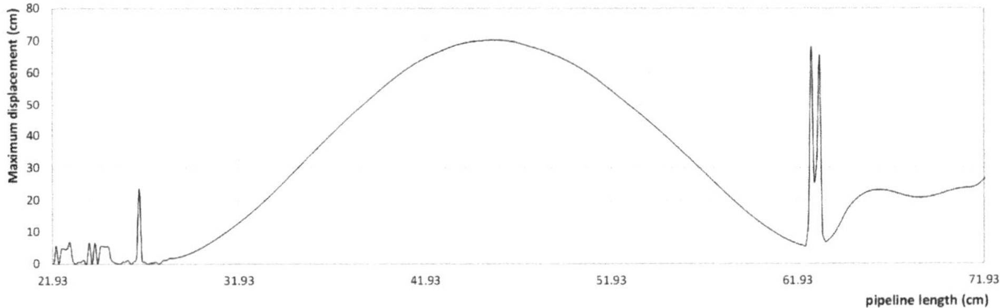

Figure 6. Maximum displacement in each point of the pipeline during a transient event.

From the deformed shape of the structure, some inconveniences can be observed, starting from the 1st to 20th m of length of the analysed pipe branch. As an overview, the deformed shape of this suspended pipeline has a point where the anchor supports, exceed their damping limit (*i.e.*, maximum expansion), permitting some elements to vibrate with a self-period. This influences the global behaviour of the entire structure leading to its fall, as it can be seen from *in situ* pictures (see Figure 1).

5. Conclusions

The purpose of this study is to ensure that in case of an extreme event, such as a water-hammer, the pipeline system will perform its intended function: position retention (the pipeline does not fall), leak tightness (the pipeline does not leak), and operability (the pipeline system delivers and regulates flow). However, when the pipeline is suspended, other constraints should be taken into account, due to the serious risk of damage by pipe buckling due to too high bending.

Based on existing pipeline damage, the structural flaws of the suspending oil pipeline were modelled using a finite element method. The suspending pipeline was calculated for two different combinations: ULS1-maximum overloading and ULS3-live. For the first, the structure fulfilled the safety conditions, unlike the ULS3 combination, where the standard design specifications failed in all aspects: (i) maximum pipeline bending stress; (ii) maximum deflection at mid-span; (iii) maximum acceptable stress; and (iv) damping limit. By using a moving point load in ULS3, the maximum mid-span achieved in the simulation results (*i.e.*, 702.40 mm) are validated with the *in situ* measurements (*i.e.*, ~750 mm).

The structural model can be used for simulating real extreme events where the maximum pressure is lower than the critical pressure (*i.e.*, when extreme conditions are guaranteed). Thus, assuming here the water–hammer event as the maximum point load that moves along the pipeline, the results obtained give a good approximation of the real case and can be adopted by engineers and designers to meet pipeline protection standards.

Acknowledgments: This research was funded by the Portuguese Foundation for Science and Technology (FCT) through the Doctoral Grant—SFRH/BD/68293/2010 of the first author. Also, thanks to CEris (Civil Engineering Research and Innovation for Sustainability) of the Department of Civil Engineering, Architecture and Georesources, Instituto Superior Técnico (IST), Universidade de Lisboa.

Author Contributions: Mariana Simão simulated the accident according to the data, kindly lent by Igor Orynyak; Helena M. Ramos and Jesus Mora-Rodriguez analyzed the data; All authors contributed reagents/materials/analysis tools; Mariana Simão wrote the paper. Helena M. Ramos and Jesus Mora-Rodriguez reviewed the manuscript.

Conflicts of Interest: The authors declare no conflict of interest.

References

1. Richie, A.M. Managing infrastructure problems that arise from earthquakes. *Mater. Perform.* **2003**, *42*, 48–51.
2. Simão, M.; Mora, J.; Ramos, H.M. Fluid-structure interaction with different coupled models to analyse an accident occurring in a water supply system. *J. Water Supply Res. Technol. Aqua* **2015**, *64*, 302–315. [CrossRef]
3. Simão, M.; Mora-Rodriguez, J.; Ramos, H.M. Mechanical interaction in pressurized pipe systems: Experiments and numerical models. *Water* **2015**, *7*, 6321–6350. [CrossRef]
4. Ramos, H. *Unsteady Flows: Water Systems Vulnerability*; Lisbon Instituto Superior Técnico: Lisbon, Portugal, 2006.
5. Ramos, H. *Vulnerability Assessment in the Operational Management of Supply Systems*; Lisbon Instituto Superior Técnico: Lisbon, Portugal, 2007.
6. Ramos, H.; Almeida, A.B.; Borga, A.; Anderson, A. Simulation of severe hydraulic transient conditions in hydro and pump systems. *Recur. Hídr. APRH* **2005**, *26*, 7–16.
7. Ramos, H.; Borga, A.; Bergant, A.; Covas, D. Analysis of surge effects in pipe systems by air release/venting. *Recur. Hídr. APRH* **2005**, *26*, 45–55.
8. Rajani, B.; Zhan, C.; Kuraoka, S. Pipe-soil interaction analysis of jointed water mains. *Can. Geotech. J.* **1996**, *33*, 393–404. [CrossRef]
9. Almeida, A.; Ramos, H. Water supply operation: Diagnosis and reliability analysis in a Lisbon pumping system. *J. Water Supply Res. Technol. Aqua* **2010**, *59*, 66–78. [CrossRef]
10. Adachi, T.; Ellingwood, B. Serviceability of earthquake-damaged water systems: Effects of electrical power availability and power backup systems on system vulnerability. *Reliab. Eng. Syst. Saf.* **2008**, *93*, 78–88. [CrossRef]
11. Lopes, M.; Leite, A. *Economic Viability of Seismic Strengthening of Buildings*; Ingenium: Greenbelt, MD, USA, 2005.
12. Tucker, M. *Bridge Crossings with Ductile Iron Pipe*; Pipeline Division-American Society of Civil Engineers (ASCE): Denver, CO, USA, 1991.
13. Gatulli, V.; Martinelli, L.; Perotti, F.; Vestroni, F. Dynamics of suspended cables under turbulence loading: Reduced models of wind field and mechanical system. *J. Wind Eng. Wind Aerodyn.* **2007**, *95*, 183–207. [CrossRef]
14. Aronowitz, D.; Steward, N.; Bradley, R. The design of a suspended concrete transport pipeline system. *J. South. Afr. Inst. Min. Metall.* **2008**, *108*, 707–713.
15. Zhang, X.; Sun, B. Aerodynamic stability of cable-stayed-suspension hybrid bridges. *J. Zhejiang Univ. Sci.* **2005**, *6*, 869. [CrossRef]
16. Teng, Z.C.; Sun, J. A New Seismic Design Methods for Suspending Oil Pipeline. In Proceedings of the 2011 International Conference on Remote Sensing, Environment and Transportation Engineering (RSETE), Nanjing, China, 24–26 June 2011; pp. 2843–2847.
17. Antaki, G.; Aiken, S. *Seismic Design and Retrofit*; American Lifelines alliance: Atlanta, GA, USA, 2002.
18. Timoshenko, S.; Goodier, J. *Theory of Elasticity*, 3rd ed.; International Student, Ed.; McGraw-Hill: New York, NY, USA, 1982.

19. Moussou, P.; Potapov, S.; Paulhiac, L.; Tijsseling, A. Industrial cases of FSI due to internal flows. In Proceedings of the 9th International Conference on Pressure Surges, Chester, UK, 24–26 March 2004; pp. 167–181.
20. Bažant, Z.P.; Cedolin, L. *Stability of Structures: Elastic, Inelastic, Fracture and Damage Theories*, 1st ed.; World Scientific Publishing: Singapore, 2010.
21. Yan, X.; Shide, H. Research on elastic dynamic stability of CFST arch model. *J. Tongji. Univ.* **2005**, *33*, 6–10. (In Chinese)
22. Xiang, H.F.; Bao, W.G.; Chen, A.R.; Lin, Z.X.; Liu, J.X. *Wind-Resistant Design Specification for Highway Bridges (JTG/TD60–01–2004)*; China Communications Press: Beijing, China, 2004.
23. Orynyak, I.; Vasylyuk, V.; Radchenko, S.; Batur, A. The modelling of the dynamical behavior of the aerial pipeline bridge at the rupture of its underground part during the water pressure testing. *Ukr. J. Oil Gas. Ind.* **2006**, *4*, 27–32. (In Ukrainian)
24. Advanced Structural Analysis Software. Available online: http://www.autodesk.com/products/robot-structural-analysis/overview (accessed on 13 June 2016).
25. Zienkiewicz, O. The birth of the finite element method and of computational mechanics. *Int. J. Numer. Method Eng.* **2004**, *60*, 3–10. [CrossRef]
26. Brenner, S.; Scott, R. *The Mathematical Theory of Finite Element Methods*, 2nd ed.; Springer-Verlag New York: New York, NY, USA, 2002.
27. American Society of Mechanical Engineers (ASME B31.3). *Engineering Standards Manual*; ASME: New York, NY, USA, 2014.
28. American Society of Mechanical Engineers (ASME B31.4). *Liquid Petroleum Transportation Piping*; ASME: New York, NY, USA, 2002.
29. Telles, P.C.S. *Industrial Pipes-Materials Projects and Assembly*, 10th ed.; Editora LTC: Rio de Janeiro, Brazil, 2001. (In Portuguese)

Article

Seismic-Reliability-Based Optimal Layout of a Water Distribution Network

Do Guen Yoo [1], Donghwi Jung [1], Doosun Kang [2] and Joong Hoon Kim [3,*]

[1] Research Center for Disaster Prevention Science and Technology, Korea University, Seoul 136-713, Korea; godqhr425@korea.ac.kr (D.G.Y.); donghwiku@gmail.com (D.J.)

[2] Department of Civil Engineering, Kyung Hee University, Yongin-si, Gyeonggi-do 446-701, Korea; doosunkang@khu.ac.kr

[3] School of Civil, Environmental and Architectural Engineering, Korea University, Anam-ro 145, Seongbuk-gu, Seoul 136-713, Korea

* Correspondence: jaykim@korea.ac.kr; Tel.: +82-2-3290-3316; Fax: +82-2-928-7656

Academic Editor: Helena Ramos
Received: 18 November 2015; Accepted: 25 January 2016; Published: 3 February 2016

Abstract: We proposed an economic, cost-constrained optimal design of a water distribution system (WDS) that maximizes seismic reliability while satisfying pressure constraints. The model quantifies the seismic reliability of a WDS through a series of procedures: stochastic earthquake generation, seismic intensity attenuation, determination of the pipe failure status (normal, leakage, and breakage), pipe failure modeling in hydraulic simulation, and negative pressure treatment. The network's seismic reliability is defined as the ratio of the available quantity of water to the required water demand under stochastic earthquakes. The proposed model allows no pipe option in decisions, making it possible to identify seismic-reliability-based optimal layout for a WDS. The model takes into account the physical impact of earthquake events on the WDS, which ultimately affects the network's boundary conditions (e.g., failure level of pipes). A well-known benchmark network, the Anytown network, is used to demonstrate the proposed model. The network's optimal topology and pipe layouts are determined from a series of optimizations. The results show that installing large redundant pipes degrades the system's seismic reliability because the pipes will cause a large rupture opening under failure. Our model is a useful tool to find the optimal pipe layout that maximizes system reliability under earthquakes.

Keywords: seismic reliability; water distribution system; optimal layout; Anytown network

1. Introduction

An earthquake occurs as a result of a sudden release of energy in the earth's crust. The released massive energy creates seismic waves, which cause deformation of the water distribution system (WDS) of which components are mainly connected beneath the ground. The damage of an earthquake is the multiple failure of system components. For example, for an earthquake occurred in Kobe, Japan in 1995, many pipes were ruptured, causing water to flow out of the system, while some of the pump stations stopped working completely due to power outage. However, the likelihood of concurrence of multiple failures within a system is low under normal conditions. Therefore, a different strategy should be adopted for WDS design of the regions with the risk of earthquake occurrence, such as Japan, Korean peninsula, and West Coast cities of the United States.

During the last two decades, many optimal design approaches have been proposed for WDSs. The early works have mainly taken into account the economic cost as a single objective [1–5]. Thereafter, some studies have included the system's performance index within the optimization framework. Lansey *et al.* [6] developed a chance-constrained least-cost model in which the capacity reliability is defined as the probability that stochastic pressures are equal to or higher than pressure requirement and are constrained at a certain level while minimizing cost. The nodal demands, pressure requirements, and pipe roughness coefficients were assumed to be uncertain while optimizing the pipe size. Several later studies also used chance-constrained optimization models for WDS design [7,8].

Kapelan *et al.* [9] was one of the earliest studies that proposed the multi-objective optimal design of WDS. Since then, most design approaches have adopted two objective functions to minimize economic cost and to maximize system reliability [10,11]. The various system reliability indices were suggested to reflect the uncertainties of pipe roughness and system demands when sizing the system. Component failures have been rarely considered in this stream of optimal WDS design.

Su *et al.* [12] is one of the few studies that considered component failure in WDS design. In their study, reliability was defined as the probability of water demand provision under pipe break (failure) conditions and is used as a constraint for the least-cost design of a WDS. The minimum cut-set (*i.e.*, the most critical set of pipes) was identified by closing the pipe individually for calculating the system reliability. Note that the pipe failure conditions were not considered in hydraulic simulation within the optimization framework, since the computational intensity is overwhelming.

The seismic hazard assessment models have been developed recently. HAZUS [13] was the first model to assess the economic losses of an infrastructure by earthquake. HAZUS is mainly intended to assess the seismic damage but not detailed simulation of the system behavior. Later, the Mid-America Earthquake Center developed a seismic impact assessment model and investigated interdependencies between water and power systems [14].

Early earthquake studies in the water domain focused on investigating individual components' physical behavior under earthquakes rather than quantifying the system-wide performance by modeling the WDS and earthquakes [15–17]. In a recent study, Fragiadakis *et al.* [18] proposed a seismic reliability assessment model of a WDS using survival curves of pipes based on general seismic assessment standards and American Lifelines Alliance (ALA) [19] guidelines. However, detailed hydraulic simulations were not conducted in this paper. Later studies began proposing methodologies to evaluate seismic reliability with hydraulic simulations using well-known hydraulic solvers and seismic simulations [20–29].

The latest and most popular assessment model is the graphical iterative response analysis for flow following earthquakes (GIRAFFE [30]) developed by a research team at Cornell University. GIRAFFE can simulate various pipe leakage and breakage conditions by using EPANET [31]. The model's graphical user interface helps to visualize the model results, which is compatible with other geographic information system tools. After the initial development in 2008, the model has been improved and validated through many case studies [27,32–36]. However, GIRAFFE uses a controversial approach to treat negative pressure that removes the nodes' negative pressure and connected pipes. This process is repeated until the negative-pressure nodes are no longer produced. This approach can be time-consuming because the system file must be revised iteratively.

Yoo *et al.* [37] developed seismic reliability assessment model under stochastic earthquake events. The model quantifies the seismic reliability of a WDS through a series of procedures: stochastic earthquake generation, seismic intensity attenuation, determination of the pipe failure status (normal, leakage, and breakage), pipe failure modeling in hydraulic simulation, and negative pressure treatment.

To the authors' best knowledge, this study is the first attempt to develop an economic, cost-constrained optimal WDS design approach that takes into account seismic reliability based on detailed hydraulic simulations. The proposed model maximizes the system's seismic reliability while satisfying the constraints on economic cost and node pressure requirements. The physical impacts of a seismic wave to WDS components are simulated to determine the failure conditions. The seismic reliability is defined as the ratio of the supplied water to the required demand under stochastic earthquake events. A well-known benchmark network, the Anytown network, is used for the applications of the optimal design and layout maximizing the seismic reliability.

2. Methodology

In this study, a seismic reliability-based WDS optimization approach is proposed. Figure 1 shows the flowchart of the proposed WDS optimal design model which consists of two sub-models: seismic reliability estimation and optimization model. Seismic reliability estimation model (SREM) first generates stochastic earthquake events by using earthquake generation module. Here, the locations and magnitude of stochastic earthquakes are determined and peak ground acceleration (PGA) reached at each pipe is calculated (described in Section 2.1). Note that no hydraulic calculation is performed from this module. This module also determines the failure mode of pipe where each pipe is classified as normal, leakage, or breakage (Section 2.2).

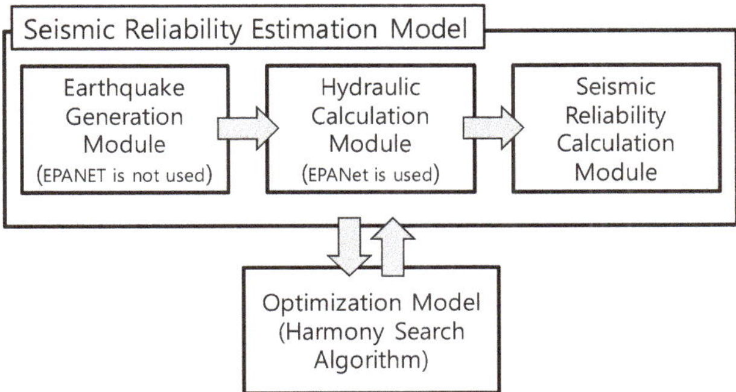

Figure 1. Flowchart of the proposed water distribution system (WDS) optimal design model.

The failure modes are transferred to hydraulic calculation module where nodal pressures under the failure conditions (*i.e.*, the boundary conditions to the WDS system equation) are calculated (Figure 1 and Section 2.3). Pipe leakage and breakage are modeled by the emitter in EPANET. In order to better simulate the conditions, a semi pressure-driven analysis (semi-PDA) is proposed. After an initial run of EPANET (a hydraulic simulator for DDA), the negative-pressure nodes' demand is set to zero, and the second run is made (Section 2.4).

The resulting hydraulics are provided to seismic reliability calculation module to quantify seismic reliability which is defined as the ratio of available supply to the required demand under stochastic earthquake events. The proposed WDS optimal design model maximizes the system's seismic reliability with constraints on economic cost and pressure requirements (Figure 1 and Section 2.5). Harmony search algorithm (HSA) is used for optimization (Figure 1) [38,39]. In the last section of the Methodology, Todini [40]'s resilience indicator is defined. This indicator is not used in the proposed model but used for a postoptimization analysis.

The following subsections describe the details of the methodology, such as earthquake simulation, objective function, optimization approach, and semi-PDA method. Note that the following subsections are in the order that Figure 1 represents.

2.1. Earthquake Intensity Attenuation

The proposed model considers the physical characteristics of the WDS's components to determine the failure modes given earthquake intensity. Therefore, supplemental information such as the types of pipes and surrounding soil characteristics should be provided. Network information such as the network layout, nodal demands and coordinates, pipe diameters and lengths, and sizes of the pump and tanks are entered from the EPANET input file. The nodal coordinates are used to calculate the location of a pipe that is used for calculating the distance from the epicenter.

Earthquake intensity is attenuated as it travels from the epicenter. In the model, three attenuation equations were adopted to quantify the damped energy of seismic waves. Kawashima *et al.* [41] relates the distance from the epicenter and earthquake magnitude to PGA (cm/s^2) based on the 90 earthquakes that have occurred in Japan as

$$PGA = 403.8 \times 10^{0.265M} \times (R + 30)^{-1.218} \tag{1}$$

where M = earthquake magnitude (e.g., M7 stands for earthquake magnitude 7), and R = the distance from the epicenter (km).

Lee and Cho [42] presented a similar PGA function that was developed using the historical data of South Korea as

$$\log PGA = -1.83 + 0.386M - \log R - 0.0015R \tag{2}$$

Instead of using the horizontal distance from the epicenter, Baag *et al.* [43] relate the Euclidean distance from focus (Δ) and earthquake magnitude to PGA as

$$\ln PGA = 0.40 + 1.2M - 0.76\ln\Delta - 0.0094\Delta \tag{3}$$

The average of the three calculated PGAs (from Equation (1) to Equation (3)) was used in our model.

2.2. Determination of Pipe Failure Mode

The fragility of the pipes was determined based on the pipe repair rate (RR, number of pipe repairs/km) and the PGA. In this study, the RR equation suggested by Isoyama *et al.* [44] and adopted by ALA [19] was used. Isoyama *et al.* related the RR to the PGA by multiplying some correction factors denoting the pipe types, diameters and soil characteristics (Table 1) as

$$RR = C1 \times C2 \times C3 \times C4 \times 0.00187 \times PGA \tag{4}$$

where C1, C2, C3, and C4 represent the correction factors according to the pipe diameter, pipe material, topography, and soil liquefaction, respectively. Because of the lack of information, this study only considered the correction factor of the pipe diameter, C1, in Equation (4), while assuming other correction factors are equal to one. As seen in Table 1, for the smaller pipe, C1 increases and results in the higher probability of damage given the same PGA.

Table 1. Correction factors (Isoyama *et al.* [44]).

	Category	Correction Factor
	D < 100	1.6
Pipe diameter (mm) (C1)	100 $\leqslant$ D < 200	1.0
	200 $\leqslant$ D < 500	0.8
	500 $\leqslant$ D	0.5
	Asbestos-Cement Pipe	1.2
	Poly-Vinyl Chloride Pipe, Vent Pipe	1.0
Pipe material (C2)	Cast Iron Pipe	1.0
	Poly-Ethylene Pipe, High Impact (3-Layer) Pipe	0.8
	Steel Pipe	0.3
	Ductile Cast Iron Pipe	0.3
	Narrow Valley	3.2
	Terrace	1.5
Topography (C3)	Disturbed Hill	1.1
	Alluvial	1.0
	Stiff Alluvial	0.4
	Total	2.4
Liquefaction (C4)	Partial	2.0
	None	1.0

The post-earthquake pipe status was classified into normal, leakage, or breakage conditions. Pipe leakage is defined as a small rupture on the pipe wall or at the joint through which continuous minor water loss occurs. Pipe breakage is defined as a complete separation of the pipe into two pieces that causes complete loss of transportation ability. ALA [19] suggested that the probability of pipe breakage ($P_{f_{breakage,i}}$) is a function of the RR and the length of the pipe (L_i, km) expressed as

$$P_{f_{breakage,i}} = 1 - e^{-RR \times L_i} \tag{5}$$

As a rule of thumb, the probability of pipe leakage ($P_{f_{leakage,i}}$) is assumed to be five times higher than the probability of pipe breakage [19]. If the cumulative probability of the two conditions is less than 1.0 (e.g., $P_{fbreakage,i} = 0.1$ and $P_{fleakage,i} = 0.5$), the complementary probability is assigned for the normal condition ($P_{normal} = 0.4$). However, if the cumulative probability of the leakage and breakage is greater than 1 (e.g., $P_{fbreakage,i} = 0.18$ and $P_{fleakage,i} = 0.90$), the model assumes that the normal condition (without any damage) is not available, and the two failure probabilities are normalized to make the sum of 1.0 ($P_{fbreakage,i} = 0.18/(0.18 + 0.90) = 0.17$ and $P_{fleakage,i} = 0.90/(0.18 + 0.90) = 0.83$).

2.3. Pipe Failure Modeling

Once the status of each pipe is determined, the information is entered into a network solver, EPANET, for hydraulic simulation (*i.e.*, solving for WDS system equations consisting of conservations of mass and energy). To simulate the pipe leakage, pressure-dependent flows (Q_l) are assigned to the closest node to the damaged pipe using the following equation:

$$Q_l = C_D p^{\alpha} \tag{6}$$

where C_D = the discharge coefficient ($= \left(\dfrac{2g}{\gamma_w}\right)^{\alpha} A$) (the emitter coefficient in EPANET), where g = gravitational acceleration and α = the exponent of the power function; γ_w = the specific weight of water; A = the opening area of the damaged pipe; and p = pressure at the closest node.

Lambert [45] conducted an experimental study to investigate the accuracy of the power function model in Equation (6) and provided a guideline on using exponent α based on the pipe material and level of leakage. It was suggested that the exponent of 0.5 would be used to simulate large detectable leakages in metal pipes; the exponent of 1.0 (linear relationship between discharge and pressure) would be used if no information on the pipe material is available. Puchovsky [46] theoretically derived a discharge coefficient and validated it using sprinkler data.

The shape and area of the opening on ruptured pipe varies depending on pipe material, origin of the pipe damage, and direction of external force. Large opening area is more likely to occur in the failure of large pipes than smaller pipes. To that end, it was assumed that the total opening area is equivalent to 10% of the entire cross-sectional pipe area during leaks. In case of breakage, the entire cross-sectional area was used as the opening area.

If a pipe was tagged as leaking, it was modeled in a hydraulic simulation, as illustrated in Figure 2a, and the discharge coefficient was assigned to the downstream node. The pipe breakage was modeled as shown in Figure 2b. The discharge coefficient was assigned to the upper node in flow direction. Then, the broken pipe was set to "closed" to disconnect the water flow. The demand of the node connected to a broken pipe was modified to consider the degraded delivery capacity due to disconnection. In the model, the nodal demands of both end junctions of the broken pipe were reduced by degrees of node (DoN). Here, DoN is defined as the number of connections/edges that a node has to other nodes in a network. As shown in Figure 2b, the base demand of the upstream node is reduced by 25% (= 1/DoN = 1/4), while the downstream node is reduced by 33.3% (= 1/DoN = 1/3).

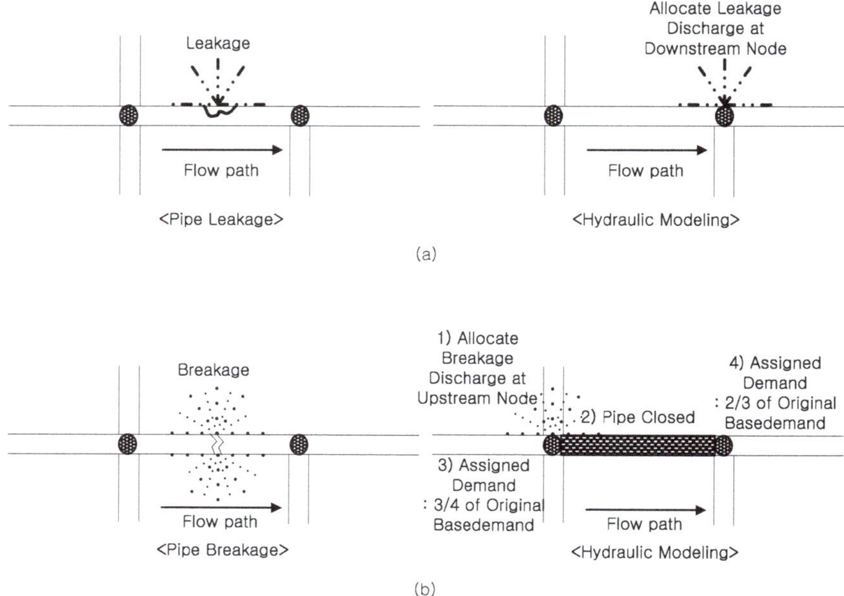

Figure 2. A schematic diagram to describe pipe damage modeling: (**a**) leakage condition and (**b**) breakage condition.

2.4. Negative Pressure Treatment

Hydraulic analysis approaches for water distribution network can be classified as a DDA and a pressure-driven analysis (PDA). DDA assumes that the demand at individual node is satisfied regardless of the associated pressure, while PDA assumes that nodal discharge is dependent on the pressure head expressed by the head-outflow relationship (HOR). DDA often generates negative pressure when simulating abnormal conditions, such as multiple pipe breaks. PDA, on the other hand, provides more realistic hydraulics avoiding negative pressures. WaterGEMS [47], WDNetXL [48], and WaterNetGen [49] are well-known programs equipped with PDA option based on HOR. However, the system-specific HOR should be provided, which is spatially and temporally inconsistent and also operational dependent. Therefore, for PDA techniques used in pipe network analyses, the analyzer should assume a HOR for the whole network. Given that hydraulic states of pipe networks can vary greatly with changes in HORs, analysis results cannot guarantee high accuracy.

Quasi-PDA methods suppress the occurrence of negative pressure through repetitive DDA analyses. In general, if negative pressure occurs from the first DDA run, better hydraulic calculation results can be derived through quasi-PDA methods by resetting the nodal demands and the components' status.

Ballantyne *et al.* [21] used the KYPIPE model [50] for hydraulic analyses. KYPIPE, a DDA program similar to EPANET, can also produce negative pressures during simulation of pipeline destruction. Ballantyne *et al.* calculated system reliability by assuming that water could not be supplied to points with negative pressures, but the hydraulic results around the negative-pressure node still did not have realistic values. The method of Shinozuka *et al.* [22] and Shi [27] completely removed negative-pressure nodes and the connected pipelines from the original network, but the process required time-consuming repetitive hydraulic analyses and regeneration of input files.

Here, a quasi-PDA approach is proposed for realistic hydraulic simulation under multiple failure conditions to avoid the occurrence of negative pressure. That is, if negative pressure occurred after DDA simulation using EPANET, the updated base demand (after modification described in Chapter 2.3) of the negative-pressure node is set to zero and DDA is repeated. If negative pressure reappears, the pressure of the relevant node is assumed to zero. Compared to Shinozuka *et al.* [22] and Shi [27] approach, the proposed approach saves overhead processing time by avoiding regeneration of the EPANET input file for the new configuration. In addition to the negative pressure treatment, pressure-dependent water supply is considered in calculating the quantity of available water at each node. The available nodal demand at node j ($Q_{avl,j}$) is estimated as

$$Q_{avl,j} = \begin{cases} 0 & \text{if } P_j \leqslant 0 \\ Q_{new,j} \times \sqrt{\dfrac{P_j}{P_{min}}} & \text{if } 0 < P_j < P_{min} \\ Q_{new,j} & \text{if } P_j \geqslant P_{min} \end{cases} \tag{7}$$

where $Q_{new,j}$ = the updated base demand with pipe breakage consideration and negative pressure treatment at node j; P_j = the pressure head at node j; and P_{min} = the minimum pressure requirement.

2.5. Seismic Reliability Indicator

Various surrogate measures of WDS reliability have been proposed including capacity reliability [6,7], resilience [40], robustness [10,11], availability [51]. Bao and Mays [52] proposed three formulations of WDS system reliability which can be calculated from nodal reliability values: minimum nodal reliability, arithmetic mean reliability, and flow-weighted mean reliability. The first one concerns on the worst nodal reliability while the latter two values indicate system-wide reliability. In order to represent post-earthquake system performance in the water supply, a seismic reliability measure should be able to reflect system-wide water availability rather than local level of system performance.

To quantify system-wide seismic reliability, a new reliability indicator is proposed herein. The system seismic reliability (S_S) is defined as the ratio of the total available system demand to the total required system demand:

$$S_S = \frac{\sum_{j=1}^{m} Q_{avl,j}}{\sum_{j=1}^{m} Q_{req,j}} \tag{8}$$

where m = total number of demand nodes; and $Q_{req,j}$ = the required demand at node j.

A water system planner would intend to design the most reliable network for earthquakes under a given budget condition. To that end, the proposed model here is a single-objective optimal design model that maximizes the system's seismic reliability with a constraint on economic cost. The optimization model is formulated as follows:

$$\begin{aligned} \text{Maximize } F &= S_s \\ \text{s.t.} \, CC &\leqslant CC_{given} \end{aligned} \tag{9}$$

where CC = the pipe construction cost; CC_{given} = the available budget for pipe system.

Equation (9) is valid for each set of available commercial pipe diameters and in conditions where the minimum pressure head is guaranteed. The pipe construction cost accounting for the pipe material and installation is expressed as:

$$CC = \sum_{i=1}^{n} \left(uc \left(D_i \right) \times L_i \right) \tag{10}$$

where $uc \left(D_i \right)$ = the unit cost of the pipe with diameter D_i determined for the ith pipe (USD/m); and L_i = the length of the ith pipe.

2.6. Resilience Indicator

In this study, Todini [40]'s resilience is quantified from a range of designs of the Anytown network and compared with seismic reliability. Todini [40] introduced a resilience index to quantify the resilience of looped network. Resilience is defined as surplus power available within the network as a percentage of net input power as:

$$\text{Resil} = \frac{\sum_{j}^{m} Q_{req,j} \left(H_j - H_{req}\right)}{\sum_{k=1}^{nr} Q_k H_k + \sum_{i=1}^{np} \dfrac{\text{Power}_i}{\gamma} - \sum_{j}^{m} Q_{req,j} H_{req}} \tag{11}$$

where H_j = total head of the jth node; Q_k = flow provided by reservoir k (m^3/s); H_k = head at reservoir k; Power$_i$ = power of the ith pump (Nm/s); γ = specific weight of water (N/m^3); nr and np = number of reservoir and pumps, respectively. Note that this indicator was used only for a postoptimization analysis.

Todini's resilience is one of the most popular and widely used surrogate measures of WDS reliability. Farmani *et al.* [53] investigated the trade-off between economic cost and the resilience for a rehabilitation problem of the Anytown network. Prasad and Park [54] proposed a multiobjective optimization approach to minimize cost and maximize modified version of Todini's resilience indicator. Recently, Gheisi and Naser [55] compared entropy-based reliability, Todini's resilience, and three modified versions of Todini's resilience in the twenty-two potential pipe layouts of a hypothetical network.

3. Study Network

The proposed optimization approach is applied for optimal design of a well-known benchmark WDS, the Anytown network. The network, which was firstly published by Walski *et al.* [56], was modified by Jung *et al.* [11] for pipe only and pipe/pump designs that minimize total cost and maximize system robustness. The original benchmark network was modified by removing the two tanks and the connected riser pipes to be solely supplied by a single reservoir with a fixed source head. The fixed source head is elevated from 3 m (10 ft) to 73.2 m (240 ft). A peaking factor of 1.8 is applied to given average-based demand (2005 average daily use) to create the daily peak condition.

Other modifications of the study network were made for application purposes. First, this study suggests new pipe layouts and sizes assuming there are no existing pipes in the system. Pump and tank design are not considered. Second, in addition to 10 commercial pipe sizes (152, 203, 254, 305, 356, 406, 457, 508, 610, and 762 mm), zero diameter can be selected suggesting no pipe installation for the potential path. The unit costs of the commercial pipes are adopted from Walski *et al.* [56].

To generate random epicenters, a 2 × 2 grid was created and laid on the study network as seen in Figure 3. The rectangular boundary of the grid is defined by the four end nodes: north, south, east, and west. Total 900 earthquakes are generated and consistent number of earthquakes are assigned at each corner of the grid (marked as "x" in Figure 3). The earthquake intensity reached at the pipe is a function of the earthquake magnitude and the Euclidean distance from the epicenter. Note the earthquake magnitude of M4 was generated from each corner of the grid. The minimum pressure requirement of 28.12 m (40 psi) should be satisfied under based demand condition and is also considered in Equation (7).

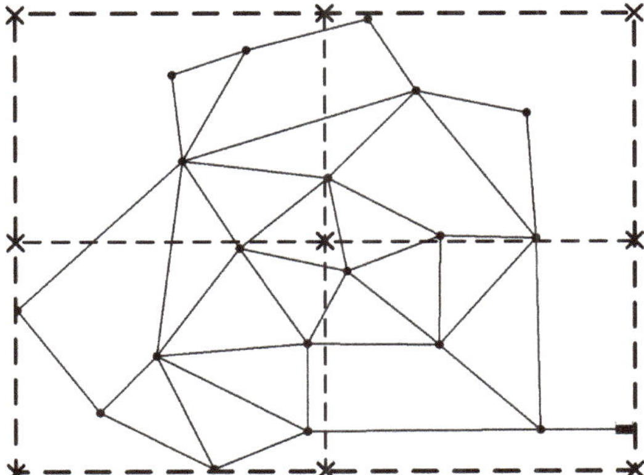

Figure 3. Study network layout (epicenter is marked as "x").

HSA was used to find an optimal solution for the pipe sizing problem. HSA was inspired by musical performance process and widely used for WDS optimizations. While the applications encompass from pipe network design to pump scheduling, HSA was proven to be generally outperform other algorithms such as genetic algorithm [57–59].

Several assumptions and simplifications were made in this study: (1) the seismic damages of the pump, tank, reservoir, and valve are not considered; (2) the earthquake's focal depth (tectonics) is assumed to be 10 km; (3) the coordinate of the center of the pipe is used as a reference point for calculating the distance from the epicenter; (4) Pipes are cast iron pipes; and (5) the Anytown network is laid on the alluvial plain with no liquefaction. Based on the assumption (4) and (5), all correction factors except C1 in the Equation (4) (C2-4) are equal to one.

4. Application Results

First, we applied two different commercial pipe sets for the pipe sizing of the study network. This analysis investigates the WDS's layout changes with respect to the seismic reliability increase. The same total cost constraint was applied for the two designs and the resultant pipe layouts were compared. Then, seven designs with a fixed layout and different redundancy levels were compared for the systems' seismic reliability. Finally, discussions and suggestions on improving WDS seismic reliability were provided at the end of this section.

4.1. Different Available Pipe Sizing Options

The first application is intended to investigate the impact of seismic reliability on the network's layout. Two sets of commercial pipe sizes are assumed to be available for two design cases. In Case 1, all commercial pipes (152, 203, 254, 305, 356, 406, 457, 508, 610, and 762 mm) except the zero size option are available. In Case 2, zero diameter is available in addition to the commercial sizes in Case 1. Considering the same cost constraint (CC_{given} = 18 million (M) USD) in Equation (9) provided a platform for consistent comparison of the resulting designs.

Figure 4 shows the optimal seismic reliability values of the two case designs. The corresponding optimal pipe layouts are presented in Figure 5. Contrary to expectations, the seismic reliability decreased with the availability of more pipe sizes. In Case 1, at least a 152 mm pipe should be installed because no pipe option is not available. By comparing Figures 5a,b and 6, we can observe that 152 mm pipes were installed in Case 1 for the link at which no pipe was constructed in Case 2. The 152 mm pipe, which is the most vulnerable pipe to earthquakes, almost always causes failure. Being able to have no pipe instead of a 152 mm pipe increased the system's seismic reliability by 0.2 (a 20% increase in the amount of available water during an earthquake). The network layout becomes smaller from Case 2 to Case 1 (Figure 5a,b), while the overall pipe diameters also increase (Figure 6).

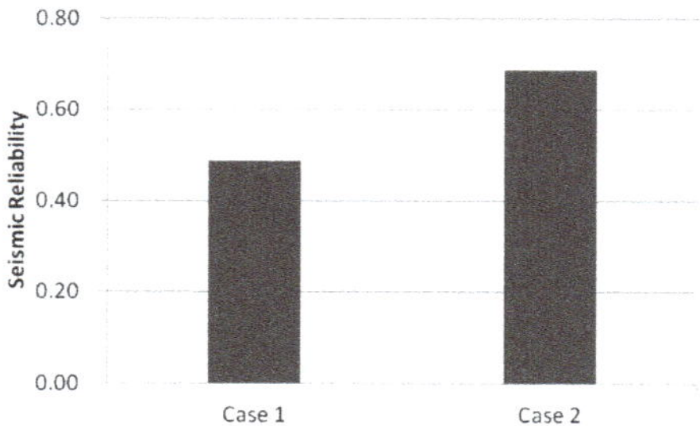

Figure 4. Maximum seismic reliability values of two case designs with different pipe sizing options.

Note that this is very different from what we observed in a traditional capacity reliability-based design. In the context of the capacity reliability field, it was believed that having additional paths and more loops would result in the increase of system reliability and redundancy [11,60]. However, the result of this study indicates that a different strategy should be available under the conditions where WDS component failures are affected by strong external forces (*i.e.*, earthquakes) and the components' physical characteristics (pipe's probability of failure as a function of RR).

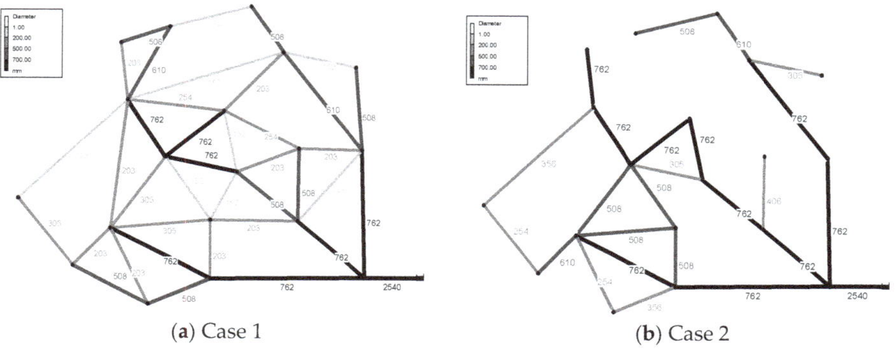

(a) Case 1   (b) Case 2

Figure 5. Pipe layout comparison for the solutions obtained from Cases 1 and 2 (Figure 4); pipe diameters are in mm; the thicker and darker pipe is larger. (**a**) Case 1; (**b**) Case 2.

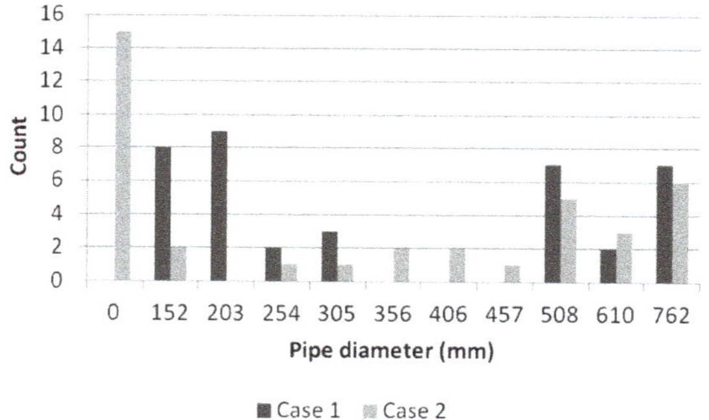

Figure 6. Histogram of pipes by the pipe diameter.

This plateau in seismic reliability was also observed in the optimal pipe designs of the Anytown network by applying different total cost constraints and using the available pipe sizes in Case 1. Figure 7 shows the Pareto optimal solutions' total cost and seismic reliability. The marginal cost becomes infinite for the designs whose cost is greater than 16.5 M USD. A reliability increase can no longer be achieved once a sufficient investment is made. Although the designs greater than 16.5 M USD have more large pipes compared to the solutions less than 16.5 M USD (the number of pipes equal to or larger than 508 mm is between 7 and 10, while the designs less than 16.5 M USD have three to five pipes), no benefit of having larger pipes was obtained with respect to seismic reliability. For effective and economical improvement of WDS seismic reliability, the threshold investment for a network should first be identified.

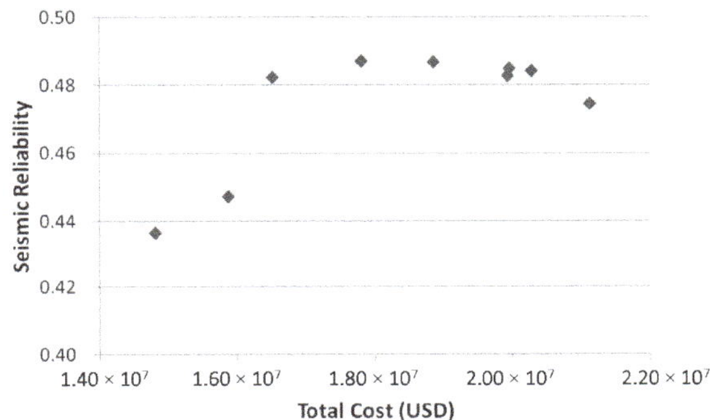

Figure 7. Trade-off relationship between total cost and seismic reliability in Case 1 where all pipes (152, 203, 254, 305, 356, 406, 457, 508, 610, and 762 mm) are available and without zero pipe option.

4.2. Constant Layout with a Single Pipe Sizing Option

The impacts of having large pipes are also investigated through the seismic reliability evaluation of seven uniform designs. The Design 1 has 305 mm for all pipes in the study network. Design 2, 3, 4, 5, 6, and 7 have 356, 406, 457, 508, 610, and 762 mm, respectively, for all pipes. The seismic reliabilities of the seven designs are shown in Figure 8. For comparison, Todini's resilience is also calculated from the seven designs and plotted in Figure 8. While there is a large increase in seismic reliability from the 457 mm design (Design 4) to the 508 mm design (Design 5), reliability decreases from the 508 mm design to the 762 mm design (Design 7). This explains why we observed a plateau in seismic reliability in Figure 7. On the other hand, resilience (a traditional reliability measure) consistently increases with increasing pipe sizes. The marginal cost of improving resilience increases substantially for a resilience value of 0.3–0.8 and stabilized for a value higher than 0.8.

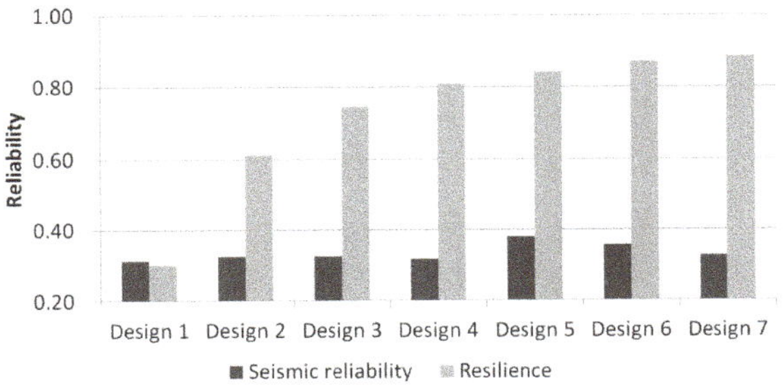

Figure 8. Seismic reliability of the seven uniform designs; all pipes are 508 mm in Case 5.

As the pipe sizes decrease, the correction factor for RR (Equation (4)) increases, finally resulting in high pipe breakage and leakage probability. However, although the pipes failed, the resulting impact is not significant to the system's seismic reliability because the calculated discharge coefficient, which is a function of the pipe's cross-sectional area (Table 2), is not large. However, as seen in Table 2, large pipes such as 508, 610, and 762 mm have a smaller failure probability compared to small pipes, but the resulting discharge coefficient is more than 10 times as big as that of small pipes. The failure effects are more significant for system seismic reliability compared to small pipes.

Table 2. Correction factors and discharge coefficients (Equation (6)) for the pipe sizes considered.

Pipe Sizes		Pipe's Cross-Sectional Area (A)	Correction Factor (C1)	Discharge Coefficient	
mm	inch			Breakage (100% of A)	Leakage (10% of A)
152	6	28	1	1074	107
203	8	50	0.8	1910	191
254	10	79	0.8	2985	298
305	12	113	0.8	4298	430
356	14	154	0.8	5850	585
406	16	201	0.8	7640	764
457	18	254	0.8	9670	967
508	20	314	0.5	11,938	1194
610	24	452	0.5	17,191	1719
762	30	707	0.5	26,861	2686

4.3. Random Designs

Finally, random designs that satisfy pressure requirements are generated from the proposed model to confirm the aforementioned conclusion. Figure 9 shows the profiles of seismic reliability and resilience of many random designs. We can clearly see that there is an inflection point around the total cost of 23 M USD and seismic reliability around 0.4, after which the overall seismic reliability decreases (Figure 9a). Because the total cost is a direct function of the pipe diameter, this plot indicates that installing large pipes does not always guarantee an increase in seismic reliability. The printed solutions are all suboptimal solutions that are dominated by the Pareto solutions found in Figure 7. On the other hand, installing large pipes resulted in an upward trend in resilience (Figure 8b).

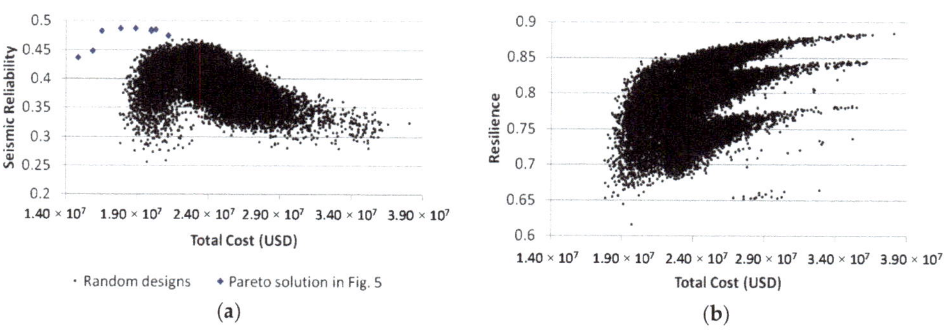

Figure 9. (a) Seismic reliability of randomly generated solutions (**black dot**) and Pareto solutions shown in Figure 7 (**blue diamond**) and (b) resilience of the same random designs.

Under normal failure condition, having more additional paths and installing large pipes throughout the system are beneficial with respect to system reliability (*i.e.*, ability to supply required quantity of water) and redundancy (*i.e.*, level of pressure redundancy). However, we observed this does not apply under earthquake failures. Earthquake deforms WDS components and their original function. Once failed under earthquakes, large pipes, which can deliver large volume of water under normal condition, help accelerate water loss out of the system. This study was the first attempt to take into account such irregular hydraulic behavior in the WDS modeling and design and highlight the need to find the optimal pipe layout considering the system's performances under the two different states.

5. Summary and Conclusions

Earthquakes can cause many simultaneous failures of components throughout WDSs, which result in very different and severe conditions compared to normal failure conditions that water communities usually deal with (e.g., single pipe failure). In this study, an economic, cost-constrained optimal design approach of a WDS was proposed to maximize seismic reliability. The network's seismic reliability is defined as the ratio of the available supply to the required water demand during stochastic earthquakes. The physical relationship between the earthquake intensity and the WDS components' vulnerability characteristics was defined and utilized for quantifying seismic reliability. To overcome the limitation of DDA in modeling earthquake failures, negative pressures were assumed to have realistic hydraulic calculations. Then, we investigated the seismic reliability improvement with respect to the economic investment in the system and the associated topological and pipe diameter changes. A traditional benchmark network, the Anytown network, was used to demonstrate the approach. Random earthquakes around the study network were generated for reliability quantification.

The results were quite different from what we generally observed from traditional reliability-based design (e.g., resilience-based design). First, allowing redundant small pipes in the system did not help improve seismic reliability. Those pipes cause frequent failures because of their low durability to earthquake forces. Second, having too-large pipes also degrades the system reliability during earthquakes. Compared to small pipes that are 152 mm to 305 mm, a big pipe (762 mm) has a lower failure probability; however, the failure's influence is very significant to the system once it fails. The large pipe's cross-sectional area will help release more water out of the system. Therefore, the first step to efficiently improve the WDS seismic reliability should be to identify the most appropriate pipe sizes for a system. For example, from the Anytown network, having a uniform 508 mm for all pipes provides the highest system reliability of 0.38 among seven uniform designs.

This study has several limitations that future research must address. First, this model is sensitive to correction factors for the pipe and the assumption of the opening area of the pipe under failure. Intensive sensitivity analysis should be conducted with experimental studies to simulate the most realistic failure behavior of WDS components. Second, this study only considers the pipe failures while pump, valve, tank, and reservoir failure can occur during an earthquake. Therefore, various failure types could be included in future studies. The proposed semi-PDA approach and EPANet can be replaced with a PDA-based network solver in order to simulate more realistic behavior of WDS under earthquake events. Third, while this study focuses on the system's reliability right after an earthquake occurs, post-earthquake recovery strategies should also be investigated to enhance the overall WDS's reliability/resilience. This could be found mostly in the context of operation and management. Fourth, the proposed semi-PDA approach and EPANet can be replaced with a PDA-based network solver in order to simulate more realistic behavior of WDS under earthquake events. Finally, interdependencies among multiple lifeline infrastructures (e.g., the water, power, transportation, and communication systems) can help improve and recover an individual infrastructure's reliability during an earthquake. Therefore, more efforts should be made to identify these interdependencies.

Acknowledgments: This work was supported by a grant from The National Research Foundation (NRF) of Korea, funded by the Korean government (MSIP) (No. 2013R1A2A1A01013886) and the Korea Ministry of Environment as "The Eco-Innovation project (GT-11-G-02-001-2)".

Author Contributions: Do Guen Yoo and Donghwi Jung carried out the survey of previous studies, analysis of proposed method, participated in the sequence alignment and drafted the manuscript. Joong Hoon Kim and Doosun Kang conceived the original idea of the study, and helped to write the final manuscript.

Conflicts of Interest: The authors declare no conflicts of interest.

References

1. Schaake, J.; Lai, D.D. *Linear Programming and Dynamic Programming Applications to Water Distribution Network Design*; Report No. 116; Department of Civil Engineering, Massachusetts Institute of Technology: Cambridge, MA, USA, 1969.
2. Alperovits, E.; Shamir, U.U. Design of optimal water distribution systems. *Water Resour. Res.* **1977**, *13*, 885–900. [CrossRef]
3. Lansey, K.; Mays, L. Optimization model for water distribution system design. *J. Hydraul. Eng.* **1989**, *115*, 1401–1418. [CrossRef]
4. Simpson, A.; Dandy, G.; Murphy, L. Genetic algorithms compared to other techniques for pipe optimization. *J. Water Resour. Plan. Manag.* **1994**, *120*, 423–443. [CrossRef]
5. Savic, D.; Walters, G. Genetic algorithms for least-cost design of water distribution networks. *J. Water Resour. Plan. Manag.* **1997**, *123*, 67–77. [CrossRef]
6. Lansey, K.; Duan, N.; Mays, L.; Tung, Y. Water distribution system design under uncertainty. *J. Water Resour. Plan. Manag.* **1989**, *115*, 630–645. [CrossRef]
7. Xu, C.; Goulter, C. Reliability-based optimal design of water distribution networks. *J. Water Resour. Plan. Manag.* **1999**, *125*, 352–362. [CrossRef]
8. Babayan, A.; Kapelan, Z.; Savic, D.; Walters, G. Least cost design of robust water distribution networks under demand uncertainty. *J. Water Resour. Plan. Manag.* **2005**, *131*, 375–382. [CrossRef]
9. Kapelan, Z.; Savic, D.; Walters, G. Multiobjective design of water distribution systems under uncertainty. *Water Resour. Res.* **2005**, *41*, W11407-1–W11407-15. [CrossRef]
10. Giustolisi, O.; Laucelli, D.; Colombo, A. Deterministic *versus* stochastic design of water distribution networks. *J. Water Resour. Plan. Manag.* **2009**, *135*, 117–127. [CrossRef]
11. Jung, D.; Kang, D.; Kim, J.H.; Lansey, K. Robustness-based design of water distribution systems. *J. Water Resour. Plan. Manag.* **2014**, *140*. [CrossRef]
12. Su, Y.; Mays, L.; Duan, N.; Lansey, K. Reliability-based optimization model for water distribution systems. mboxemphJ. Hydraul. Eng. **1987**, *113*, 1539–1556. [CrossRef]
13. Federal Emergency Management Agency (FEMA). *HAZUS97 Technical Manual*; FEMA: Washington, DC, USA, 1997.
14. Kim, Y.S.; Spencer, B.F.; Song, J.; Elnashai, A.S.; Stokes, T. *Seismic Performance Assessment of Interdependent Lifeline Systems*; MAE Center: Urbana, IL, USA, 2007.
15. Hall, W.; Newmark, N. Seismic design criteria for pipelines and facilities. *J. Tech. Counc. ASCE* **1978**, *104*, 91–107.
16. Wright, J.P.; Takada, S. *Earthquake Response Characteristics of Jointed and Continuous Buried Lifelines*; Grant Report No. 15; National Science Foundation: New York, NY, USA, 1980.
17. Hwang, R.N.; Lysmer, J. Response of buried structures to traveling waves. *J. Geotech. Eng. Div.* **1981**, *107*, 183–200.
18. Fragiadakis, M.; Christodoulou, S.E.; Vamvatsikos, D. Reliability assessment of urban water distribution networks under seismic loads. *Water Resour. Manag.* **2013**, *27*, 3739–3764. [CrossRef]
19. American Lifelines Alliance. *Seismic Fragility Formulations for Water Systems Part 1 Guideline*; American Lifeline Alliance: Washington, DC, USA, 2001.
20. Eguchi, R.T.; Taylor, C.E.; Hasselman, T.K. *Earthquake Vulnerability Models for Water Supply Components*; Technical Report No. 83-1396-2c; J.H. Wiggins Company: Redondo Beach, CA, USA, 1983.
21. Ballantyne, D.B.; Berg, E.; Kennedy, J.; Reneau, R.; Wu, D. *Earthquake Loss Estimation Modeling of the Seattle water System*; Technical Report; Kennedy/Jenks/Chilton: Federal Way, WA, USA, 1990.
22. Shinozuka, M.; Tan, R.Y.; Toike, T. Serviceability of Water Transmission Systems under Seismic Risk. In *Lifeline Earthquake Engineering, the Current State of Knowledge*; American Society of Civil Engineers: New York, NY, USA, 1981.
23. Shinozuka, M.; Hwang, H.; Murata, M. Impact on water supply of a seismically damaged water delivery system. In *Lifeline Earthquake Engineering in the Central and Eastern U.S.*; American Society of Civil Engineers: New York, NY, USA, 1992; pp. 43–57.

24. Shinozuka, M.; Rose, A.; Eguchi, R.T. *Engineering and Socioeconomic Impacts of Earthquakes*; Monograph Series 2; Multidisciplinary Center for Earthquake Engineering Research: Buffalo, NY, USA, 1998.

25. Markov, I.; Mircea, G.; O'Rourke, T. *An Evaluation of Seismic Serviceability of Water Supply Networks with Application to the San Francisco Auxiliary Water Supply System*; Technical report NCEER 94-0001; National Center for Earthquake Engineering Research University of Buffalo, State University of New York: Buffalo, NY, USA, 1994.

26. Hwang, H.; Lin, H.; Shinozuka, M. Seismic performance assessment of water delivery systems. *J. Infrastruct. Syst.* **1998**, *4*, 118–125. [CrossRef]

27. Shi, P. Seismic Response Modeling of Water Supply Systems. Ph.D. Thesis, Cornell University, Ithaca, NY, USA, 1 January 2006.

28. Liu, G.Y.; Chung, L.L.; Yeh, C.H.; Wang, R.Z.; Chou, K.W.; Hung, H.Y.; Chen, S.A.; Chen, Z.H.; Yu, S.H. *A Study on Pipeline Seismic Performance and System Post-Earthquake Response of Water Utilities (1/2)*; Technical Report MOEA-WRA-0990095; Water Resource Agency, MOEA: Taipei, Taiwan, 2010.

29. Liu, G.Y.; Chung, L.L.; Huang, C.W.; Yeh, C.H.; Chou, K.W.; Hung, H.Y.; Chen, Z.H.; Chou, C.H.; Tsai, L.C. *A Study on Pipeline Seismic Performance and System Post-Earthquake Response of Water Utilities (2/2)*; Technical Report MOEA-WRA-1000090; Water Resource Agency, MOEA: Taipei, Taiwan, 2011.

30. GIRAFFE. *GIRAFFE User's Manual*; School of Civil and Environmental Engineering, Cornell University: Ithaca, NY, USA, 2008.

31. Rossman, L.A. *EPANET 2 User's Manual*; U.S. Environmental Protection Agency (EPA): Cincinnati, OH, USA, 2000.

32. Wang, Y. Seismic Performance Evaluation of Water Supply Systems. Ph.D. Thesis, Cornell University, Ithaca, NY, USA, 1 January 2006.

33. Shi, P.; O'Rourke, T.D.; Wang, Y. Simulation of earthquake water supply performance. In Proceedings of the 8th National Conference on Earthquake Engineering, Paper No. 8NCEE-001295, EERI, Oakland, CA, USA, 18–22 April 2006.

34. Wang, Y.; O'Rourke, T.D. Characterizations of seismic risk in Los Angeles water supply system. In Proceedings of the 5th China-Japan-US Symposium on Lifeline Earthquake Engineering, Haikou, China, 26–28 November 2007.

35. Bonneau, A.L. Water Supply Performance during Earthquakes and Extreme Events. Ph.D. Thesis, Cornell University, Ithaca, NY, USA, 2008.

36. Bonneau, A.L.; O'Rourke, T.D. *Water Supply Performance during Earthquakes and Extreme Events*; Technical Report MCEER-09-0003; University of Buffalo, State University of New York: Buffalo, NY, USA, 2009.

37. Yoo, D.G.; Jung, D.; Kang, D.; Kim, J.H.; Lansey, K. Seismic hazard assessment model for urban water supply networks. *J. Water Resour. Plan. Manag.* **2015**. [CrossRef]

38. Geem, Z.W.; Kim, J.H.; Loganathan, G.V. A New Heuristic Optimization Algorithm: Harmony Search. *Simulation* **2001**, *76*, 60–68. [CrossRef]

39. Kim, J.H.; Geem, Z.W.; Kim, E.S. Parameter Estimation of the Nonlinear Muskingum Model Using Harmony Search. *J. Am. Water Resour. Assoc.* **2001**, *37*, 1131–1138. [CrossRef]

40. Todini, E. Looped water distribution networks design using a resilience index based heuristic approach. *Urban Water* **2000**, *2*, 115–122. [CrossRef]

41. Kawashima, K.; Aizawa, K.; Takahashi, K. Attenuation of peak ground motion and absolute acceleration response spectra. In Proceedings of the 8th World Conference on Earthquake Engineering (WCEE), San Francisco, CA, USA, 24–28 July 1984; pp. 257–264.

42. Lee, K.; Cho, K.H. Attenuation of peak horizontal acceleration in the Sino-Korea Craton. In Proceedings of the Annual Fall Conference of Earthquake Engineering Society of Korea, Cheonan, Korea, 27 September 2002; pp. 3–10.

43. Baag, C.E.; Chang, S.J.; Jo, N.D.; Shin, J.S. Evaluation of seismic hazard in the southern part of Korea. In Proceedings of the International Symposium on Seismic Hazards and Ground Motion in the Region of Moderate Seismicity, Seoul, Korea, 1 November 1998; pp. 31–50.

44. Isoyama, R.; Ishida, E.; Yune, K.; Shirozu, T. Seismic damage estimation procedure for water supply pipelines. In Proceedings of the 12th World Conference on Earthquake Engineering (WCEE), Auckland, New Zealand, 1–4 January 2000; p. 1762.

45. Lambert, A. What do we know about pressure-leakage relationships in distribution systems? In Proceedings of the IWA Conference on Systems Approach to Leakage Control and Water Distribution System Management, Brno, Czech Republic, 16 May 2001.

46. Puchovsky, M.T. *Automatic Sprinkler Systems Handbook*; National Fire Protection Association (NFPA): Quincy, MA, USA, 1999.

47. Bentley Systems. *WaterGEMS*; Bentley Systems Incorporated: Exton, PA, USA, 2006.

48. Giustolisi, O.; Savic, D.A.; Berardi, L.; Laucelli, D. An Excel-based Solution to Bring Water Distribution Network Analysis Closer to Users. In Proceedings of the Computer and Control in Water Industry, Exeter, UK, 5–7 September 2011; Volume 3, pp. 805–810.

49. Muranho, J.; Ferreira, A.; Sousa, J.; Gomes, A.; Marques, A.S. Pressure-dependent Demand and Leakage Modeling with an EPANET Extension–WaterNetGen. *Procedia Eng.* **2014**, *89*, 632–639. [CrossRef]

50. Wood, D. *KYPipe Reference Manual*; Civil Engineering Software Center, University of Kentucky: Lexington, KY, USA, 1995.

51. Cullinane, M.; Lansey, K.; Mays, L. Optimization-availability-based design of water-distribution networks. *J. Hydraul. Eng.* **1991**, *118*, 420–441. [CrossRef]

52. Bao, Y.; Mays, L. Model for water distribution system reliability. *J. Hydraul. Eng.* **1990**, *116*, 1119–1137. [CrossRef]

53. Farmani, R.; Walters, G.; Savic, D. Trade-off between total cost and reliability for Anytown water distribution network. *J. Water Resour. Plan. Manag.* **2005**, *131*, 161–171. [CrossRef]

54. Prasad, T.D.; Park, N. Multiobjective genetic algorithms for design of water distribution networks. *J. Water Resour. Plan. Manag.* **2004**, *130*, 73–82. [CrossRef]

55. Gheisi, A.; Naser, G. Multistate reliability of water-distribution systems: Comparison of surrogate measures. *J. Water Resour. Plan. Manag.* **2015**, *141*. [CrossRef]

56. Walski, T.; Brill, E.; Gessler, J., Jr.; Goulter, I.; Jeppson, R.; Lansey, K.; Lee, H.; Liebman, J.; Mays, L.; Morgan, D.; Ormsbee, L. Battle of the Network Models: Epilogue. *J. Water Resour. Plan. Manag.* **1987**, *113*, 191–203. [CrossRef]

57. Geem, Z.W. Harmony search in water pump switching problem. In *Advances in Natural Computation*; Springer: Berlin, Germany; Heidelberg, Germany, 2005; pp. 751–760.

58. Geem, Z.W. Optimal cost design of water distribution networks using harmony search. *Eng. Optim.* **2006**, *38*, 259–277. [CrossRef]

59. Geem, Z.W. Harmony search optimisation to the pump-included water distribution network design. *Civ. Eng. Environ. Syst.* **2009**, *26*, 211–221. [CrossRef]

60. Lansey, K. Sustainable, robust, resilient, water distribution systems. In Proceedings of the 14th Water Distribution Systems Analysis Conference, Adelaide, South Australia, 24–27 September 2012; pp. 1–18.

Article

Initial Provincial Water Rights Dynamic Projection Pursuit Allocation Based on the Most Stringent Water Resources Management: A Case Study of Taihu Basin, China

Min Ge [1,2,*], Feng-Ping Wu [3] and Min You [3]

1   School of Business Administration, Hohai University, Changzhou 213022, China
2   School of Business, Jiangsu University of Technology, Changzhou 213013, China
3   School of Business, Hohai University, Nanjing 211100, China; wfp@hhu.edu.cn (F.-P.W.); youmin@126.com (M.Y.)
*   Correspondence: ge_min19@126.com; Tel.: +86-519-8697-8206

Academic Editors: Helena Margarida Ramos and Davide Viaggi
Received: 24 August 2016; Accepted: 3 January 2017; Published: 10 January 2017

Abstract: Clarification of initial water rights is the basis and prerequisite for a water rights trade-off market and also an effective solution to the problem of water scarcity and water conflicts. According to the new requirements for the most stringent water resources management in China, an initial provincial water rights allocation model is proposed. Firstly, based on analysis of multiple principles for initial provincial water rights allocation including total water use, water use efficiency, water quality of water function zones, regional coordination and sharing, an index system of initial provincial water rights allocation is designed. Secondly, according to dynamic projection pursuit technique, an initial provincial water rights allocation model with the total water use control is set up. Moreover, the self-adaptive chaotic optimization algorithm is applied to tackle the model. Finally, a case study of Taihu Basin is adopted. Considering the multiple scenarios of three different water frequencies (50%, 75% and 90%) and planning year 2030, the empirical results show Jiangsu Province always obtains the most initial water rights. When the developing situation of provinces are given more consideration, Shanghai should acquire more initial water rights than Zhejiang Province; but when the dynamic increment evolving trend of provinces is taken more into account, Shanghai should obtain less initial water rights than Zhejiang Province. The case about Taihu Lake further verifies the feasibility and effectiveness of the proposed model and provides a multiple-scenarios decision making support for entitling the initial water rights with the most stringent water resources management constrains in Taihu Basin.

Keywords: most stringent water resources management; initial provincial water rights; dynamic projection pursuit

1. Introduction

China is facing a great challenge as water is still a major constraint on the nation's economic and social development due to the growing population, urbanization, and industrialization at a time when there are serious water shortages, growing pollution, severity of droughts, and declining aquatic ecosystems. Although China is rich in water resources and ranks sixth in the world in terms of water volume, the available water in China is only 1998.6 m^3 per capita in 2014 and less than one-quarter of the world average which can be shown by the data from China Statistical Yearbook (2015). China's amount of water use reached 609.5 billion m^3 (Figure 1) in 2014. However, the annual water scarcity across the country has exceeded 50 billion m^3 on average according to the data from Global Water Partnership.

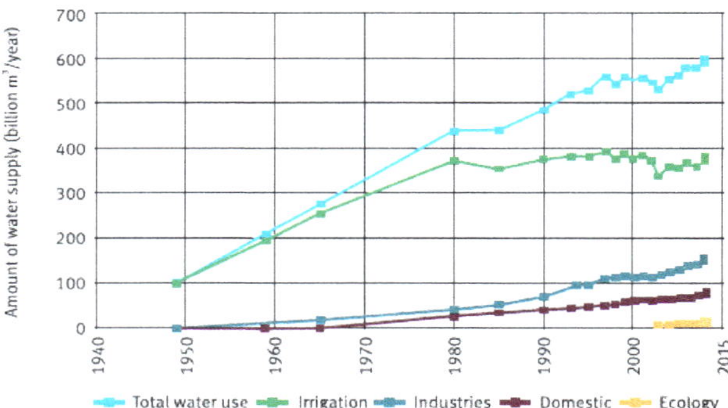

Figure 1. Increasing water use in China. Data source: China Water Resources Bulletin (1997–2000) and China Statistical Yearbook (2000–2015).

With China's previous water resources management system, the cost of using water was nearly zero which caused a similar external diseconomy of "tragedy of the commons". In that situation, injudicious development and overuse of water cannot be avoided and the negative impacts on environment came out including rivers drying up, ecosystem degradation, and environment deterioration.

At present, under the new scenarios of the most stringent water resources management system, in order to address these water problems, China is establishing a legal framework including initial water rights, water licensing, and a water rights trade-off to reform water allocation mechanisms and property rights systems to fit more closely with China's developing and maturing market economy. In fact, the final distribution of wealth, however, does depend on how property rights are initially assigned. Just as the Coase Theorem presented, the initial allocation of property rights often matters for efficiency due to transaction costs cannot be neglected [1]. Therefore, assignment of initial water rights is particularly important for fairness of water use and water efficiency. Clarification of initial water rights in China is not only the basis and prerequisite for a water rights trade-off market but also an effective solution to the problem of water scarcity and water conflicts.

Initial provincial water rights are an important part of river basin initial water rights. Existing literature about initial provincial water rights optimal allocation is in pursuit of a more scientific and effective allocation plan. However, as the major reforms to the implementation of the new most stringent water resources management system in China, the existing theory and practice of provincial initial water rights allocation should meet the new requirements of this system. Therefore, how to assign a reasonable allocation of initial provincial water rights, following the benchmark of the new most stringent water resources management system, is not only the major element and core aspect of basin initial water rights allocation but also an important approach to enhancing the efficiency of water resources utilization and significant for achieving harmony between economic and social development and available water resources in China. Recently, initial provincial water rights optimal allocation has been an active research field both in foreign and domestic areas. A majority of literature in this area can be divided into four categories.

The first category focuses on using hybrid allocation method to solve the initial provincial water rights allocation problems. For instance, Wang Hao et al. [2] found water resources allocation in China should follow a "natural-artificial" dual knowledge and perception mode for modern river basin water cycling process. Wu Dan et al. [3] established the bi-level optimization model of the compound system for basin initial water rights allocation to realize the reasonable allocation of water rights in different regions and different industries. Considering the pollution limits of water functional zones, which is the one "red line" of the most stringent water resources management, Ge Min et al. [4] proposed a provincial initial water rights incentive allocation model with total pollutant discharge constraints.

The second category pays close attention to multi-participation allocation model. For instance, Ralph [5] set up a generalized simulation model called Water Right Analysis Package (WRAP) which was implemented in the state of Texas for the state's 23 river basins. Yan-ping Chen et al. [6] divided the regions of the same basin into a disadvantaged group and an advantaged group, based on which they founded an evolutionary game model. They also adjusted the water rights of different regions according to the evolutionary stable strategy. Read et al. [7] established the model of negotiation with multiple-decision makers over water rights allocation.

The third category concerns an allocation model of multi-objective optimization. For instance, Feng-ping Wu et al. [8] established a multi-objective and semi-structural fuzzy optimization model in order to solve the problem of initial allocation of water rights of the first hierarchy. Xian-feng Huang et al. [9] created a multi-objective chaotic optimization algorithm. Condon et al. [10] presents the development of a water allocation model (WAM) for an integrated physical hydrology model. His management model uses linear optimization to maximize satisfaction of demand.

The fourth category focuses on applying interactive or harmonious allocation method to tackle the initial provincial water rights allocation problems. For instance, Feng-ping Wu and Min Ge et al. [11] put forward principles for harmonious and disharmonious judgment of initial water rights allocation, and they also came up with an interactive method of initial water rights allocation which could fully assimilated the ideas of different apartments from different regions about the initial allocation of water rights. Furthermore, Wang et al. [12] built a harmonious model for water rights allocation, and Feng-ping Wu et al. [13] proposed the harmonious method of allocating basin water rights.

According to the existing literature, despite different perspectives, much research has applied a considerable amount of initial provincial water rights allocation methods in the pursuit of a more scientific and effective water rights allocation plan. However, along with the evolution of the social-economic-ecological complex system, water rights management in China is now facing many new constraints. The first one is that China is implementing the most stringent mechanism of water resources management which contains "three red lines", namely a "red line" for controlling the total water use, a "red line" for improving the water use efficiency and a "red line" for controlling the pollution of water function zones. Initial provincial water rights allocation must meet the new requirements of the water resources management system in China. The second one is that China advocates the construction of an ecological civilization for which the improvement of water quality is an important guarantee. The third one is that China has put forward five concepts for development and policies for poverty alleviation, and how to realize the balanced and efficient use of water with initial provincial water rights allocation really needs further discussion.

For the new scenarios of water resources management presented above, the existing studies of relevant aspects are inadequate as follows:

- Firstly, current models of initial provincial water rights allocation still lack a comprehensive consideration of the key elements, namely the "three red lines"—including total water use, water efficiency, and water quality. Therefore, in contrast to the existing models of initial provincial water rights allocation, one contribution of this paper is to propose new comprehensive insights for the initial provincial water rights allocation method following the benchmark of the all "three red lines" under the new scenarios of the most stringent water resources management in China.

- Secondly, the social-economic-ecological complex system of the provinces is always in the process of dynamic evolution. Most existing allocation methods are limited to analysis of static cross-section data. Seldom do they consider both developing situation of a system and its dynamic evolving trend. Therefore, the other contribution of this paper is to consider initial provincial water rights allocation based on the total water use control as a three-dimensional dynamic decision question for time, indicator, and allocation plan. We are concerned not only about the static analysis of the system, but also the dynamic development of the system.

- Thirdly, the existing indicator system studies are inadequate for meeting the new most stringent water resource management requirements. Hence, the third contribution of this paper is that we design an index system of initial provincial water rights allocation by absorbing the concepts of "three red lines" and the idea of coordinated and sharable development of provinces. Based on the designed index system, we set up an initial provincial water rights allocation model with the total water use control by dynamic projection pursuit technique. Furthermore, we tackle the model with a self-adaptive chaotic optimization algorithm to acquire the solutions for province initial water rights allocation, and study the case of Taihu Basin. The outline of the paper can be depicted in Figure 2.

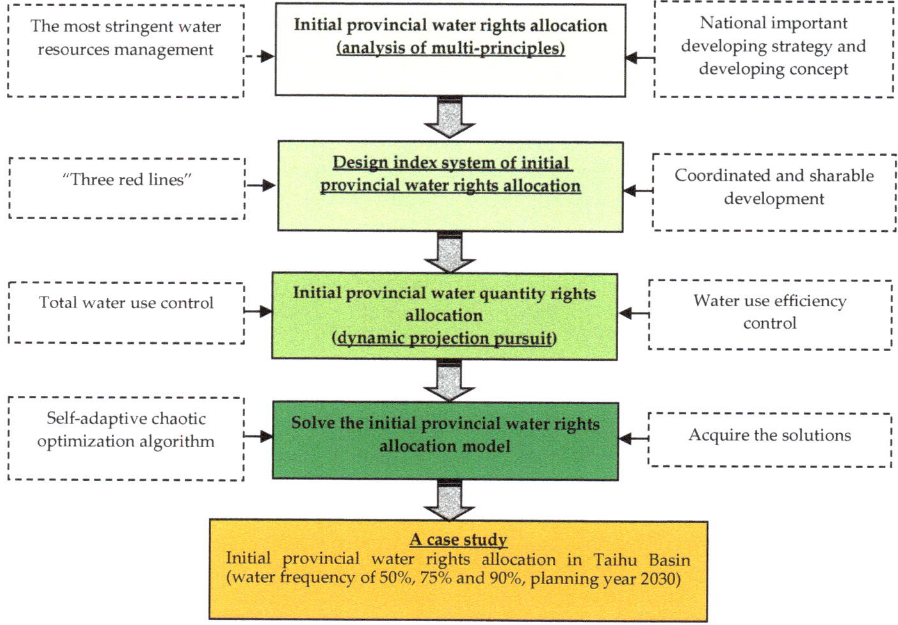

Figure 2. Outline of initial provincial water right allocation based on most stringent water resources management.

The rest of this study is organized as follows. Section 2 designs an index system of initial provincial water rights allocation with multiple principles. Section 3 establishes a dynamic projection pursuit allocation model of initial provincial water rights. Section 4 details a case study of Taihu Basin. Section 5 provides the conclusions.

2. Design of the Index System

2.1. The Principles of Establishing Index System

The principles of the establishing index system are important to ensure this method and conclusion are accepted by the administration and the public [13]. The existing literature presents different principles of initial provincial water rights allocation. In order to meet the new requirements of the most stringent water resources management in China, we follow the principles of "three red lines". Furthermore, we follow the principle of "coordinate development, share the future" to design the indicator system for the initial provincial water rights allocation.

2.1.1. Principle of Controlling the Total Water Use

As the key content of the most stringent water resources management system, control gives dual attention to two dimensions of water quantity and quality. The principle of total water use is shown on the foundation of water resources capacity which means a macroscopic control over the water use of basins or provinces. It sets a bearing capacity of water resources within which people can ensure sustainable social and economic development through rational and efficient water use. As the main controlling factor, it aims to protect ecological environment, and also strengthen the constraining force of water resources management and promote the optimal use of water resources.

2.1.2. Principle of Improving the Water Use Efficiency

The principle of improving the water use efficiency has water conservation as its main target. It can reflect controlling the process of water saving, in addition, it can reveal the high utilization rate of water resources and lower pollutant discharge, which is a benefit for water quality protection. At a macro level, this red line can control the water total amount of a basin or a province, furthermore, at a micro level, it is the quota management to the total water use of industries, enterprises, and water users, which can directly influence not only the water quantity but also the water quality.

2.1.3. Principle of Controlling Pollution of Water Function Zones

The principle of controlling the pollution of water function zones, which is mainly used for limiting the total pollutant discharge, safeguarding the proportion of drinking water sources meeting the required standard, and keeping the ecological base flow for eco-environmental needs. It protects water environment bearing capacity including water quality and water ecology. Furthermore, it restores the damaged water ecology, improves water quality, and decreases the amount of pollutants in the basin.

2.1.4. Principle of "Coordinate Development, Share the Future"

The principle of "coordinate development, share the future" leads a profound reform idea in China. At the present stage, initial water rights allocation research should satisfy bearable balanced development of environmental resources between regions, and also should include the idea of coordinated and sharable development. In fact, there is a natural difference in water resources endowment between different provinces including bad geographical location, backward economic development, and a worsening ecological environment. We should give full consideration to the contribution of different provinces for basin water resources management in water pollution control and flood discharging. Moreover, in water rights management, we must ensure coordination and equality within initial provincial water rights allocation and protect the basic rights of disadvantaged groups. Then we are able to implement the poverty-reducing policy and share the achievements of development.

2.2. Index System Design

According to the characteristics of initial provincial water rights allocation, combined with the principle for "three red lines" control and absorbing the idea of coordinated and sharable development, considering the representativeness and availability of related data, we designed the index system of initial provincial water rights allocation which is shown in Table 1.

In fact, because of the specificities of water resources, initial provincial water rights allocation proves to be a complicated problem including the aspects of social development, economic growth, and the construction of ecological civilization. Consequently, the index system in Table 1 attempts to achieve the following basic aims:

- Coordinate the general interests and the partial interests, so as to maximize the overall satisfaction in a basin.
- Clearly depict, describe, and measure the developing state and development trends in the whole river basin system.
- Equally reflect the requirements of every stakeholder.

Table 1. The index system of initial provincial water rights allocation.

Object Layer	Principle Layer	Index Layer	No	Attribute
Initial provincial water rights allocation	Total water use control	Current water use (billion m^3)	P_1	Profit type
		Water yield of province (billion m^3)	P_2	Profit type
		Population of province (10 thousand person)	P_3	Profit type
		Area of province (km^2)	P_4	Profit type
	Water use efficiency control	Water use per capita (m^3/person)	P_5	Profit type
		Volume of water used for irrigation per unit area (10 thousand m^3/km^2)	P_6	Profit type
		Ten thousand Yuan GDP water use (Yuan)	P_7	Cost type
		Per capita GDP (Yuan)	P_8	Profit type
	Water pollution control	The amount of pollutant discharge (billion m^3)	P_9	Cost type
		Discharge standard-meeting rate of wastewater (%)	P_{10}	Profit type
	Coordination and sharing of provinces	Contribution rate of the flood control for provinces (%)	P_{11}	Profit type
		Protection level of disadvantaged groups	P_{12}	Profit type

The meaning of indexes can be illustrated as follows: (1) Current water use (P_1) reflects the current water use situation and also expresses the respect for the provincial different history of water use; (2) Water yield of province (P_2) is decided by annual average runoff. It represents the respect of priority for local water sources. Specifically, those with more water yields should be given more initial water rights; (3) Population of province (P_3) reflects people who lives in the same river basin should own the same initial water rights; (4) Provincial area (P_4) shows province should be given more initial water rights if it has more area; (5) Water use per capita (P_5) is the ratio between provincial current water use and its population; (6) The volume of water used for irrigation per unit area (P_6) is the ratio between provincial current water use and its irrigation area; (7) Ten thousand Yuan GDP water use (P_7) reflects water resources use efficiency. A smaller indicator value means higher water resources use efficiency; (8) Per capita GDP (P_8) is the ratio between provincial GDP and its population which can reflect the general economic development level; (9) The amount of pollutant discharge (P_9) represents the idea that more provincial pollutant discharge should result in less initial water rights; (10) Discharge standard-meeting rate of wastewater (P_{10}) means the water function area with low water quality standard should result in less initial water rights; (11) Contribution rate of the flood control for provinces (P_{11}) reflects the ratio between provincial flood discharge and river basin flood discharge. The higher ratio means the higher contribution and they should be assigned more initial water rights, consequently; (12) Protection level of disadvantaged groups (P_{12}) reflects the disadvantaged provinces with bad geographical location or a backward economic development level and worsening ecological environment should be effectively protected for the assignment of initial water rights. The protection level of disadvantaged groups P_{12} can be represented by grades scaling 1–10 [13]. The higher grades of the protection level of the disadvantaged groups mean that more initial water rights should be assigned.

3. Dynamic Projection Pursuit Allocation Model of Initial Provincial Water Rights

Initial provincial water rights allocation based on water use control is a three-dimensional dynamic decision making problem for time, index, and allocation plans. Taking full account of the primitive values of indexes which reflect developing situation of allocation system and annual increment values of indexes which present its dynamic increment evolving situation, we will adopt dynamic projection pursuit (DPP) technology to acquire the weights of the times and indexes. Furthermore, by calculating the best projection value, we can finally obtain the optimal allocation ratios and plans for the initial provincial water rights. During this process, we choose the dynamic projection pursuit method, because it is different from the existing initial water rights allocation methods. It can reflect the dynamics of data and overcome the difficulties of determining the weights of times and indexes.

3.1. Basic Data Processing

(1) S_k denotes the province, $k = 1, 2, ..., q$; T_i denotes the year, $i = 1, 2, ..., n$; P_j denotes the index, $j = 1, 2, ..., m$.

(2) Let a_{kij} denote the primitive value of index P_j for province S_k at year T_i. The analysis matrix of year T_i can be described by Equation (1).

$$A_i = \left(a_{kij}\right)_{q \times m} \tag{1}$$

where $k = 1, 2, ..., q$; $i = 1, 2, ..., n$; $j = 1, 2, ..., m$.

Let a_{kij}' denote the absolute increment value of indexes P_j for province S_k at year T_i, which can be written as follows:

$$a_{kij}' = a_{kij} - a_{k(i-1)j} \tag{2}$$

where $i \geq 2$, where $k = 1, 2, ..., q$; $i = 2, ..., n$; $j = 1, 2, ..., m$. Thus, we can get the absolute increment value matrix for the initial provincial water rights allocation at year T_i, which can be presented by Equation (3).

$$A_i' = \left(a_{kij}'\right)_{q \times m} \tag{3}$$

where $k = 1, 2, ..., q$; $i = 2, ..., n$; $j = 1, 2, ..., m$.

(3) Adopting the improved efficiency coefficient method to remove dimensions of indexes, then matrix A_i and A_i' are normalized.

Let b_{kij} denote the non-dimensional value of a_{kij}. When a_{kij} represents a profit type, b_{kij} can be obtained by Equation (4).

$$b_{kij} = \left\{ \left(a_{kij} - \min_k \min_i a_{kij}\right) / \left(\max_k \max_i a_{kij} - \min_k \min_i a_{kij}\right) \right\} \times 40 + 60 \tag{4}$$

Moreover, when a_{kij} represents cost type, b_{kij} can be obtained by Equation (5).

$$b_{kij} = 100 - \left\{ \left(a_{kij} - \min_k \min_i a_{kij}\right) / \left(\max_k \max_i a_{kij} - \min_k \min_i a_{kij}\right) \right\} \times 40 \tag{5}$$

where $k = 1, 2, ..., q$, $i = 1, 2, ..., n$, $j = 1, 2, ..., m$. $b_{kij} \in [60, 100]$.

Let c_{kij} denote the non-dimensional value of a_{kij}'. When a_{kij}' is a profit type, c_{kij} can be obtained by Equation (6).

$$c_{kij} = \left\{ \left(a_{kij}' - \min_k \min_i a_{kij}'\right) / \left(\max_k \max_i a_{kij}' - \min_k \min_i a_{kij}'\right) \right\} \times 40 + 60 \tag{6}$$

Moreover, when a_{kij}' is cost type, c_{kij} can be obtained by Equation (7).

$$c_{kij} = 100 - \left\{ \left(a_{kij}' - \min_k \min_i a_{kij}'\right) / \left(\max_k \max_i a_{kij}' - \min_k \min_i a_{kij}'\right) \right\} \times 40 \tag{7}$$

where $k = 1, 2, ..., q$, $i = 2, ..., n$, $j = 1, 2, ..., m$. $c_{kij} \in [60, 100]$.

(4) The normalized matrix A_i can be expressed by Equation (8).

$$B_i = \left(b_{kij}\right)_{q \times m} \tag{8}$$

where $k = 1, 2, ..., q; i = 1, 2, ..., n; j = 1, 2, ..., m$.

Moreover, the normalized matrix A_i' can be estimated by Equation (9).

$$C_i = (c_{kij})_{q \times m} \tag{9}$$

where $k = 1, 2, ..., q; i = 2, ..., n; j = 1, 2, ..., m$.

(5) Comprehensively considering b_{kij} and c_{kij}, the comprehensive index analysis matrix can be expressed by Equation (10).

$$E_i = (e_{kij})_{q \times m} \tag{10}$$

where $i = 2, ..., n, j = 1, 2, ..., m$. The element of matrix E_i can be measured by Equation (11).

$$e_{kij} = \alpha b_{kij} + \beta c_{kij} \tag{11}$$

Here, e_{kij} means the weighted sum of b_{kij} and c_{kij}, $k = 1, 2, ..., q, i = 2, ..., n, j = 1, 2, ..., m$. Where, α and β respectively represent the weights of matrix B_i and matrix C_i. Moreover, $0 \leq \alpha \leq 1$, $0 \leq \beta \leq 1, \alpha + \beta = 1$.

When $\alpha > \beta$, it reflects that we give more consideration to the developing situation of indexes themselves; when $\alpha < \beta$, it means the increment or growth situation of indexes should be given more consideration; when $\alpha = \beta = 0.5$, it illustrates the situation of the indexes themselves and increment (or growth) situation of indexes are equally important during the allocation for initial water rights.

(6) Transfer E_i into the provincial index analysis matrix of initial water rights allocation:

$$E_k = (e_{kij})_{n \times m} \tag{12}$$

where $k = 1, 2, ..., q$. Then positive ideal matrix E^+ and negative ideal matrix E^- can be written as follows:

$$e_{ij}^+ = \max\{ e_{kij} \big| k = 1...q\} \tag{13}$$

$$e_{ij}^- = \min\{ e_{kij} \big| k = 1...q\} \tag{14}$$

3.2. Descriptions of Decision Variables

After basic data processing, the decision variables of our model can be described as follows: Let $d(k)$ denote the one-dimensional projection value; Let $\theta = (\omega_1, ...\omega_m, \lambda_2, ...\lambda_n)$ denotes projection direction, where ω_j is the weight of index P_j, λ_i is the weight of year T_i; Let $d^*(k)$ denote the best projection value; Let ω_{S_k} denote the ratio of initial provincial water rights allocation for the province S_k; Let w_{S_k} denote quantity of the initial water rights of the province S_k.

3.3. Construct Objective Function of Projection

The purpose of constructing projection objective function is to decrease high dimensional data into one-dimensional projection data. Each step is presented in the following:

Step 1: Synthesize matrix E_k into the one-dimensional projection value $d(k)$ with projection direction $\theta = (\omega_1, ...\omega_m, \lambda_2, ...\lambda_n)$.

$$d(k) = [\sum_{i=2}^{n} \lambda_i (\sum_{j=1}^{m} \omega_j (e_{kij} - e_{ij}^-)^2)]^{0.5} / \{[\sum_{i=2}^{n} \lambda_i (\sum_{j=1}^{m} \omega_j (e_{kij} - e_{ij}^+)^2)]^{0.5} + [\sum_{i=2}^{n} \lambda_i (\sum_{j=1}^{m} \omega_j (e_{kij} - e_{ij}^-)^2)]^{0.5}\} \tag{15}$$

Furthermore, $d(k)$ is the proximity between the positive and negative ideal solution for province S_k.

Step 2: Based on the principle of dispersing the projection value $d(k)$ as far as possible, we can establish projection index function $f(\theta)$ which can be shown in Equation (16).

$$f(\theta) = s_d = \left[\sum_{k=1}^{q} \left(d(k) - \overline{d(k)} \right)^2 / (q-1) \right]^{0.5} \tag{16}$$

where $\overline{d(k)}$ is the mean value of $d(k)$, $k = 1, 2, ..., q$.

Step 3: For getting the best projection value $d(k)$, the maximum value of projection indexes function should be found which can be calculated by Equation (17).

$$\begin{cases} \max f(\theta) = s_d \\ S.T. \begin{cases} c_1(\theta) = \sum_{j=1}^{m} \omega_j - 1 = 0 \\ c_2(\theta) = \sum_{i=2}^{n} \lambda_i - 1 = 0, \omega_j > 0, \lambda_i > 0 \end{cases} \end{cases} \tag{17}$$

By solving Equation (17), the best projection value $d^*(k)$ of the province S_k can be measured. The bigger $d^*(k)$ means more advantages can be obtained by province S_k during the initial water rights allocation.

3.4. Solutions of Initial Provincial Water Rights Allocation

Normalize the best projection value $d^*(k)$, then the ratio of initial provincial water rights allocation ω_{S_k} which can be presented by Equation (18).

$$\omega_{S_k} = d^*(k) / \sum_{k=1}^{q} d^*(k) \tag{18}$$

Set W as the total amount of initial water rights allocated by the river basin. The quantity of initial water rights of the province S_k can be calculated by Equation (19).

$$w_{S_k} = W \times \omega_{S_k} = W \times \left(d^*(k) / \sum_{k=1}^{q} d^*(k) \right) \tag{19}$$

Here, Equation (17) is non-linear programming (NP). In fact, many approaches can solve this kind of nonlinear optimization problem including chaotic optimization algorithm (COA), genetic algorithm (GA), and simulated annealing algorithm [14]. We will choose self-adoptive chaotic optimization algorithm [15] to solve Equation (17).

4. A Case Study of Taihu Basin

4.1. Data Sources

The administration areas in Taihu Basin mainly include Jiangsu province, Zhejiang province, and Shanghai city. Taihu Basin is one of the most developed and most modernized region in China. By far, the amount of annual water use has reached 36.5 billion m^3, while the amount of the perennial water resources of the basin is only 19.6 billion m^3 Therefore, the gap between supply and demand is great.

At present, in Taihu Basin, water shortage, worsening water pollution, and increasing demand of water resources have imposed undue pressure on the carrying capacities of the water resources and water environment. As a result, in order to promote the sustainable development of the basin, it is an inevitable requirement to clarify the initial provincial water rights and implement the allocation plan of Taihu Basin.

According to the materials of Taihu Basin and Southeast Rivers Water Resources Bulletin [16], some values of indexes for Jiangsu province, Zhejiang province, and Shanghai can be presented respectively as follows (we adopted the annual average values during 2011–2014 as the values of P_2, P_4, P_{11}, and P_{12}):

(1) The index of water yield of province P_2: Jiangsu province is 6.6 billion m^3, Zhejiang province is 7.27 billion m^3 and Shanghai is 2.01 billion m^3.

(2) The area of province P_4: Jiangsu province is 19,399 km^2, Zhejiang province is 12,093 km^2 and Shanghai is 5178 km^2.

(3) The contribution rate of flood control for province P_{11}: Jiangsu province is 34.5%, Zhejiang province is 15.7%, and Shanghai is 49.8%.

(4) The protection level of disadvantaged groups P_{12}: Jiangsu province is graded 8, Zhejiang province is graded 9, and Shanghai is graded 7.

The other specific relevant data of Taihu Basin during 2011–2014 are summarized in Table 2.

Table 2. Provincial index values of Taihu Basin during 2011–2014.

Province	Year	Index Value							
		P_1	P_3	P_5	P_6	P_7	P_8	P_9	P_{10}
Jiangsu	2011	184.2	2939	627	94.95	57	8.21	29.0	28.9
	2012	188.2	2960	636	97.02	60	8.97	29.1	13.2
	2013	193.8	2985	649	99.9	58	9.84	29.2	26.3
	2014	193.5	2996	646	99.75	45	10.68	28.3	19.5
Zhejiang	2011	51.9	1763	294	42.92	43	6.54	12.1	35.7
	2012	51.4	1775	289	42.50	39	6.97	12.1	28.6
	2013	52.6	1791	294	43.50	36	7.53	12.6	35.7
	2014	50.1	1797	279	41.43	32	7.98	12.9	46.2
Shanghai	2011	91.5	1175	779	176.71	77	12.01	22.6	16.7
	2012	109.7	1184	927	211.86	62	12.53	23.1	0.00
	2013	117.7	1194	986	227.31	64	13.44	22.9	0.00
	2014	99.7	1198	832	192.55	53	14.47	22.9	66.7

Notes: Data sources: Taihu Basin and Southeast Rivers Water Resources Bulletin (2011–2014) and Water Quality Bulletin of Taihu Basin and Important Water Function Areas of Southeast Rivers (2011–2014).

4.2. Calculation of Initial Provincial Water Right Alloction

Using the initial data of indexes in Table 2, selecting three typical parameter combinations of parameter α and β ($\alpha = 0.4, \beta = 0.6$; $\alpha = 0.5, \beta = 0.5$; $\alpha = 0.6, \beta = 0.4$), according to the initial provincial water dynamic projection pursuit allocation model based on the self-adaptive chaotic algorithm in this paper, we can obtain related solutions by MATLAB 7.01 including the value of optimal projection direction $\theta = (\omega_1, ... \omega_{12}, \lambda_2, \lambda_3, \lambda_4)$ and optimal projection value $d^*(k)$. Using Equation (18), we can acquire the ratios of the initial provincial water rights allocation which are shown in Table 3.

According to the materials of Taihu Basin and Southeast Rivers Water Resources Bulletin, under the three different water frequencies including 50%, 75% and 90%, aiming for the planning year 2030, the distributable total amounts of initial water rights in Taihu Basin are 329.2 hundred million m^3, 355.4 hundred million m^3, and 392.6 hundred million m^3 respectively.

Table 3. Results of initial provincial water rights allocation in Taihu Basin.

Related Parameter Selecting		Initial Population Size B = 200, Chaotic Iteration M = 100, Initial Temperature T_0 = 100
$\alpha = 0.6$ $\beta = 0.4$	Optimal projection direction value $\theta = (\omega_1, ...\omega_{12}, \lambda_2, \lambda_3, \lambda_4)$	(0.1624, 0.0741, 0.0535, 0.0088, 0.0778, 0.0678, 0.0492, 0.2125, 0.0907, 0.0339, 0.0808, 0.0884, 0.1345, 0.1841, 0.6814)
	Optimal projection value $d^*(k)$, $k = 1, 2, 3$	Jiangsu $d^*(1)$ = 0.5803; Zhejiang $d^*(2)$ = 0.4298; Shanghai $d^*(3)$ = 0.5039
	Ratio of province initial water rights allocation	Jiangsu ω_{S_1} = 38.33%; Zhejiang ω_{S_2} = 28.39%; Shanghai ω_{S_3} = 33.28%
$\alpha = 0.5$ $\beta = 0.5$	Optimal projection direction value $\theta = (\omega_1, ...\omega_{12}, \lambda_2, \lambda_3, \lambda_4)$	(0.1147, 0.1022, 0.0011, 0.0473, 0.0922, 0.0866, 0.0310, 0.1037, 0.0953, 0.3102, 0.0513, 0.0045, 0.2074, 0.2843, 0.5083)
	Optimal projection value $d^*(k)$, $k = 1, 2, 3$	Jiangsu $d^*(1)$ = 0.6045; Zhejiang $d^*(2)$ = 0.4032; Shanghai $d^*(3)$ = 0.4794
	Ratio of province initial water rights allocation	Jiangsu ω_{S_1} = 40.65%; Zhejiang ω_{S_2} = 27.11%; Shanghai ω_{S_3} = 32.24%
$\alpha = 0.4$ $\beta = 0.6$	Optimal projection direction value $\theta = (\omega_1, ...\omega_{12}, \lambda_2, \lambda_3, \lambda_4)$	(0.0941, 0.2306, 0.0899, 0.0333, 0.1185, 0.0636, 0.0042, 0.0724, 0.1725, 0.0331, 0.0145, 0.0733, 0.1222, 0.4322, 0.4456)
	Optimal projection value $d^*(k)$, $k = 1, 2, 3$	Jiangsu $d^*(1)$ = 0.6812; Zhejiang $d^*(2)$ = 0.5772; Shanghai $d^*(3)$ = 0.3987
	Ratio of initial provincial water rights allocation	Jiangsu ω_{S_1} = 41.11%; Zhejiang ω_{S_2} = 34.83%; Shanghai ω_{S_3} = 24.06%

According to the results of initial provincial water rights allocation in Table 3, we can acquire the larger weight of index P_j and year T_i, and also the ranking results of ratios for provincial initial water rights allocation which can be expressed in Table 4.

Table 4. Ranking results of initial provincial water rights allocation in Taihu Basin, China.

Typical Parameter Combinations for Parameter α and β	The Larger Weight of Index P_j	The Larger Weight of Year T_i	The Ranking Results of Ratios for Provincial Initial Water Rights Allocation
$\alpha = 0.6, \beta = 0.4$	Current water use ($\omega_1 = 0.1624$); Per capita GDP ($\omega_8 = 0.2125$)	2014 ($\lambda_4 = 0.6814$)	*Jiangsu ≻ Shanghai ≻ Zhejiang*
$\alpha = 0.5, \beta = 0.5$	Current water use ($\omega_1 = 0.1147$); discharge standard-meeting rate of wastewater ($\omega_{10} = 0.3102$)	2014 ($\lambda_4 = 0.5083$)	*Jiangsu ≻ Shanghai ≻ Zhejiang*
$\alpha = 0.4, \beta = 0.6$	The water yield of province ($\omega_2 = 0.2306$); The amount of pollutant discharge ($\omega_9 = 0.1725$)	2013 ($\lambda_3 = 0.4322$); 2014 ($\lambda_4 = 0.4456$)	*Jiangsu ≻ Zhejiang ≻ Shanghai*

Consequently, according to the ratio of initial provincial water rights allocation which is shown in Table 3, using Equation (19), we can obtain multiple scenarios of the solutions for initial provincial water rights allocation under three different water frequencies in the planning year 2030 which can be presented in Table 5.

Table 5. Multiple scenarios of solutions for provincial initial water rights allocation in Taihu Basin, China (water frequencies of 50%, 75% and 90%, planning year 2030, Taihu Basin).

Solutions of Water Rights Allocation		Jiangsu Province			Zhejiang Province			Shanghai		
		Different Water Frequency			Different Water Frequency			Different Water Frequency		
		50%	75%	90%	50%	75%	90%	50%	75%	90%
Solutions of dynamic projection pursuit allocation model (100 million m³)	$\alpha = 0.6, \beta = 0.4$	126.18	136.22	150.48	93.45	100.89	111.45	109.57	118.29	130.67
	$\alpha = 0.5, \beta = 0.5$	133.82	144.47	159.59	89.26	96.36	106.45	106.12	114.57	126.56
	$\alpha = 0.4, \beta = 0.6$	135.33	146.10	161.39	114.67	123.79	136.75	79.21	85.51	94.46

4.3. Analysis of Result

In Table 4, the empirical results of the larger weight of index P_j and year T_i, and the ranking results of provincial initial water rights allocation can be described as follows:

- When $\alpha \geq 0.5$, as is clearly indicated in Table 4: (1) According to the larger weight of year T_i, the ranking results of provincial initial water rights are all mainly influenced by the data in 2014; (2) Considering the larger weight of indexP_j: ① when $\alpha > 0.5$, current water use (P_1) and per capita GDP (P_8), which respectively belong to the principles of the total water use control and water use efficiency, are two important indexes for the ranking results; ② when $\alpha = 0.5$, the current water use (P_1) and discharge standard-meeting rate of wastewater (P_{10}), which respectively belong to the principles of the total water use control and controlling pollution have considerable impacts on the ranking results; (3) When $\alpha > 0.5$ and $\alpha = 0.5$, these two ranking results show if we are more concerned about the primitive values of indexes themselves, namely in the short term. The developing situation of provinces are given more consideration, Jiangsu Province should obtain the largest ratio of initial provincial water rights allocation all the time, followed by Shanghai and Zhejiang Provinces. Furthermore, from the short-term perspective, the principles of total water use control and water use efficiency should be taken into account. In addition, along with the increasing value of β, the principle of controlling pollution is vital for the initial water rights allocation.

- When $\beta > 0.5$, the empirical results can be expressed as follows: (1) According to the larger weight of year T_i, the data from 2013 and 2014 are crucial to the ranking results; (2) Based on the larger weight value of index P_j, the water yield of province (P_2) and the amount of pollutant discharge (P_9) which respectively belong to the principles of the total water use control and controlling pollution will mainly influence the allocation results of initial water rights; (3) When $\beta > 0.5$, the ranking results show if we give more weight to the annual increment values of indexes, namely in the long term, we give greater consideration to the dynamic increment evolving trend of provinces, Jiangsu Province still acquire the largest ratio of initial provincial water rights allocation, but Zhejiang Province will obtain more ratio than Shanghai. Furthermore, from the long term perspective, the principles of the total water use control and controlling pollution will be considerable for the allocation of the provincial initial water rights.

5. Conclusions

Under the new background of the most stringent water resources management in China, an effective and reasonable allocation of initial provincial water rights following the benchmark of the three red lines is the basis and prerequisite for a water rights trade-off market. It is an effective solution to the problem of water scarcity and a significant approach for achieving harmony between economic and social development and available water resources in China. The dynamic projection pursuit allocation method proposed to obtain the multiple scenarios of solutions for initial provincial water rights has met the requirements of controlling the total water use, improving water use efficiency, and controlling pollution of water function zones. Furthermore, it absorbed the idea of coordinated and sharable development. The empirical research of Taihu Basin further verified the feasibility and effectiveness of the model in this paper. The empirical results shown under the different water frequencies of 50% 75% and 90%, for the planning year 2030, where the developing situation of provinces is given more consideration, Jiangsu Province obtains the most initial water rights, followed by Shanghai City and Zhejiang Province. If we are more concerned about the dynamic increment evolving trend of provinces, Jiangsu Province will acquire the most quantity of initial provincial water rights, but Zhejiang Province should obtain more than Shanghai City. Further research can focus on improving the index system of initial provincial water rights allocation to satisfy the requirements of the most stringent water resources management better.

Acknowledgments: The authors would like to thank the financial support provided by the National Natural Science Foundation of China (Grant No. 71301063; Grant No. 41271537), sponsorship provided by "Jiangsu Overseas Research and Training Program for University Prominent Young and Middle-aged Teachers and Presidents" (2014) and "333" high-caliber talents cultivation project of Jiangsu Province (No. (2013)III-2110).

Author Contributions: Min Ge designed the research and drafted the manuscript; Feng-Ping Wu conducted the model simulation, reviewed, commented on, and revised the manuscript; Min You collected the data. All authors have read and approved the final manuscript.

Conflicts of Interest: The authors declare no conflict of interest.

References

1. Coase, R.H. The Problem of Social Cost. *J. Law Econom.* **1960**, *3*, 1–44. [CrossRef]
2. Wang, H.; Wang, J.; Qin, D. Research advances and direction on the theory and practice of reasonable water resources allocation. *Adv. Water Sci.* **2004**, *15*, 123–128.
3. Wu, D.; Wu, F.; Chen, Y. The bi-level optimization model of the compound system for basin initial water right allocation. *Syst. Eng. Theory Pract.* **2012**, *32*, 196–202.
4. Ge, M.; Wu, F.; You, M. A provincial initial water rights incentive allocation model with total pollutant discharge control. *Water* **2016**, *8*, 525. [CrossRef]
5. Ralph, W.A. Modeling river-reservoir system management, water allocation, and supply reliability. *J. Hydrol.* **2005**, *300*, 100–113.
6. Chen, Y.; Wu, F.; Zhou, Y. Analysis of evolutionary game between strong group and vulnerable group in initial water rights allocation. *Soft Sci.* **2011**, *25*, 11–15.
7. Read, L.; Madani, K.; Inanloo, B. Optimality versus stability in water resource allocation. *J. Environ. Manag.* **2014**, *133*, 343–354. [CrossRef] [PubMed]
8. Wu, F.; Ge, M. Initial allocation model for water right of the first hierarchy. *J. Hohai Univ. (Nat. Sci.)* **2005**, *33*, 216–219.
9. Huang, X.; Shao, D.; Gu, W.; Dai, T. Optimal water resources deployment based on multi-objective chaotic optimization algorithm. *J. Hydraul. Eng.* **2008**, *39*, 183–188.
10. Condon, L.E.; Maxwell, R.M. Implementation of a linear optimization water allocation algorithm into a fully integrated physical hydrology model. *Adv. Water Resour.* **2013**, *60*, 135–147. [CrossRef]
11. Wu, F.; Ge, M. Method for interactive water right initial allocation based on harmoniousness judgment. *J. Hohai Univ. (Nat. Sci.)* **2006**, *34*, 104–107.
12. Wang, Z.J.; Zheng, H.; Wang, X.F. A harmonious water rights allocation model for Shiyang river basin, Gansu province, China. *Int. J. Water Resour. Dev.* **2009**, *25*, 355–371.
13. Wu, F.; Chen, Y. *Research on the Harmonious Allocation Method of Initial Water Rights in a Basin*; China Water Power Press: Beijing, China, 2010.
14. Liu, Z.; Yang, D. Research advances of chaos optimization algorithms for engineering global optimization. *Chin. J. Comput. Mech.* **2016**, *33*, 269–286.
15. Wang, S.; Cheng, C.; Wu, X.; Li, B. Application of self-adaptive chaos whole annealing genetic algorithm to optimal operation of hydropower station groups. *J. Hydroelectr. Eng.* **2014**, *33*, 63–71.
16. Taihu Basin Authority of Ministry of Water Resources. *Taihu Basin and Southeast Rivers Water Resources Bulletin*; Taihu Basin Authority of Ministry of Water Resources: Shanghai, China, 2011–2014.

Article

Flood Reduction in Urban Drainage Systems: Cooperative Operation of Centralized and Decentralized Reservoirs

Eui Hoon Lee [1], Yong Sik Lee [1], Jin Gul Joo [2], Donghwi Jung [3] and Joong Hoon Kim [1,*]

[1] School of Civil, Environmental and Architectural Engineering, Korea University, 306, Eng. Building, Anam-dong, Seongbuk-gu, Seoul 02841, Korea; hydrohydro@naver.com (E.H.L.); lsiky88@korea.ac.kr (Y.S.L.)

[2] Department of Civil Engineering, Dongshin University, Naju 520-714, Korea; civilguy97@hanmail.net

[3] Research Center for Disaster Prevention Science and Technology, Seoul 02841, Korea; donghwiku@gmail.com

* Correspondence: Jaykim@korea.ac.kr; Tel.: +82-02-3290-3316

Academic Editor: Helena Margarida Ramos
Received: 5 July 2016; Accepted: 17 October 2016; Published: 22 October 2016

Abstract: Failure of drainage systems leads to urban flooding; therefore, structural measures such as the installation of additional drainage facilities, including pump stations and detention reservoirs, have been adopted in the past to prevent and mitigate urban flooding. These measures, however, are costly and time consuming. To maximize flood mitigation efficiency, it is essential to also implement non-structural measures such as effective operation of drainage facilities. In this study, we propose a new cooperative operation scheme for urban drainage systems that involves linking centralized reservoir (CR) and decentralized reservoir (DR) operations by sharing water level information at monitoring nodes. Additionally, we develop a resilience index to assess the system's ability to mitigate, restore, and recover from inundation (i.e., failure). Most results show that flood reduction and resilience in cooperative operations are better than the current operation. However, the results of CR operation for 2010 are worse than the current operation at high monitoring node levels (1.4 m–1.5 m), and the results of DR operation for 2011 are worse than the current operation at low monitoring node levels (0.8 m–0.9 m). All results related to flood reduction and resilience in cooperative operation are superior to the current operation.

Keywords: urban flooding; centralized reservoir; decentralized reservoir; cooperative operation

1. Introduction

In urban areas, various drainage facilities such as pump stations and detention reservoirs have been constructed. Pump stations can prevent the backwater effect in urban streams that leads to flooding in drainage systems. Detention reservoirs can reduce the peak discharge in urban drainage systems. It is difficult, however, to prevent flooding due to extreme rainfall, even though all drainage facilities are designed and constructed using the concept of design flood frequency [1]. Both structural and non-structural measures can be implemented to prevent and reduce flooding. Structural measures are any physical construction used to reduce or avoid possible impacts of flooding, as well as engineering techniques used to achieve flood-resistance and resilience in urban drainage systems. Non-structural measures are any measure not involving physical constructions, which use knowledge or practical operation to reduce flood risks.

Currently, however, drainage facilities are not as effective as expected due to a greater amount of extreme rainfall, impervious areas, and runoff in urban areas. Non-structural measures such as rainfall prediction, flood forecasting, and drainage facility operation have emerged as a new alternative that can support structural measures [2].

Recently, significant attention has been paid to studies on the effectiveness of pump stations and detention reservoirs for disaster prevention. Studies of detention reservoirs have included: optimization of detention facilities using multi-objective genetic algorithms (MOGA) [3], determination of reservoir detention capacity [4], and stochastic rainfall analysis for storm tank performance evaluation in urban drainage systems [5,6]. Studies on reservoir location, parameters, and optimal design capacity have drawn a lot of attention [7,8]. In addition, pump station studies have focused on: the operation of prediction-based rainwater pump stations in urban basins [9], generalized methods for storm water pumping station design [10], and the real time operation of rainwater pump stations [11]. There have also been studies on drainage systems in urban basins, including the assessment of urban drainage system resilience using a hydraulic assessment model [12], and an investigation of the relationships between precipitation and floods [13].

Many studies have investigated the design/operation of rainwater pump stations and the capacity/location of detention reservoirs; however, there have been no previous studies on the cooperative operation of pump stations and detention reservoirs. Cooperative operation is advantageous as pump stations and detention reservoirs are a part of the catchment system and their operation may drastically affect water levels and flooding in the drainage network. This study investigates the cooperative operation of pump stations and detention reservoirs, which are non-structural measures that can compensate for the time-consuming and costly nature of structural measures. Drainage pumps at pump stations and detention reservoirs are operated on the basis of centralized reservoirs (CR) and decentralized reservoirs (DR), with monitoring nodes that consider the drainage system situation and implement cooperative operation when necessary. To maximize flood sharing between each drainage facility, cooperative operation is performed on the basis of the water levels in the pump stations, drainage system, and detention reservoirs. CR operation can improve rapid drainage effectiveness and DR operation can increase detention effectiveness. Cooperative operation between CR and DR combines the advantages of CR/DR operation and effective flood mitigation in urban drainage areas. In addition, we propose and apply the resilience index as a standard for determining the condition of urban drainage systems.

2. Methods

The methodology used in this study consists of five parts. First, monitoring nodes for cooperative operation of drainage facilities are selected. Second, we perform CR operation based on monitoring node levels and review the resulting flood volume. The aim of CR operation is rapid urban area drainage and increased drainage effectiveness. Third, we perform DR operation based on monitoring node levels and review the resulting flood volume. The aim of DR operation is to obtain additional capacity in DR and improve detention effectiveness. Fourth, we review the operation of CR and DR based on monitoring node levels and compare the resulting flood volume with that of current drainage facility operations. The objective of cooperative operation between CR and DR combines the rapid drainage effectiveness of CR and the detention effectiveness of DR. Fifth, we use the resilience index to quantify how well each operation method mitigates the magnitude of failure (i.e., flooding) of the drainage system.

2.1. Overview

Two procedures are used in the selection of monitoring nodes for cooperative operation of drainage facilities. The first step identifies the node where overflow first occurs, and the second identifies nodes where the maximum overflow occurs. Each procedure is described below.

The first flooding node is determined using rainfall probability, as proposed by Huff [14], with selection of an appropriate quartile for the study basin. The Huff formula distributes rainfall in a selected quartile once the event duration is selected. The Storm Water Management Model (SWMM) is run with uniform rainfall depths, starting at 1 mm and increasing by 1 mm until flooding occurs. The node where the first flooding occurs is designated as one of the monitoring nodes. We continue to review the first flooding node as rainfall duration changes. We also identify the maximum flooding node using the Huff distribution.

There are four quartiles in the Huff distribution. Equations (1)–(4) are different from other regions because these are the ones available in Seoul (Korea Precipitation Frequency Data Server, www.k-idf.re.kr) [15]. The quartiles in the Huff distribution represent the peak values in the rainfall data. The appropriate quartile is selected for the study area from among the four quartiles of the Huff distribution, each of which has a different timing of peak rainfall. For example, the maximum value of the first quartile is located between 0% and 25% of the total rainfall duration. The maximum values of the second, third, and fourth quartiles are then located between 25% and 50%, 50% and 75%, and 75% and 100% of the total rainfall duration.

When an appropriate Huff quartile is selected for the basin, synthetic precipitation occurs according to frequency and duration times. Rainfall-runoff simulation depends on rainfall probability. The node with the highest flood volume is selected as the second monitoring node. In the study area of Seoul, the first, second, third and fourth Huff quartiles are determined using Equations (1)–(4), respectively.

$$y = 29.289x^6 - 95.64x^5 + 119.7x^4 - 70.768x^3 + 18.176x^2 + 0.2426x - 0.0007 \tag{1}$$

$$y = -38.505x^6 + 118.93x^5 - 132.67x^4 + 60.815x^3 + 8.3001x^2 + 0.7296x + 0.0005 \tag{2}$$

$$y = 37.835x^6 - 106.21x^5 + 105.18x^4 - 44.549x^3 + 9.1084x^2 - 0.3603x + 0.0005 \tag{3}$$

$$y = -25.498x^6 + 63.755x^5 - 57.196x^4 + 22.882x^3 - 3.4377x^2 + 0.4955x - 0.0002 \tag{4}$$

where y indicates the accumulated rainfall ratio and x indicates the accumulated time ratio. The time ratio x is cumulative time divided by total rainfall duration. The rainfall ratio y is cumulative rainfall at time ratio x divided by total rainfall. The third quartile of the Huff distribution is selected to review the first and maximum flooding nodes in the target basin. Yoon et al. (2013) suggested that the third quartile in the Huff distribution is appropriate for South Korea [16]. Figure 1 describes the complete process for obtaining rainfall data using the Huff distribution.

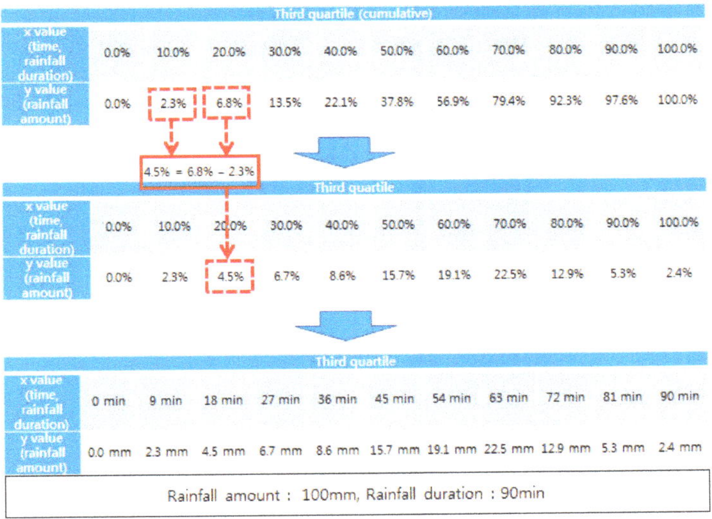

Figure 1. Process for obtaining rainfall data using the Huff distribution.

Using the process in Figure 1, each rainfall event can be obtained using the Huff distribution. If the total rainfall amount is 100 mm and the duration is 90 min in the third quartile, we can obtain rainfall data using the Huff distribution and use it as input data for the rainfall-runoff model. The total rainfall amount is set to 11 mm, in 1 mm increasing intervals, if there is no flooding after simulation of the rainfall-runoff model. This process is repeated until there is flooding in one node. We call this node the first flooding node.

SWMM, developed by the United States Environmental Protection Agency (US-EPA) in 1971, was used to simulate the rainfall-runoff process and to investigate the first and maximum flooding nodes. SWMM has been developed to simulate the flow volume and water quality within urban drainage systems. SWMM can be used for planning, analysis, and design related to stormwater runoff, combined/sanitary sewers, and other drainage systems in urban areas. Furthermore, it can be used in rural area simulations. It is a comprehensive model that can be used to simulate runoff volume caused by rainfall events and runoff of pollutants in surface and sewer pipes, to trace runoff volume in sewer distribution networks, and to calculate retained volume [17]. Users can select either steady-flow routing, kinematic wave routing, or dynamic wave routing as flow routing options in SWMM.

We analyze the flooding that is contingent on CR operation on the basis of monitoring node levels. When the level at the monitoring node indicates possible flooding of sewer conduits, the backwater effect caused by high CR water level is minimized by operation of drainage pumps in order to reduce the burden on sewer conduits. Thus, flooding is reduced through early operation of pump station drainage pumps to minimize CR levels.

We applied the method of Lee et al. to DR operation analysis based on monitoring node levels [18]. One of the weaknesses of urban detention reservoirs is a vulnerability to continuous rainfall. To counter this, we secure additional space in the DR by operating drainage pumps if the sewer conduit levels, identified by the monitoring node, are low. However, if the level of the DR is higher than the limit (discharge level), discharge in the DR is initiated whilst current drainage facility operations are performed.

We also evaluate cooperative operation of CR and DR based on monitoring node levels. Discharge released in order to secure additional space in the DR is transported to the CR at the basin outlet through the drainage system. Backwater effects of the CR are reduced by early discharge of reserved water. The ultimate goal of cooperative operation is to reduce inundation in urban areas. This is achieved through efficient sharing of flood discharge by linking the operation of drainage facilities, such as upstream detention reservoirs in drainage areas, with other drainage facilities, such as downstream pump stations. In addition, we develop a resilience index to quantify the ability of the system to mitigate inundation (i.e., system failure), and then quantitatively compare current and new operational approaches of drainage facilities. Resilience of drainage systems can be calculated on the basis of patterns from flooding to recovery. Figure 2 summarizes the cooperative operation approach used in our study.

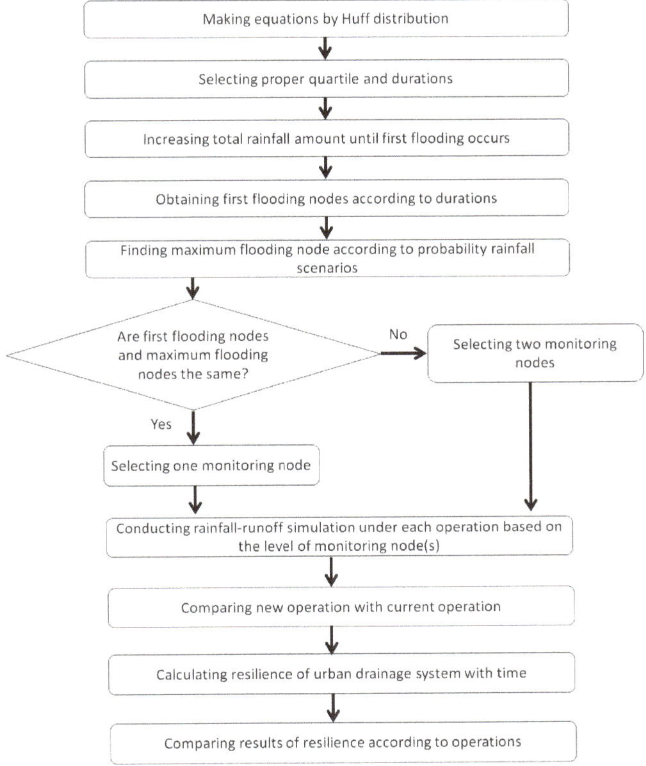

Figure 2. Flowchart for cooperative operation of urban drainage systems.

2.2. Components of Urban Drainage Systems in Korea

The pump stations that are currently installed and operational in Korea are divided into two types; those with and without CR. Pump stations without CR are usually installed in small basins that have low maximum inflow, and pump stations with CR are installed in large basins with high maximum inflow. In addition, gate pump stations have recently been introduced, but these are largely limited to basins without CR. The structure of a typical pump station with CR is shown in Figure 3.

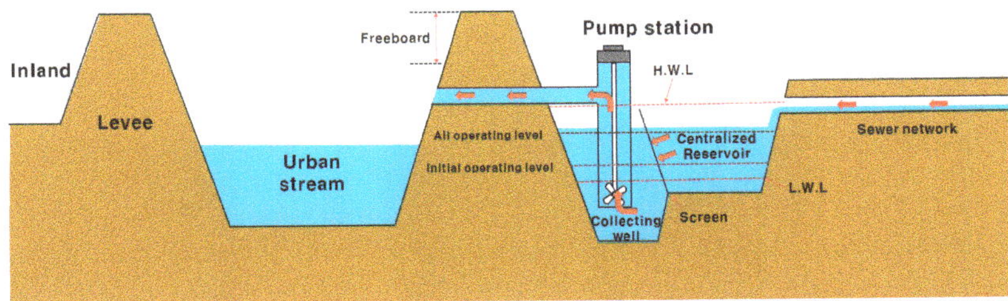

Figure 3. Structure of a typical pump station with a centralized reservoir.

For each CR in pump stations, a High Water Level (HWL) and a Low Water Level (LWL) are defined. The initial and total operating levels for drainage pumps are determined between the LWL and the HWL (Figure 3). The LWL is above the base of the CR and the initial operating level is higher than the LWL in order to prevent cavitation of drainage pumps, which may occur when there is no CR inflow.

The DR is located in the midstream and upstream; it is connected to sewer conduits or rivers, and is designed to reduce peak discharge (Figure 4). There are two types of DR, online and offline. An online DR is generally larger and is effective in continuous storms. However, due to its large capacity, it is difficult to install in urban areas. An offline DR is generally smaller, and is not so effective in continuous storms, but its small capacity makes it easy to install in urban areas.

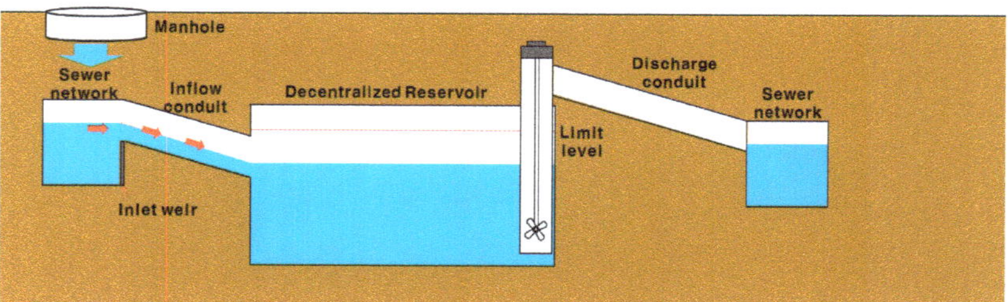

Figure 4. Structure of a typical decentralized reservoir.

Operation of the offline DR is controlled by inlet and outlet components. The height of the inlet weir directly affects DR operation and its capacity to retain runoff. However, once weir height is determined, there is no way to control inflow volume. With consideration of current design standards (only inflow which goes over the weir at a fixed height can be delivered), it is impossible to control inflow volume during operation. Thus, it is important to focus on controlling outflow volume in order to secure additional space within the DR. Using outflow control, the DR retains rainfall until the rainfall event ends, and then initiates runoff within a designed exclusion time or until the limit for drainage is reached. However, this operation has some disadvantages. As detention or drainage is performed without consideration of the condition of the sewer conduits, it is vulnerable to flooding. In addition, if it continues to rain, the DR retains the first peak inflow and is then full so cannot store additional inflow. To compensate for this, runoff is regulated according to the condition of the sewer distribution network in order to secure space for the first peak inflow as well as additional inflow.

2.3. Operation of Urban Drainage Facilities

Pump stations operate drainage pumps based on the level of the CR. The initial operating level of drainage pumps is set by taking into account cavitation and flood volume. When rainfall occurs, drainage pumps are activated if the CR reaches the initial operating level. After that, the operating level is determined by considering the capacity, number of drainage pumps, CR area, and effective depth. As the CR is operated on a fixed level basis, without considering sewer conduit conditions, it is not effective for reducing inundation.

For operation of the CR proposed in this study, we select monitoring nodes representative of the sewer distribution network (Figure 5). An early operation of drainage pumps at the pump station is determined by water levels at the monitoring nodes in order to maintain the minimum level required to prevent cavitation. Like the CR, the DR is operated independently of other urban drainage facilities. Offline DRs are installed within drainage systems in urban areas and drain only in two cases: first, when the level of the DR reaches full capacity (the limit level), discharge starts by operating drainage pumps; and second, when rainfall is deemed to have ended, runoff is initiated. However, as discussed in Section 2.2, the system has some disadvantages. It is not straightforward to determine whether rainfall has ended. Moreover, when the discharge is released, the condition of the sewer conduits is not considered.

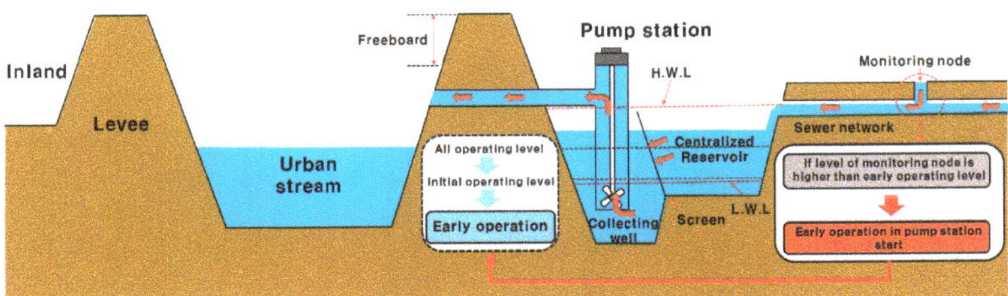

Figure 5. New operation method for a pump station with a centralized reservoir.

As explained above, current CR and DR operations run independently of each other and of the sewer conduits, and are not effective in reducing urban flooding or adapting during a heavy rainfall event. Cooperative operation of the facilities, involving connecting the DR to the sewer conduits and to the CR, is essential in order to address urban flooding. In order to operate them as one system, the condition of the sewer conduits is examined in real time on the basis of levels at the monitoring nodes (Figure 6).

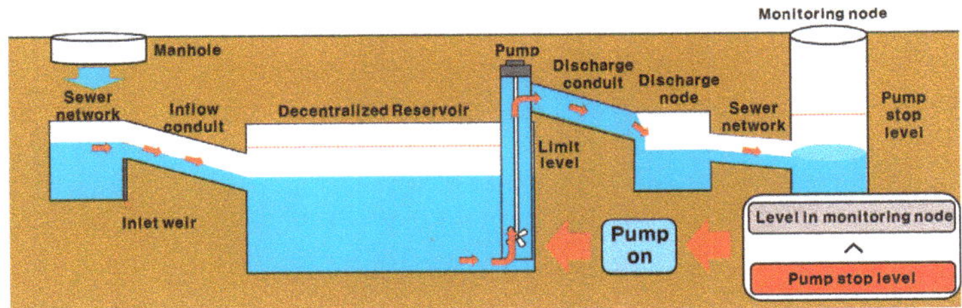

Figure 6. New operation method for a decentralized reservoir.

2.4. Cooperative Operation of Urban Drainage Facilities

For the cooperative operation of urban drainage facilities, a cooperative operation of both the main and the sub pump station is performed. When the capacity of the main pump station is not sufficient to drain the outlet, the sub pump station is added. When natural drainage is carried out, the sewer is directed to the main pump station with a large CR. When drainage is forced using drainage pumps, flooding is mitigated by considering the capacity of each pump station (pump capacity + CR capacity). Such cooperative pump station functions are available when pump stations are connected with a DR through sewer conduits. Cooperative operation with a DR installed in the basin is not considered in this type of cooperative operation between pump stations.

For the cooperative operation of urban drainage facilities proposed in this study, the water level at monitoring nodes is selected on the basis of the early operating level of the CR and the pump stop level of the DR. When sewer conduit levels are elevated and reach the early operating level of the CR, pumps are operated and rainwater is rapidly drained. When the level reaches the pump stop level of the DR, drainage pumps are stopped to remove the burden on sewer conduits.

In this section, we describe three characteristics of cooperative operations between CR and DR in detail. These are: additional space in the DR; reduction of the backwater effect by CR; and flood mitigation of the DR/sewer conduits/CR. First, we can secure additional space in the DR by drainage, based on monitoring node levels. Moreover, we reduce the risk of floods caused by CR backwater effects by performing early drainage in pump stations based on monitoring node levels. In addition, controlling flow delivery based on performance of the DR, sewer conduits, and CR provides efficient flood-mitigation for the entire urban drainage network. As low water levels are maintained due to rapid drainage of urban drainage facilities, including pump stations and detention reservoirs, the CR and DR can be equipped with sufficient capacity to prevent floods caused by continuous rainfall (Figure 7). Cooperative operation is more effective for flood mitigation than both current and new drainage facility operations because of the detention effectiveness of DR, the rapid drainage effectiveness of CR, and the cooperation effectiveness of the monitoring nodes.

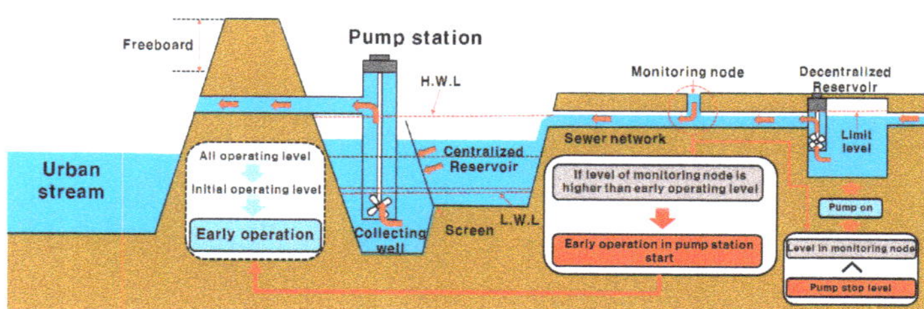

Figure 7. Cooperative operation of a centralized and a decentralized reservoir.

2.5. Resilience Index for Urban Drainage Systems

Many indicators have been proposed for quantifying the ability for urban drainage systems to mitigate flooding [19–21]. For example, House et al. (1993) suggested an index for receiving water impacts and making better judgments regarding the acceptable performance of CSO (Combined Sewer Overflows). Cembrano et al. (2004) used a performance index for reducing flooding and discharge pollution in containing gates and detention tanks of urban drainage systems. Mitchell (2005) proposed an urban catchment wetness index (UCWI) for mapping hazards related to non-point pollution in urban drainage system.

There are few studies applying the concept of a resilience index for the evaluation of urban drainage systems. Mugume et al. (2014) is the only study to propose a resilience index according to flood depth in urban drainage systems [12]. The resilience index for urban drainage systems evaluates their ability to recover from floods, and flood (failure)–restoration–recovery is considered as one process.

The resilience of urban drainage systems can be defined as the ability to prepare for, recover, and restore after facility malfunction and flooding (failure). For example, failure is defined as the occurrence of flooding in urban areas due to malfunctions of equipment, such as sudden power outages in pump stations and remote control system shut-down of water gates in pump stations. Figure 8 conceptually defines the resilience of a system as a function of failure depth and recovery time.

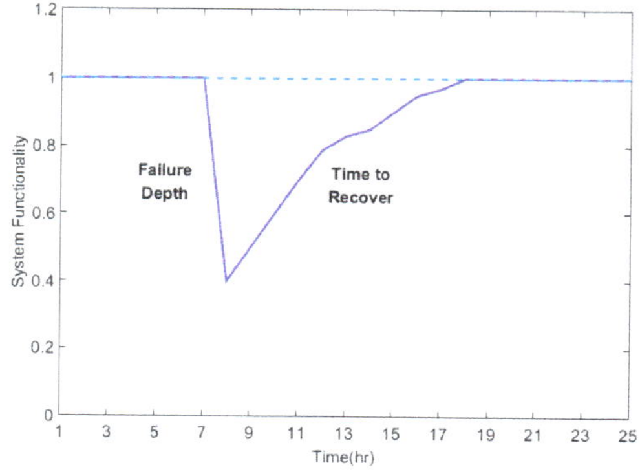

Figure 8. Conceptual illustration of resilience.

Resilience is calculated using the performance evaluation function, which is based on the variables of flood volume per minute, accumulated rainfall for a certain time until the present, and basin area. Values of the performance evaluation function range from 0 to 1. When there is no flooding in urban drainage facilities, the value is 1. Equation (5) gives the performance evaluation function $u(T)_i$ at the ith minute (arbitrary time).

$$u(T)_i = \max \left(0, \quad 1 - \frac{F_i}{R_d \times A_u} \right) \qquad \left(R_d = \sum_{i-t_c}^{i} R_i \right) \tag{5}$$

where F_i is flood volume (m^3), t_c is time of concentration (min), R_d is accumulated rainfall during current time (t) and i $- t_c$ (10^{-3} m = mm), and A_u is the basin unit area (10^{-8} m^2 = 0.01 km^2). High values of the performance evaluation function mean that the urban drainage system has a high ability to recover after flooding (failure) or minor system failure, and low values indicate a low ability to recover. The resilience of the entire urban drainage system is calculated as shown in Equation (6).

$$R_s = \frac{1}{T_n} \int_{T_0}^{T_n} u(T) dT \tag{6}$$

where R_s is resilience of the entire system, T_0 is the start time and T_n is the end time. In this study, we calculate and compare resilience of the current urban drainage system with the DR system and the cooperative system, including pump stations and detention reservoirs. For validation with experimental/measured in-situ data, measurement of the flooding area and flooding depth by aerial photography is essential. Nowadays, drones are employed for this; however, there are several associated problems such as high costs and inaccuracy of measured data.

3. Applications and Results

3.1. Study Area

Seoul is one of the most urbanized cities in the world and consists of 25 administrative districts. The Han River flows through the center of Seoul. It has several tributaries, including Banpo stream, Wangsuk stream, Anyang stream, and Jungnang stream. In this study, we select Dorim stream as a target. Dorim stream runs through the Guro-gu, Yeongdeungpo-gu, and Gwank-gu administrative districts, and its tributaries include Daebang-cheon and Bongcheon-cheon.

As shown in Figure 9, there are 10 pump stations, 2 detention reservoirs, and 2 tributaries in the Dorim stream basin. The basin above Daerim 3 pump station is selected for the application of cooperative operation, as proposed in this study. A detention reservoir is installed in the upstream drainage system connected to Daerim 3 pump station, which is located at the basin outlet. Since the DR was installed in 2009, there have been two episodes of flooding, in 2010 and 2011.

For this study, we select the largest events in the recent flood record in order to compare the different operating systems using disaster year books issued by the Ministry of Public Safety and Security [22–27], which contain information on disaster type, area, duration, statistical data, and typhoon trajectories. Table 1 lists rainfall periods from 2008 to 2014, as identified in the disaster year books [22–27]. The target basin was damaged in significant events on 23 September 2010 and 27 July 2011. Rainfall duration and frequency, respectively, is estimated as 240 min and 100 years for the 2010 event, and 1080 min and 100 years for the 2011 event.

Table 1. Large rainfall events between 2008 and 2014 (data from Korea Meteorological Administration).

Year	2008	2009	2010	2011	2012	2013	2014
Date	23/7–25/7	9/7	20/9–21/9	26/7–28/7	20/8–22/8	12/7–14/7	23/8–27/8
Total Precipitation (mm)	280	312.5	256	386	405.2	429	384.5
Representative Return Period (year)	80	100	100	100	150	150	200

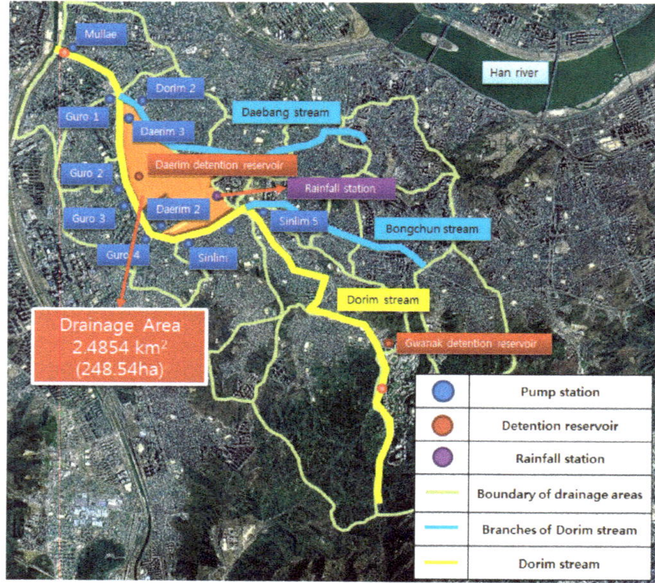

Figure 9. Dorim stream study area (from Google Earth Imagery © Google Inc., 2015).

The design return period for the Daerim 3 pump station is 30 years. A total of 12 drainage pumps (1 pump at 300 kW, 7 at 413 kW, 2 at 450 kW, and 2 at 900 kW) are installed, with a total discharge capacity of 57.02 m^3/s. The CR has a total capacity of 33,650 m^3, a HWL at 9.0 m, and a LWL at 6.8 m. The first drainage pump at the CR is operated at the initial operating level and all drainage pumps at the CR are operated at all operating level. The time of concentration in the Daerim 3 pump station basin is 30 min [28].

The detention reservoirs in the Daerim basin have a design return period of 20 years. The 5-year-flow is designed to go over the weir and into the reservoir. The weir has a width of 2 m and a height of 0.4 m, and is located on the right side of a rectangular type sewer 2 m wide and 1.5 m high. As a result, the detention reservoir has a total height of 3.2 m, a freeboard of 1 m, and a total detention volume of 2447 m^3. It is impossible for the Daerim detention reservoirs to drain naturally; however, they can also drain by operation of a submersible pump at 22 kW, which has a capacity of 9.0 m^3/min with a total head of 6.0 m [29].

The model is manually calibrated. We focus on four aims when calibrating the model: matching the peak discharge and peak time total runoff volume, and minimizing the Root Mean Square Error (RMSE) between the observed inflow data and the simulated inflow results for 2010 and 2011. First, rainfall data for 21 September 2010 and 27 July 2011 are used for calibration of the rainfall-runoff model. Inflow is estimated using pump operation records and the variance of CR level because CR in the pump station does not measure inflow. Figure 10 describes the observed/simulated inflow hydrograph in the target basin and the accuracy of calibration/validation.

Table 2 shows the types and values of the calibrated parameters in SWMM. For calibration and validation of the target basin, three parameters are used: percentage of the impervious area, width of the subcatchment, and percentage of slopes in the subcatchment. The total number of parameters is 4728 because the total number of subcatchments is 1576. In this study, Manning's roughness coefficient in the conduits is assumed to be uniformly 0.014 following a technical report written after the recent rehabilitation and maintenance of the Daerim 3 pump station and the design of the Daerim detention reservoir.

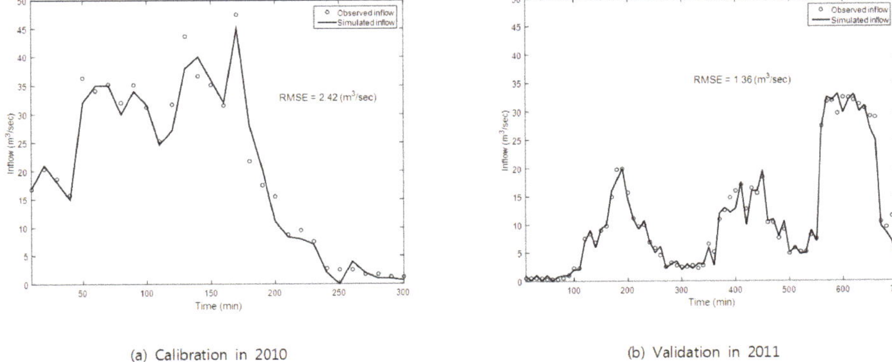

(a) Calibration in 2010

(b) Validation in 2011

Figure 10. Calibration and validation of 2010 and 2011 data.

Table 2. Summary of calibration parameters.

Type	Parameter	Range
Subcatchment	Percentage of impervious area (%)	0–100
	Width (m)	0–60
	Percentage of slope (%)	0–100

The "control" module in EPA-SWMM is used for various simulations of CR and DR operations. It is a simple code whereby, for example "If node DB-1 depth $\geq$ 2.33 then pump ps1 status = on". On the basis of GIS data, we construct and simulate sewage networks in Seoul, including the Daerim detention reservoir (Figure 11).

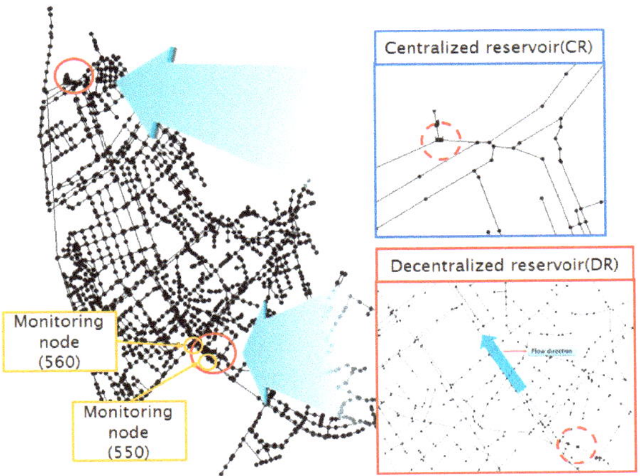

Figure 11. Urban drainage systems in the target area.

3.2. Selection of Monitoring Nodes for Cooperative Operation of Urban Drainage Facilities

Monitoring nodes are selected on the basis of rainfall probability based on the third quartile of the Huff distribution, as proposed by Yoon et al. Different quartiles of the Huff distributions have different peak locations, and in the case of the third quartile, the peak tends to be in the latter half of the event. We use two methods to select monitoring nodes, based on first and maximum flooding.

First, we select three rainfall durations (30, 60, and 90 min) for simulation of the first flooding node, up to three times as long as the time of concentration in the target basin (30 min). For each rainfall duration, we increase total rainfall volume in 1 mm increments until the first flooding occurs, and identify the first flooding node. Sewer conduits in urban areas can be divided into main and branch conduits according to the associated basin area runoff coefficient (C) and drainage area (A). The first flooding generally occurred at branch conduits for which design frequency (CA < 0.12 km^2) is only 5–10 years (that of the main conduit is 30 years) and thus the conditions of the whole basin cannot be effectively determined by only considering the branch conduits. Therefore, we increase total rainfall volume until the first flooding occurs in the main conduits (CA $\geq$ 0.12 km^2).

To identify the maximum flooding node, we review a total of nine conditions of rainfall probability frequency and duration (a combination of 30, 50, and 100 year frequencies, and 30, 60, and 90 min durations) in the Huff third quartile distribution. However, the maximum flooding nodes identified in this process are not consistent across all nine conditions, except at Daerim reservoir, and are located in branch conduits making it problematic to use them as monitoring nodes. The maximum flooding nodes that are located in main conduits are all at location 550. Thus, we review maximum flooding nodes on the basis of rainfall on 23 September 2010 and 27 July 2011. The maximum flooding node occurred in the Daerim detention reservoir. However, as the interior level of the detention reservoir is considered in the current operation of drainage facilities, we only count nodes in the main conduits. Therefore, the highest flood volume in both events is observed at location 550. Details of the selected monitoring nodes are shown in Table 3. We select locations 560 and 550, the first and maximum flooding nodes, respectively, as the monitoring nodes for operation of the CR, DR, and cooperative CR/DR. The nodes at 560 and 550 are located within the 2-line main conduits with a height of 1.5 m and width of 2 m.

Table 3. First and maximum flooding nodes in the drainage system as potential monitoring nodes.

First Flooding Nodes		Maximum Flooding Nodes	
Duration (min)	First flooded node in main conduits (Total precipitation)	Date	Maximum flooded node in main conduits (flood volume)
30	560 (79 mm)	23/9/2010	550 (1597 m^3)
60	560 (104 mm)	27/7/2011	550 (143 m^3)
90	575 (154 mm)	-	-

3.3. Operation of CR Based on Monitoring Node Levels

The CR is operated as follows. When stream level is lower than the base of the drainage gate, water in the CR naturally drains through the drainage gate. When stream level is higher than the base of the drainage gate, the gate is closed and water is forced to drain using drainage pumps. When no water enters the CR, drainage pump operation is stopped.

As previously mentioned, drainage pumps in the pump station are activated according to the level of the CR. High CR water levels increase the water level in conduits of the upstream drainage system, resulting in a backwater effect. Low CR water levels decrease the level of upstream sewer conduits. Initial early operating levels for preventing cavitation in the Daerim 3 pump station are calculated using the following procedure. The initial early operating level is calculated as the sum of the required depth (A), the screen head loss (B), and the freeboard for mechanical operation (C). The entire calculation procedure for the initial operating level in the new operation shown in Table 4 is explained below.

The required depth is the minimum depth for the pump operating cycle (A), calculated by dividing the required volume in CR (A1) by the average area in CR (A2). The required volume in CR (A1) is calculated by multiplying the initial pump discharge (3.88 m^3/s = 233 m^3/min) and the preparation time (30 min), and dividing by four. The required volume in CR (A1) is 1711 m^3. The required depth (A, 0.15 m) is the value of the required volume in CR (A1, 1711 m^3) divided by the average area in CR (A2, 11,400 m^2).

Screen head loss (B) refers to the variance of the level in the collecting well of CR when the initial pump is operated. It generally has a value between 0.1 m and 0.3 m. In this study, 0.1 m is selected as the screen head loss (B). The freeboard for mechanical operation (C) is the freeboard used for mechanical work such as valve opening and pump operation when an initial pump is operated. It is generally from 0.0 m to 0.2 m. In this study, 0.05 m is the freeboard for mechanical operation (C). The last calculation for initial operating depth in the new drainage system operation is the sum (0.3 m) of the required depth (A, 0.15 m), (B, 0.1 m), and (C, 0.05 m), which is added to the elevation in the bed of the CR (6.5 m). The initial operating elevation is set at 6.8 m.

Determination of the initial operating level is complicated because it is closely related to cavitation of the initial operating pump. Excluding calculation of the initial operating level, calculation of the other operating levels is based on the required depth (A). At low elevation, between 6.5 m and 7.3 m, the operating interval is larger than at high elevation because preparation time of longer than 10 min is required. Additionally, we follow the order of pump discharge in the current operation because the order of pump discharge at each pump station is already fixed by the local government. It is determined by considering the pump rotation cycle for maintenance control of the pumps.

Table 4. Current and new operation of drainage pumps in the centralized reservoir (CR).

Elevation (m)	Pump Discharge (Current Operation, m^3/s)	Pump Discharge (New Operation, m^3/s)	Comments
6.5	-	-	Bed level of CR
6.8	-	3.88	LWL
7.2	-	8.05	
7.3	3.88	15.48	Initial operating level in current operation
7.5	8.05	19.65	-
7.6	15.48	23.36	-
7.7	19.65	27.08	-
7.8	23.36	30.80	-
7.9	27.08	57.02	-
8.0	30.80	57.02	-
8.1	57.02	57.02	-
8.3	57.02	57.02	All operating level
9.0	57.02	57.02	HWL

We set the pump discharge level at the monitoring node to 0.5 D (about 0.8 m), where D is total height, increasing to 1.0 D (1.5 m) in increments of 0.1 D, and simulate rainfall-runoff processes and pump operation independently for each case, in order to obtain flood volume. When the monitoring node level is below 0.5 D, the flood volume results are the same as for 0.5 D because reducing the level in CR (reducing the backwater effect) is not related to reducing flooding when the monitoring node level is low. We assess the results by applying rainfall from the two events of 23 September 2010 and 27 July 2011 (Figure 12).

For the 2010 event, the new CR operation resulted in lower flood volume than with the current operation, except at high monitoring node levels (1.4 m–1.5 m) (Figure 12a). Anomalous results at 1.4 m and 1.5 m by CR operation are based on current DR operation. Higher flood volumes than the current operation can be achieved if pumps are operated when monitoring node levels are high because pumps are operated according to DR levels in the current DR operation. For the 2011 event, all flood volumes were lower than those of the current CR operation (Figure 12b) because the CR level was low due to early operation of pumps, and the level decreased in upstream nodes, including flooding nodes. Figure 12b also shows that the largest flood volumes occur in the high and low monitoring node levels, but the smallest flood volumes occur in the medium monitoring node levels.

Figure 12c,d show the pumpage of CR at the time corresponding to the level of monitoring nodes in 2010 and 2011, respectively, and Figure 12e,f show the level of CR in 2010 and 2011, respectively. The results of the new CR operation were compared to that of the current operation. It was confirmed that the pumpage of CR decreases (Figure 12c,d) and the level of CR increases (Figure 12e,f) as the level of monitoring nodes increases. Therefore, total flooding volume is affected by the pumpage of CR triggered by the level of monitoring nodes under the new CR operation in which proactive pumping for the reduction of the backwater effect is capable. On the other hand, the pumpage and level of CR under the current operation do not affected by the level of monitoring nodes.

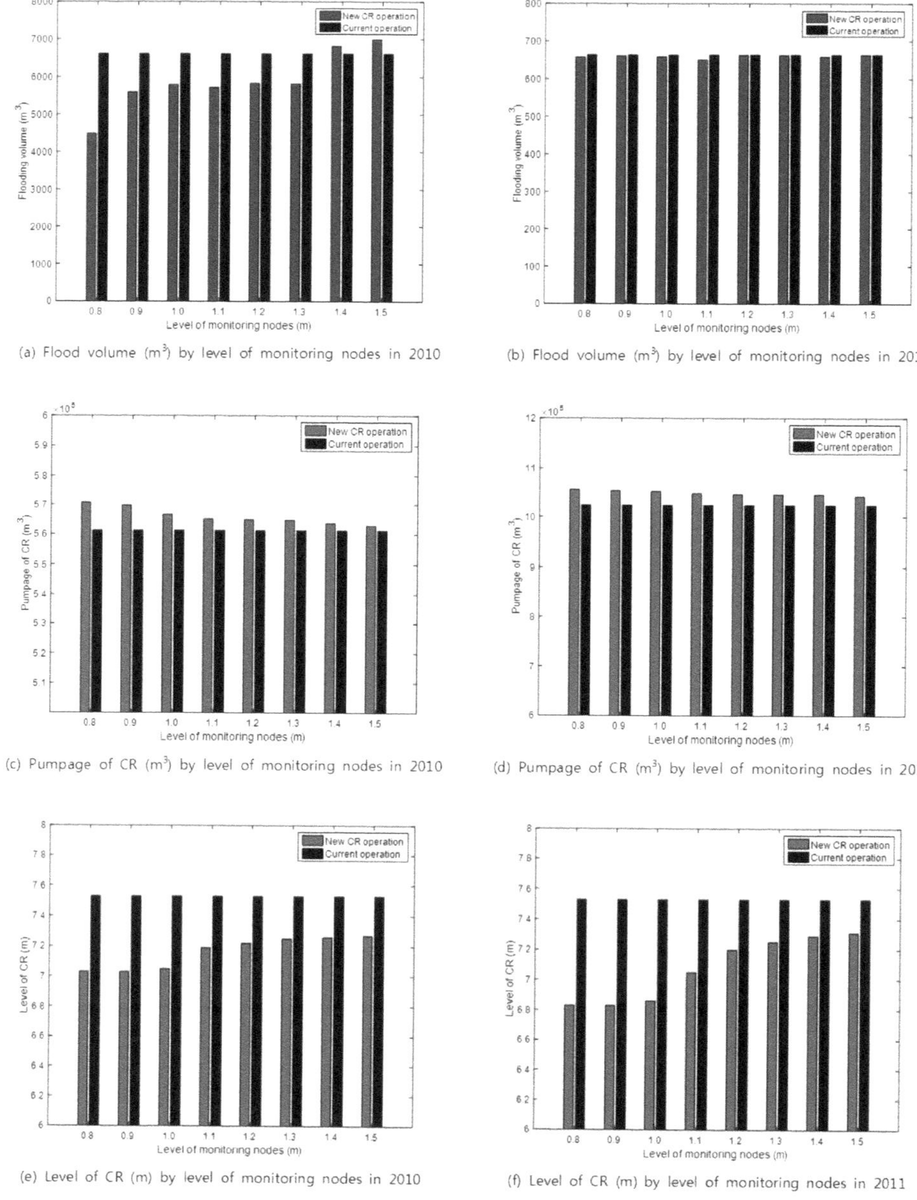

(a) Flood volume (m^3) by level of monitoring nodes in 2010

(b) Flood volume (m^3) by level of monitoring nodes in 2011

(c) Pumpage of CR (m^3) by level of monitoring nodes in 2010

(d) Pumpage of CR (m^3) by level of monitoring nodes in 2011

(e) Level of CR (m) by level of monitoring nodes in 2010

(f) Level of CR (m) by level of monitoring nodes in 2011

Figure 12. Results of new centralized reservoir operation.

Flood volume is small if there is enough storage space in DR when the level in the sewer conduits is high. Flood volume is large if there is scant storage space in DR when the level in the sewer conduits is high. However, the CR operation results vary depending on monitoring node levels due to the impact of DR operation; current DR operation is used to compare the current and new CR operations. Current operation of the DR is not related to the level of the sewer conduits, but depends on the interior level of the DR. Thus, if DR discharge coincides with the peak level in the sewer conduits, it augments the flood volume.

Flood volume patterns corresponding to various monitoring node levels for the 2010 and 2011 events differ in Figure 12. Flood volume in the new CR operation is higher than the current operation despite the low level of CR when the level of monitoring nodes are high. It is based on the capacity limitation of CR. There are irregular flooding patterns because the frequency of 2010 and 2011 rainfall events are over 100 years though the design frequency of CR is 30 years. Additionally, it is due to the locations of peak rainfall events and the CR level at the time of the peak rainfall events.

3.4. Operation of DR Based on Monitoring Node Levels

Current operation of the DR is performed as follows. Inflow from sewer conduits enters through the inlet weir. If it is not possible to drain naturally, drainage pumps are used in the detention reservoir. DR operation is controlled by discharge conditions. Discharge from the DR starts when the interior level reaches the limit level, or when the rainfall is determined to have ended. The DRs in urban areas are generally installed to make up for the lack of water delivery capacity in the sewer conduits. Even with low intensity rainfall, water overtops the weir of the inflow conduit, and storage in the DR is initiated. However, if rainfall continues, the DR is already full after the low intensity rainfall and cannot take additional water from the sewer conduits.

In this section, the new operation is applied to determine DR discharge depending on monitoring node levels (Figure 13). For the 2010 event, results of the new DR operation are improved compared to those of the current CR operation (Figure 13a). Figure 14a also shows that the largest flood volume occurs in high and low monitoring node levels, but the smallest flood volume occurs in medium monitoring node levels, both for the same reason as for the results shown in Figure 12b. Flood volume is small if there is enough storage space in the DR when the level in the sewer conduits is high, and flood volume is large if there is scant storage space in the DR when the level in the sewer conduits is high. For the 2011 event, results at monitoring nodes with levels of more than 0.9 m are better than those of the current operation, but those for the 0.8 m and 0.9 m levels are inferior because pump discharge in the new operation is smaller than pump discharge in the current operation when discharge is ceased at low monitoring node levels. This means that the new operation obtains a smaller additional DR capacity than the current operation at low pump stop levels. The variation in new DR operation results with monitoring node level is due to the design capacity of the DR. Rainfall associated with the 2010 and 2011 events has a maximum return period of up to 100 years, while DR design capacity is only 20 years. Thus, it is difficult to prevent widespread flooding and the operational effects are inconsistent.

Figure 13c,d show the pumpage of DR at the time corresponding to the level of monitoring nodes in 2010 and 2011, respectively, and Figure 13e,f show the level of DR in 2010 and 2011, respectively. The pumpage of DR was significantly increased (Figure 13c,d) and the level of DR were dramatically decreased (Figure 13e,f) if the level of monitoring nodes is higher than or equal to 1.1 m which is identified as the critical level for flooding. The consequence of such operation is low flooding volume (Figure 13a,b).

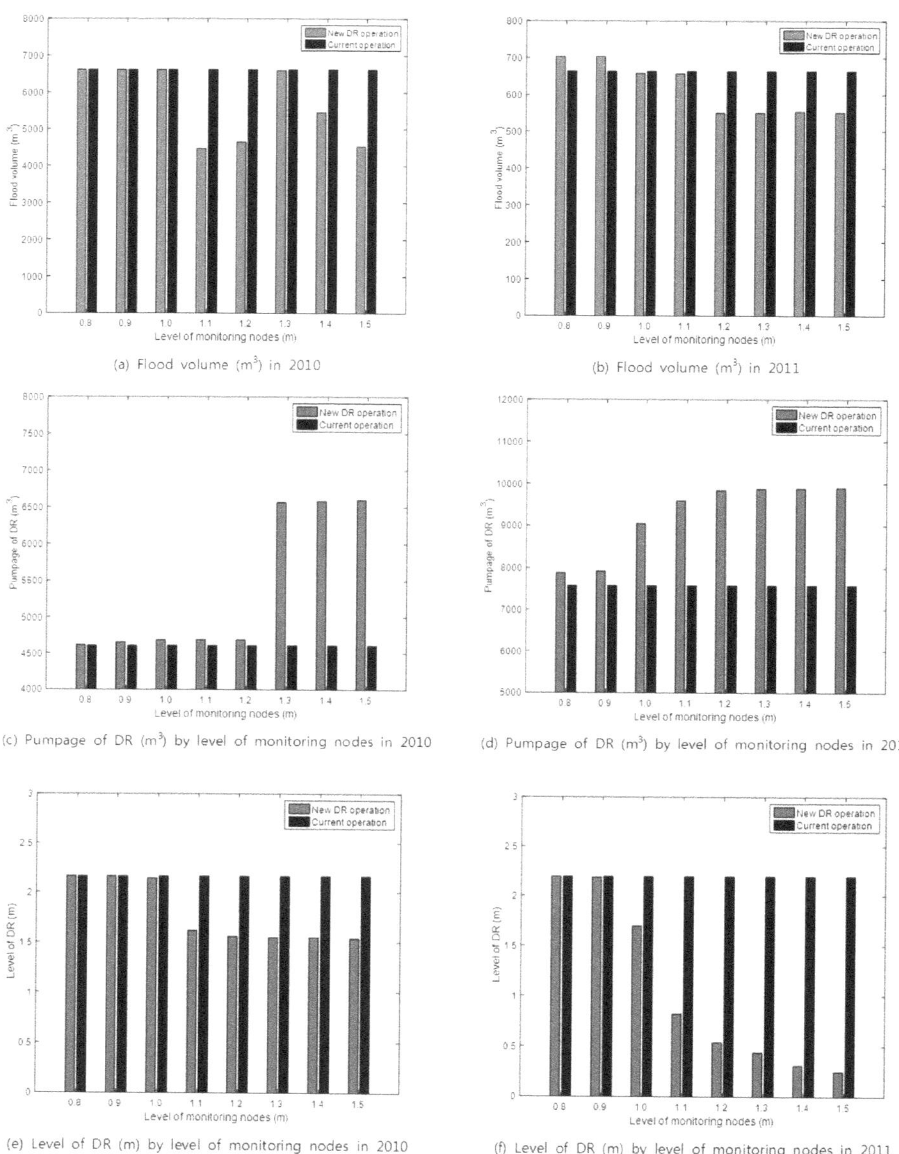

Figure 13. Results of the new decentralized reservoir operation.

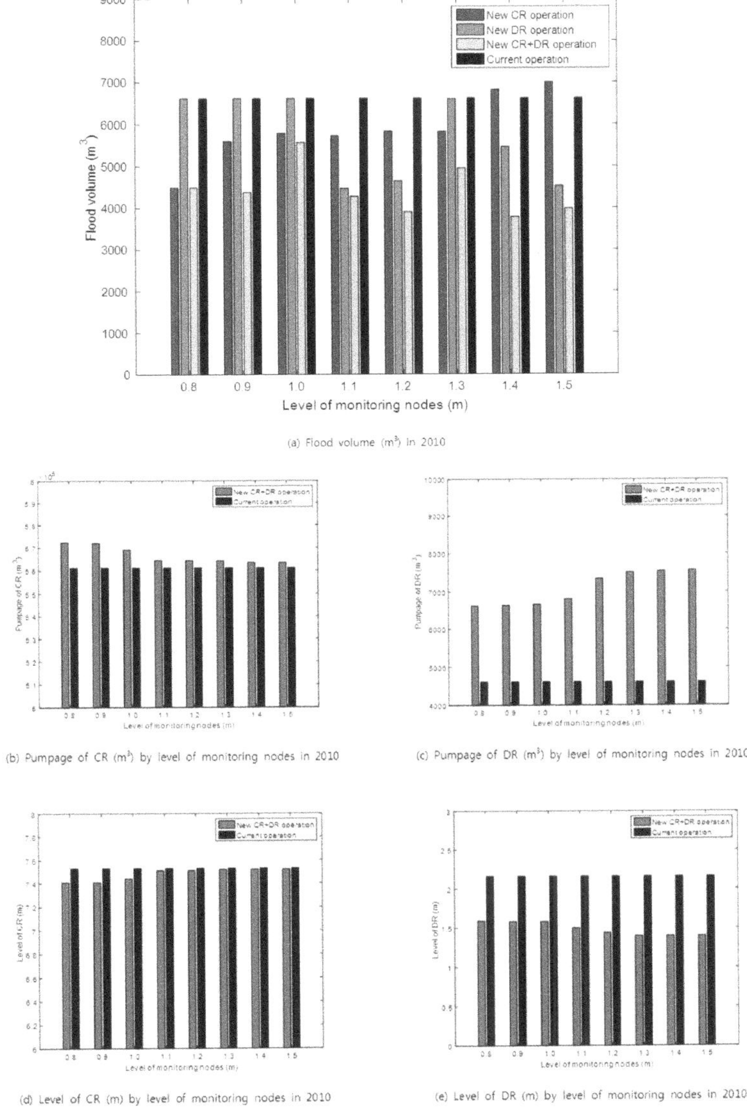

(a) Flood volume (m³) in 2010

(b) Pumpage of CR (m³) by level of monitoring nodes in 2010

(c) Pumpage of DR (m³) by level of monitoring nodes in 2010

(d) Level of CR (m) by level of monitoring nodes in 2010

(e) Level of DR (m) by level of monitoring nodes in 2010

Figure 14. Results of cooperative operation in 2010.

3.5. Cooperative Operation of Urban Drainage Facilities

Various urban drainage facilities are operated in each area, often regardless of the sewer conduit conditions, and independent of other urban drainage facilities. However, other drainage facilities within the same drainage area should be considered in order to meet the goal of urban area drainage.

For proper operation of urban drainage facilities, including the CRs and DRs installed in urban areas, it is necessary to prevent flooding, because a lack of capacity increases the flood risk in the drainage system. However, when rainfall is continuous, it is necessary to efficiently share the flood volume across urban drainage facilities. In this section, cooperative operation of the CR and DR is applied at the monitoring nodes, and the results are shown in Figures 14 and 15. For both 2010 and 2011, results of cooperative operation are superior to those of the current operation because CR/DR cooperative operation results in a smaller flood volume than all monitoring node levels of the current operation for 2010 and 2011 (Figures 14a and 15a, respectively). This is due to the detention effectiveness of DR operation and the rapid drainage effectiveness of CR. However, the same variability as the CR and DR operations is observed across the monitoring node levels, due to the way the DR operation is included.

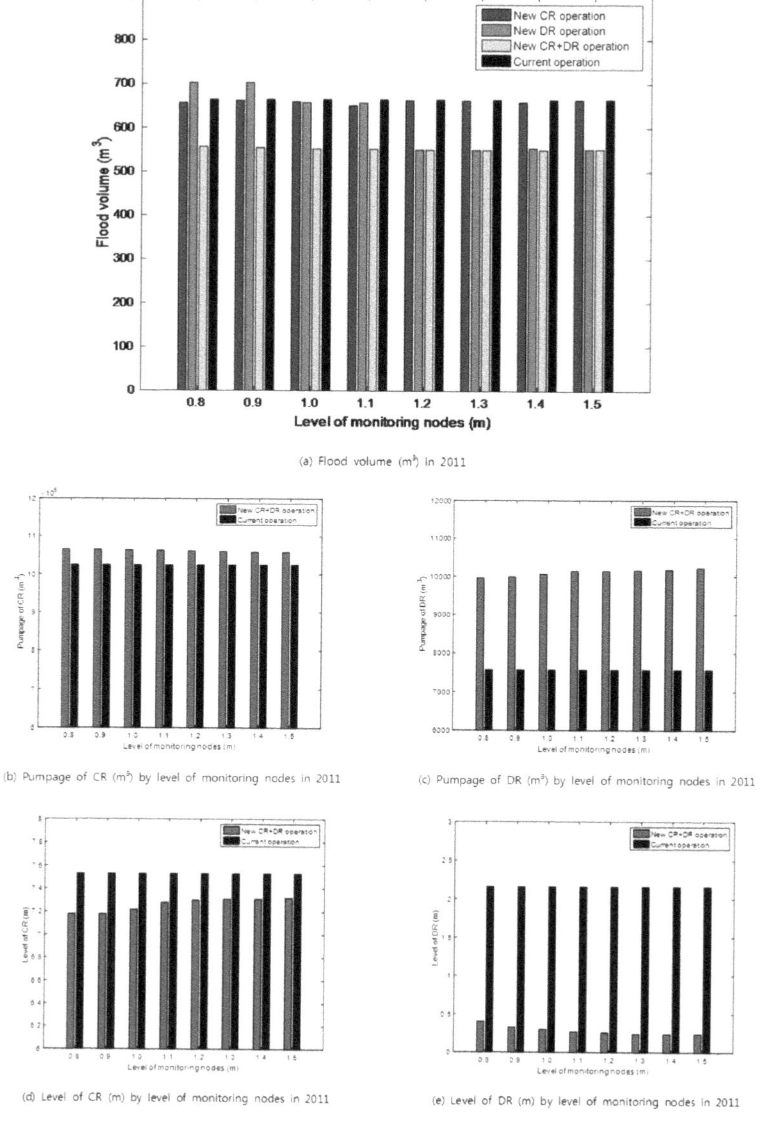

Figure 15. Results of cooperative operation in 2011.

The pumpage of CR and DR and their level under the cooperative operation were compared to those under the current operation. Figure 14b,c show the pumpage of CR and DR, respectively, with respect to the level of monitoring nodes under two operations in 2010. Figure 14d,e show the level of CR and DR in 2010, respectively. Figure 15b,c show the pumpage of CR and DR in 2011, respectively, and Figure 15d,e show the level of CR and DR in 2011, respectively. Regardless of the year of rainfall event, overall pumpage of CR and DR is higher than that of the current operation (Figures 14b,c and 15b,c) and the level of CR and DR is lower than that of the current operation (Figures 14d,e and 15d,e). The pumpage and level of CR and DR in both events show similar patterns (as seen in Figures 12 and 13, although attenuated increase and decrease were observed for the pumpage and level of DR, respectively (Figure 13c–f)). The notable point is that the CR/DR cooperative operation results in high pumpage and low level of CR and DR at all levels of monitoring nodes, compared to the current operation. As we confirmed from the previous two cases (the new CR operation and the new DR operation), it is originated from rapid drainage of CR and detention effectiveness of DR.

3.6. Evaluation of Urban Drainage System Resilience

Results of resilience evaluations for each operation method are shown in Table 5. We assess resilience at a monitoring node level of 1.1 m for each operation using the 2010 and 2011 rainfall event. It is found that the resilience of the urban drainage system increases in each new operation compared to the current. The highest resilience is obtained under cooperative operation (CR + DR in Table 5).

Table 5. Urban drainage system resilience for real rainfall data (2010, 2011) by each operation method.

Event	Current Operation	CR Operation	DR Operation	CR + DR Operation	Increase in System Resilience
2010	0.980042 (98.0042%)	0.985141 (98.5141%)	0.985166 (98.5166%)	0.986758 (98.6758%)	0.006716 (0.6716%)
2011	0.999236 (99.9236%)	0.999241 (99.9241%)	0.999380 (99.9380%)	0.999444 (99.9444%)	0.000208 (0.0208%)

The results of system resilience vary with different monitoring node level, as shown as Figure 16. In the results of 2010, CR operation shows somewhat less resilience than the current operation, though the DR operation shows better resilience than the current operation. When the monitoring node level is 1.1 m, all operations, including CR operation, DR operation, and CR/DR cooperative operation, show better resilience than the current operation.

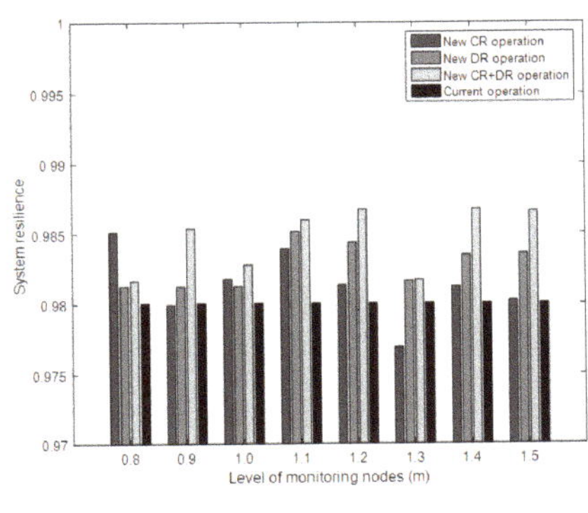

(a) System resilience in 2010

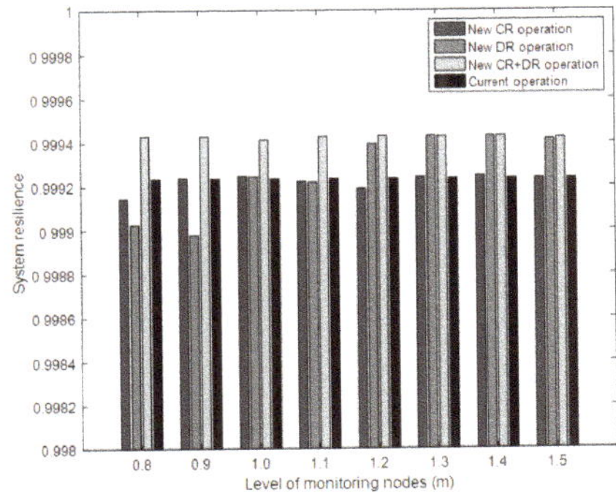

(b) System resilience in 2011

Figure 16. Results of system resilience.

The irregular patterns related to the monitoring node levels for both 2010 and 2011 are due to the discharge conditions in the DR. Discharge is conducted when the DR level reaches the limit level. The current DR operation and the DR operation in this study include this condition. Furthermore, there are several conduits with a reverse slope in the target basin. Generally, almost all sewer conduits have a slope because flow is induced by gravity; however, the design is a kind of inverted siphon due to the presence of a variety of obstacles, including Internet cables. This can result in irregular flow in the sewer network and therefore irregular patterns of system resilience.

In the results of 2011, individual operations, including CR operation and DR operation, show better results than the current operation in some parts and worse results in other parts, although CR/DR cooperative operation always shows better results than the current operation. The inferior resilience of individual CR and DR operation compared to historical operation is related to the current operation of DR and the available storage capacity in DR during the peak rainfall period. Current DR operation in CR operation starts to discharge based on the DR level, without considering the monitoring node levels. If current DR operation in individual CR operation starts to discharge pumps when monitoring node levels are high, the flood risk will be higher than for both current and cooperative operation, and system resilience will be lower than both current and cooperative operation.

Additionally, individual DR operation is inferior to the current operation in 2011 when the monitoring node levels are 0.8 m and 0.9 m. In DR operation, the DR reserves water if the monitoring node levels are high, though the DR operates the drainage pumps in the current operation when the DR level reaches the limit level. DR operation can show higher flood volumes than the current operation if the peak rainfall period occurs when the monitoring node levels are high, and the DR reserves water in DR operation. The system resilience of DR operation can be lower than both current and cooperative operation because of the available DR storage capacity when during the peak rainfall period. Furthermore, the individual operation of CR and DR over 30 and 20 years, respectively, is insufficient because the maximum return periods for 2010 and 2011 rainfall events are as long as 100 years. Therefore, it is difficult to overcome the drawbacks related to shortages in design size in individual CR and DR operation. For the 2010 rainfall event, the minimum increase for CR/DR cooperative operation compared with the current operation is 0.001612, and the maximum increase is 0.006716. For the 2011 rainfall event, the minimum increase is 0.000177 and the maximum increase 0.000194.

Figure 17 shows variations in rainfall, runoff, flood volume, and resilience over time for the 2010 event. Rainfall and runoff volume has two peaks, at 450 min and 650 min, respectively. Peak runoff volume in the CR, DR, and CR + DR operations is very similar, but is slightly higher under the current operation. Patterns of flood volume over time are similar to those of the resilience index. Considering the differences between the current and new operation in terms of rainfall and runoff relationships, the flood volumes associated with the first rainfall peak are not substantially different, but there is a significant difference associated with the second rainfall peak because the flood volume and resilience index differ greatly between the current operation and the CR/DR cooperative operation. The flood volume at the 650 min peak is 611.4 m^3 in the current operation but 100 m^3 in the CR/DR cooperative operation, which is the greatest difference for the whole period. The resilience index in the current operation is 0, but the resilience index in the CD/DR cooperative operation is 0.65. One of the reasons for the good results of the cooperative operation is due to draining the CR prior to subsequent storms, which reduces the CR level and mitigates the backwater effects of the CR.

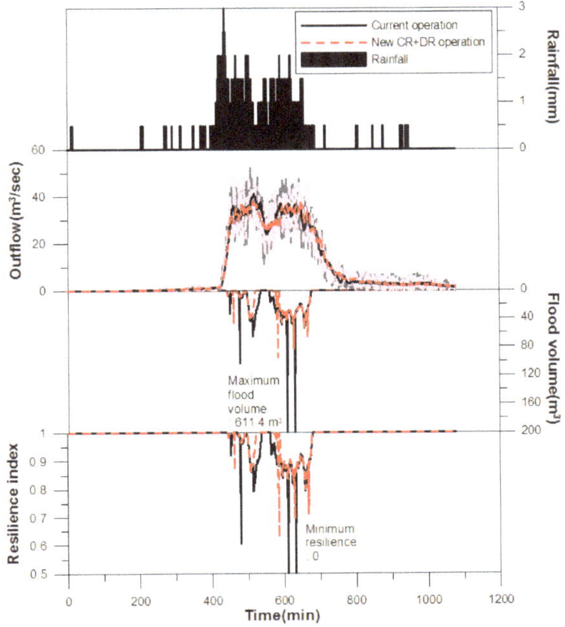

Figure 17. Rainfall, runoff, flood volume and resilience in the target watershed (2010).

4. Conclusions

The first result is the difference between the current and new operation of the CR. The current operation is controlled by CR level alone, while the new operation is more flexible and controlled by sewer conduit monitoring node levels. If the levels are high at the first and maximum flooding nodes, drainage pumps in the pump station are operated early. This process was applied to the rainfall events of 21 September 2010 and 27 July 2011. The best result was achieved when early operation occurred at a monitoring node level of 0.8 m. CR operation can improve the drainage effectiveness of CR in most cases.

The second notable result is the difference between the current and new DR operation. The current operation is controlled by DR level. The new DR operation is flexibly controlled through the sewer conduit monitoring node levels. When the level is low, discharge from the DR ensures sufficient space for additional inflow. When the level at monitoring nodes is high, drainage pumps in the DR are stopped. However, when the DR level is the same as the current operation, discharge by drainage pumps starts again. When monitoring node levels are around 1.1 to 1.5 m, the new DR operation shows better results than the current operation. DR operation can increase the detention effectiveness of DR in most cases. In some cases, DR operation and CR operation show worse results than the current operation, respectively. To improve flood mitigation effectiveness, cooperative operation between CR and DR is simulated and checked.

The third notable result is the difference between the current operation and the cooperative operation of urban drainage facilities (CR + DR operation). The current operation is performed depending on CR or DR level, but the condition of sewer conduits is not considered. The newly proposed cooperative operation determines storage or discharge at each urban drainage facility, depending on sewer conduit levels (determined at the monitoring nodes). When monitoring node levels are high, storage in the CR prevents flooding associated with backwater effects, and discharge from the DR prevents flooding by obtaining additional capacity. When monitoring node levels are low, the CR operates normally and the DR ensures additional space by discharge, depending on monitoring node levels. The results of cooperative operation for urban drainage facilities are better than those of the current operation. For the 2010 and 2011 events, the best results were observed when the monitoring node level was 1.4 m. Cooperative operation between CR and DR is superior to the current operation in all cases because cooperative operation has the advantages of both CR and DR operation.

Finally, we analyzed resilience as a new index for evaluating urban drainage systems. We calculated the resilience of the entire system using drainage volume and flood volume based on unit rainfall, time of concentration to target basin, and unit basin area. We assessed the resilience results for two significant rainfall events in 2010 and 2011, and found that cooperative operation of urban drainage facilities showed superior performance compared to the current operation.

All urban drainage systems are operated in drainage areas, but this study has shown that it is necessary to control CR drainage pump operation by taking into account urban stream levels. Moreover, it is necessary to conduct further studies on DR inlets, in order to allow the effective control of inflow volume.

Acknowledgments: This research was supported by a grant (13AWMP-B066744-01) from the Advanced Water Management Research Program funded by the Ministry of Land, Infrastructure, and Transport of the Korean government.

Author Contributions: Eui Hoon Lee and Yong Sik Lee carried out the survey of previous studies. Eui Hoon Lee and Donghwi Jung wrote the draft of the manuscript and revised the draft to the final manuscript. Eui Hoon Lee simulated CR operation, DR operation and CR/DR operation. Eui Hoon Lee, Joong Hoon Kim, Donghwi Jung and Jin Gul Joo conceived the original idea of the proposed method.

References

1. Tingsanchali, T. Urban flood disaster management. *Procedia Eng.* **2012**, *32*, 25–37. [CrossRef]
2. Andjelkovic, I. *Non-Structural Measures in Urban Flood Management*; IHP-V Technical Documents in Hydrology No. 50; UNESCO: Paris, France, 2001.
3. Chung, J.H.; Han, K.Y.; Kim, K.S. Optimization of detention facilities by using multi-objective genetic algorithms. *J. Korea Water Resour. Assoc.* **2008**, *41*, 1211–1218. [CrossRef]
4. Al-Hamati, A.A.N.; Ghazali, A.H.; Mohammed, T.A. Determination of storage volume required in a sub-surface stormwater detention/retention system. *J. Hydro-Environ. Res.* **2010**, *4*, 47–53. [CrossRef]
5. Andrés-Doménech, I.; Montanari, A.; Marco, J.B. Stochastic rainfall analysis for storm tank performance evaluation. *Hydrol. Earth Syst. Sci.* **2010**, *14*, 1221–1232.
6. Andrés-Doménech, I.; Montanari, A.; Marco, J.B. Efficiency of storm detention tanks for urban drainage systems under climate variability. *J. Water Resour. Plan. Manag.* **2012**, *138*, 36–46. [CrossRef]
7. Chill, J.; Mays, L.W. Determination of the optimal location for developments to minimize detention requirements. *Water Resour. Manag.* **2013**, *27*, 5089–5100.
8. Tao, T.; Wang, J.; Xin, K.; Li, S. Multi-objective optimal layout of distributed storm-water detention. *Int. J. Environ. Sci. Technol.* **2014**, *11*, 1473–1480. [CrossRef]
9. Tamoto, N.; Endo, J.; Yoshimoto, K.; Yoshida, T.; Sakakibara, T. Forecast-based operation method in minimizing flood damage in urban area. In Proceedings of the 11th International Conference on Urban Drainage, Edinburgh, Scotland, UK, 31 August–5 September 2008.
10. Graber, S.D. Generalized method for storm-water pumping station design. *J. Hydrol. Eng.* **2010**, *15*, 901–908. [CrossRef]
11. Hsu, N.-S.; Huang, C.-L.; Wei, C.-C. Intelligent real-time operation of a pumping station for an urban drainage system. *J. Hydrol.* **2013**, *489*, 85–97. [CrossRef]
12. Mugume, S.; Gomez, D.E.; Butler, D. Quantifying the Resilience of Urban Drainage Systems Using a Hydraulic Performance Assessment Approach. In Proceedings of the 13th International Conference on Urban Drainage, Sarawak, Malaysia, 7–11 September 2014; International Association for Hydro-Environment (IAHR): Perugia, Italy; International Water Association (IWA): Lisbon, Portugal, 2014.

13. Gaudio, R.; Penna, N.; Viteritti, V. A Combined methodology for the hydraulic rehabilitation of urban drainage networks. *Urban Water J.* **2015**, *13*, 644–656. [CrossRef]

14. Huff, F.A. Time distribution of rainfall in heavy storms. *Water Resour. Res.* **1967**, *3*, 1007–1019. [CrossRef]

15. Korea Precipitation Frequency Data Server. Available online: www.k-idf.re.kr (accessed on 3 March 2016).

16. Yoon, Y.N.; Jung, J.H.; Ryu, J.H. Introduction of design flood estimation. *J. Korea Water Resour. Assoc.* **2013**, *46*, 55–68.

17. United States Environmental Protection Agency (USEPA). *Storm Water Management Model User's Manual Version 5.0*; EPA: Washington, DC, USA, 2010.

18. Lee, E.H.; Lee, Y.S.; Joo, J.G.; Kim, J.H. Development of operation in urban offline detention reservoirs. *J. Korean Soc. Hazard Mitig.* **2016**, *16*, 227–236. [CrossRef]

19. House, M.A.; Ellis, J.B.; Herricks, E.E.; Hvitved-Jacobsen, T.; Seager, J.; Lijklema, L.; Aalderink, H.; Clifforde, I.T. Urban drainage—Impacts on receiving water quality. *Water Sci. Technol.* **1993**, *27*, 117–158.

20. Cembrano, G.; Quevedo, J.; Salamero, M.; Puig, V.; Figueras, J.; Martı, J. Optimal control of urban drainage systems. A case study. *Control Eng. Pract.* **2004**, *12*, 1–9. [CrossRef]

21. Mitchell, G. Mapping hazard from urban non-point pollution: A screening model to support sustainable urban drainage planning. *J. Environ. Manag.* **2005**, *74*, 1–9. [CrossRef] [PubMed]

22. Ministry of Public Safety and Security. *The Disaster Year Book*; Ministry of Public Safety and Security: Seoul, Korea, 2008.

23. Ministry of Public Safety and Security. *The Disaster Year Book*; Ministry of Public Safety and Security: Seoul, Korea, 2009.

24. Ministry of Public Safety and Security. *The Disaster Year Book*; Ministry of Public Safety and Security: Seoul, Korea, 2010.

25. Ministry of Public Safety and Security. *The Disaster Year Book*; Ministry of Public Safety and Security: Seoul, Korea, 2011.

26. Ministry of Public Safety and Security. *The Disaster Year Book*; Ministry of Public Safety and Security: Seoul, Korea, 2012.

27. Ministry of Public Safety and Security. *The Disaster Year Book*; Ministry of Public Safety and Security: Seoul, Korea, 2013.

28. Seoul Metropolitan Government. *Report on Design and Expansion of Daerim 3 Pump Station*; Seoul Metropolitan Government: Seoul, Korea, 2010.

29. Yeongdeungpo-Gu. *Report on Design of Daerim Detention Reservoir*; Yeongdeungpo-Gu: Seoul, Korea, 2007.

Article

Creation of an SWMM Toolkit for Its Application in Urban Drainage Networks Optimization

F. Javier Martínez-Solano [1,*], Pedro L. Iglesias-Rey [1], Juan G. Saldarriaga [2] and Daniel Vallejo [2]

[1] Departamento de Ingeniería Hidráulica y Medio Ambiente, Universidad Politécnica de Valencia, Valencia 46022, Spain; piglesia@upv.es

[2] Departamento de Ingeniería Civil y Ambiental, Universidad de los Andes, Bogotá Cra 1 Este # 19A-40, Colombia; jsaldarr@uniandes.edu.co (J.G.S.); d.vallejo49@uniandes.edu.co (D.V.)

* Correspondence: jmsolano@upv.es; Tel.: +34-96-387-7000

Academic Editor: Helena Margarida Ramos
Received: 4 May 2016; Accepted: 15 June 2016; Published: 18 June 2016

Abstract: The Storm Water Management Model (SWMM) is a dynamic simulation engine of flow in sewer systems developed by the USEPA. It has been successfully used for analyzing and designing both storm water and waste water systems. However, despite including some interfacing functions, these functions are insufficient for certain simulations. This paper describes some new functions that have been added to the existing ones to form a library of functions (Toolkit). The Toolkit presented here will allow the direct modification of network data during simulation without the need to access the input file. To support the use of this library, a testing protocol was performed in order to evaluate both calculation time and accuracy of results. Finally, a case study is presented. In this application, this library will be used for the design of a sewerage network by using a genetic algorithm based on successive iterations.

Keywords: SWMM Toolkit; sewer system; design; optimization

---

1. Introduction

The Storm Water Management Model (EPA SWMM) is a hydrologic and hydraulic model used to simulate flows in both storm water runoff and sanitary sewers. It was developed by the United States Environmental Protection Agency (USEPA), and it is a program that is able to solve the hydraulic equations of a network by using three different algorithms: steady flow, kinematic wave, and dynamic wave [1]. While the first is insensitive to the time interval, the last one requires very short time intervals (about less than one minute). This leads a high computational time.

Research regarding the optimal design of sewerage networks and drainage is becoming more common nowadays. Most of these optimization processes require numerous simulations, which is why the speed at which the calculations and the simulations are done becomes a fundamental aspect. Network design [2,3], real time control [4], and contamination source identification [5] are some examples of applications that require a hydraulic simulation of the sewerage network. Because such applications require an elevated number of simulations, the optimization process may become very slow. This is why previous studies have worked with constant discharges entering each network node, and steady state models have been used [2,6,7]. Krebs *et al.* [8] performed a sensitivity analysis to find the balance between the simulation time step through the dynamic wave model and the computation time to find a calibration application of a rain water evacuation network.

Although it might be possible to use an external program to connect with the EPA SWMM engine, it is still necessary to handle files either to modify the data or to access the results of the simulation. In order to do this, the project file must be modified or the results binary file read, respectively. Usually, file handling requires longer times than accessing data directly from the application. Hence, EPA SWMM requires a complete Toolkit that allows the manipulation of data from the application itself.

Next, a library of functions that increases the flexibility of EPA SWMM in terms of exchanging information during run-time is presented. This library will be hereinafter referred to as the Toolkit. This Toolkit notably extends those tools already present, thereby going from 9 to 22 functions. The new functions help with simulation execution, result reading, and network characteristic modification. All these steps may be done without the need for handling archives, which means a considerable reduction in calculation times.

Furthermore, a testing protocol has been assembled in order facilitate the use and implementation of this library. This protocol consists of two phases. First, savings in calculation times due to the use of the library are measured. As a result of this first phase, only new functions that imply time saving were considered. In the second phase, the results obtained by using the library are compared to those obtained directly with EPA SWMM. This comparison was made for every new function and only those presenting an exact coincidence were validated.

A typical application where computational efforts are high is the design of hydraulic networks by means of heuristic optimization algorithms. Therefore, the work is completed by using this library for the development of an optimized design model of a sewerage network by using a genetic algorithm.

2. Development of the SWMM Toolkit

In the field of water distribution networks, there are many examples of tools developed in order to facilitate the incorporation of calculating engines in optimization algorithms. EPANET might be considered as the main example of this. EPANET is a program that tracks the flow of water in each pipe within a drinking water system and was also developed by the USEPA. Within its published version, EPANET includes a programming library (EPANET Programmer's Toolkit) that allows the program to be used as typical programming commands [9]. Several authors have adapted this library in order to incorporate functions that originally were not thought about being included. This is the case of Kandiah *et al.* [10], who partially modified the EPANET Programmer's Toolkit in order to include functions for control adaptation in EPANET. Another example is the CWSNet Library [11]. CWSNet expands this library so that new elements such as variable speed pumps or specific valves could be incorporated, as well as new algorithms. Savić *et al.* [12] continued to develop CWSNet until the creation of the GANet application. Other authors, instead, chose to internally modify EPANET in order to adjust it to their particular needs. Marchi and Simpson [13], for example, modified the energetic calculation in order to correct the energy calculation for including the efficiency variation due to affinity laws when using pumps with variable speed drives. In other cases, EPANET has been continued by other institutions that have included programming libraries in their commercial programs. Bearing all this in mind, when it comes to potable water supply, there exists a library that allows the speeding up of calculation processes that require constant change or update in the network in which the simulation is being done.

However, that is not the case for sewage systems. The internal structure of EPA SWMM is derived from the structure presented by EPANET. However, unlike EPANET, EPA SWMM has not offered a set of interfacing functions as EPANET does. It only includes some basic functions allowing running of a single event simulation.

EPA SWMM provides several interface tools [14] that allow using the model by an external application. This works as long as the network characteristics have been previously defined. The performance of the EPA SWMM program is sequential. First, an input module reads the files needed for the simulation. The network data are contained in an input or project file (INP file). These data might be complemented with auxiliary files for hot start conditions, weather data, etc. Then, if the analysis contains rain water data, a hydrologic calculation must be carried out in order to transform the rain in a runoff hydrograph. A hydraulic simulation follows, and then, if considered necessary, a follow-up on contaminant substances may be done in order to produce a quality model. All results are stored in a binary results file (OUT file). Finally, a report file (RPF file) containing some results statistics is written. The whole process is summarized on the right hand side of Figure 1.

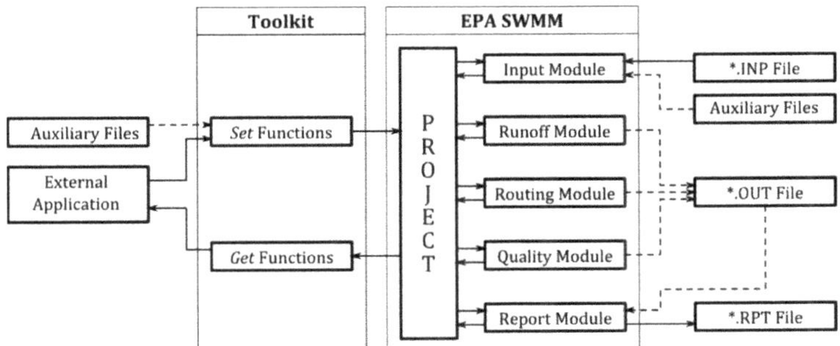

Figure 1. Block diagram of EPA SWMM and the Toolkit.

The performance of the interfacing functions provided with EPA SWMM follows the same structure, that is, as soon as the project is open, the input file must be read. Afterwards, the calculation is executed, thus being able to perform a step by step analysis. Finally, the results are written in a binary file that must eventually be read. This tool has 9 functions that can be used in order to manage the project. These functions are listed in Table 1 as group 1. However, it is not possible to obtain partial results from the said simulation, or to perform data modification until the process is complete.

Table 1. Functions included in the Toolkit.

Function Name	Description
Group 1. Project management functions (already available in EPA SWMM)	
swmm_getVersion	Retrieves the version number of SWMM engine
swmm_run	Runs a complete simulation with SWMM
swmm_open	Opens the project for a new execution
swmm_start	Initializes SWMM engine
swmm_step	Executes next time step
swmm_end	Ends the SWMM engine when the simulation has ended
swmm_getMassBalErr	Retrieves the continuity errors when the simulation has ended
swmm_report	Writes results in text format in the report file
swmm_close	Closes the project when the simulation has ended
Group 2. Get functions	
swmm_getCount	Retrieves the number of elements of the specified type
swmm_getNodeIndex	Gets the index of a node from its identifier
swmm_getNodeId	Gets the identifier of a node from its index
swmm_getLinkIndex	Gets the index of a link from its identifier
swmm_getLinkId	Gets the identifier of a link from its index
swmm_getNodeType	Retrieves the type of a node from its index
swmm_getLinkType	Retrieves the type of a link from its index
swmm_getNodeValue	Retrieves the value of a specified parameter of a node from its index
swmm_getLinkValue	Retrieves the value of a specified parameter of a link from its index
swmm_getLinkNodes	–
Group 3. Set functions	
swmm_setNodeValue	Sets the value of a specified parameter of a node from its index
swmm_setLinkValue	Sets the value of a specified parameter of a link from its index
swmm_setLinkGeom	Sets the geometry parameters of a link

This performance mode was conceived in order to work with a different user interface, developed "ad-hoc" and well-integrated in other programs. Its most evident application would be the integration of the EPA SWMM engine within a geographic information system. However, the process is slowed down because of file access. On the one hand, it is required that the input file is written for every execution. On the other, in order to process results, EPA SWMM must write the results file and the external application should read it. These two operations require a great deal of programming efforts, and they may consume a significant amount of time.

An alternative to this performance mode is the development of a new set of communication functions. These new functions, integrated as a Dynamic Link Library (DLL), could address a wide array of problems present in applications and programming languages. The final objective is that, eventually, both EPA SWMM and any other application based on the EPA SWMM engine use the same library. An application that is external to EPA SWMM can therefore replace file use for information exchange, thereby ensuring time savings.

The Toolkit replaces the file reading and writing operations with two new groups of functions, as shown on the left hand side of Figure 1. The first group–referred to as *Get Functions*–is composed of functions associated with retrieving information either from project data or simulation results. The second group (*Set Functions*) includes all the data modification functions. All the functions within this second group are simultaneously the most relevant and the most sensitive, since these functions will allow successive simulations by modifying data without having to access the files. For instance, by using the function *swmm_setLinkValue,* it is possible to change the slope or the Manning roughness coefficient of a pipe during run time. Table 1 summarizes the functions included in the Toolkit.

The optimal sizing of a storm water network might be taken as a first example. This issue is typically addressed by testing different combinations of diameters and slopes, which are ranked by using an objective function. The objective function includes the cost function for the new conduits and penalization terms when the constraints of the problem are not met. However, the whole problem has seldom been studied, even more so when considering the transient flow. Most authors have usually tackled this problem with oversimplified hypotheses. For example, in order to avoid having to perform a runoff analysis for every case, some authors use a constant inflow at network nodes instead of performing the hydrological model. Another example consists of using a steady flow model based on uniform flow. In this way, no duration is defined so that the calculation process is faster. However, it leads to the appearance of inaccuracies that increase with the size of the network and the discharge increase. Therefore, the steady flow model based on uniform flow is evidently insufficient when addressing situations for which surcharge or flooding risks exist.

If the temporal distribution of the rain water is included, then it becomes unnecessary to execute the hydrologic calculation every time. First, given the modular character of EPA SWMM, it is possible to execute a preliminary hydrologic calculation to obtain runoff hydrographs that enter in each element. With these inflows to the nodes, the hydraulic calculation can be performed as many times as it is required. The repeated modification of various network parameters will be necessary, but by using the Set functions, this no longer implies the constant rewriting of the input file. Because this operation requires a significant amount of computational time, the parameters will be modified directly in memory without accessing the writing of the files.

Finally, the binary results file EPA SWMM produces is not easy to manipulate. This file presents basic statistical summaries of the results (these summaries include items such as maximum discharges, maximum levels, volume balances, etc.). If more detailed information on a specific object must be obtained, the [REPORT] section on the project file must be modified to include this information in the result file, and then filter this latter file in order to look for the specific information needed. Since this process might be complex, it has become preferable to read the results obtained after each itineration. It is no longer necessary to read the result file, and the calculation process may be done faster.

Figure 2 shows the flow chart of an optimization process that is based on the evaluation of an objective function. The left flow chart represents the process that is based on the EPA SWMM native interfacing tool. On the other hand, the right flow chart, instead, represents the same process but implemented using the Toolkit. It can be observed that several operations can be avoided. As the number of iterations increases so will the amount of time that is saved. Bear in mind that if evolutionary algorithms are preferred, the number of evaluations of the object function might be highly elevated.

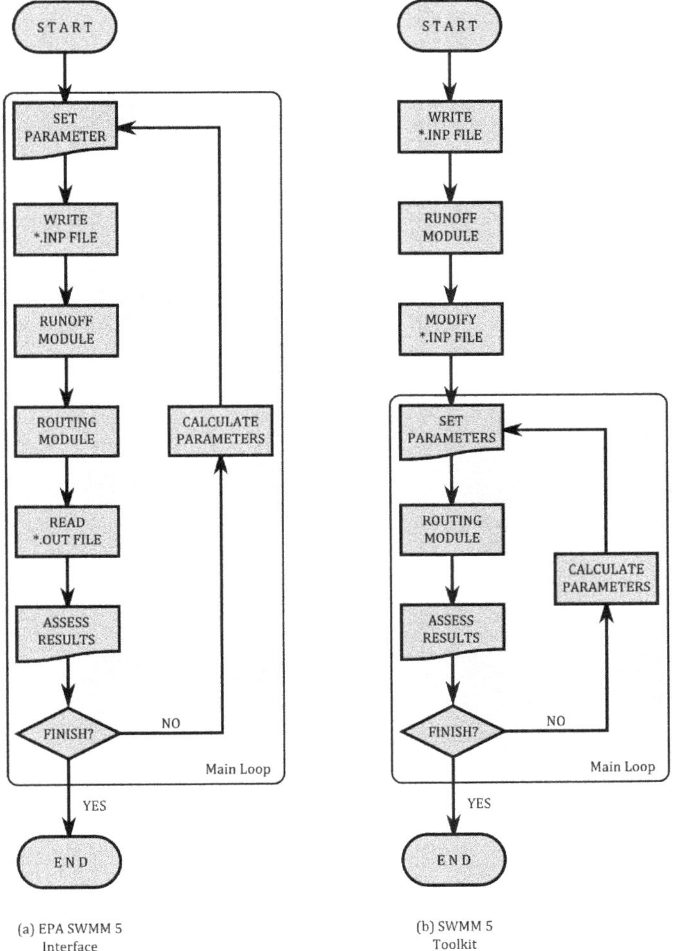

Figure 2. Comparison scheme of optimization algorithm performance using EPA SWMM 5 (**a**) and the Toolkit (**b**).

3. Validation Protocol

3.1. Velocity Tests

Given that for simulations over extended periods, the calculation times can be relevant; this aspect has always constituted a preoccupation for the program's developers. The program was originally developed in FORTRAN [15], with its successive adaptations. This way, when the original FORTRAN routines were converted to C language, the developers proceeded to make comparisons in terms of execution times [16] as well as result precision [17].

A series of tests will be done in order to support the previously mentioned statements. On the one hand, velocity tests have been carried out in order to measure calculation times for both options

(EPA SWMM 5 Interface and SWMM Toolkit). A total of 9 cases in which network size, rain duration and rain-runoff simulation are modified have been performed for each option. Just as it was done during the SWMM 4-SWMM 5 transition [16], and with the hopes of ensuring equality of conditions, every case was processed with the same computer.

In order to see the effect that is obtained as a result of modifications in network size, three specific networks have been chosen and they are shown in Figure 3. The first network (Net1) corresponds to the network that is available in the program's manual [14]. It is rather simple, as it consists of 5 nodes and 4 pipelines. The second network (Net2) is obtained by simplifying the sewer network of a district of Bogotá (Colombia) and consists of 82 nodes and 83 links. Finally, the third case corresponds to the Xirivella network (Spain). This network possesses all the elements that are typically found in a network, such as pumping stations, sewer overflows, waste water treatment plant, etc. It has 916 nodes and 931 links.

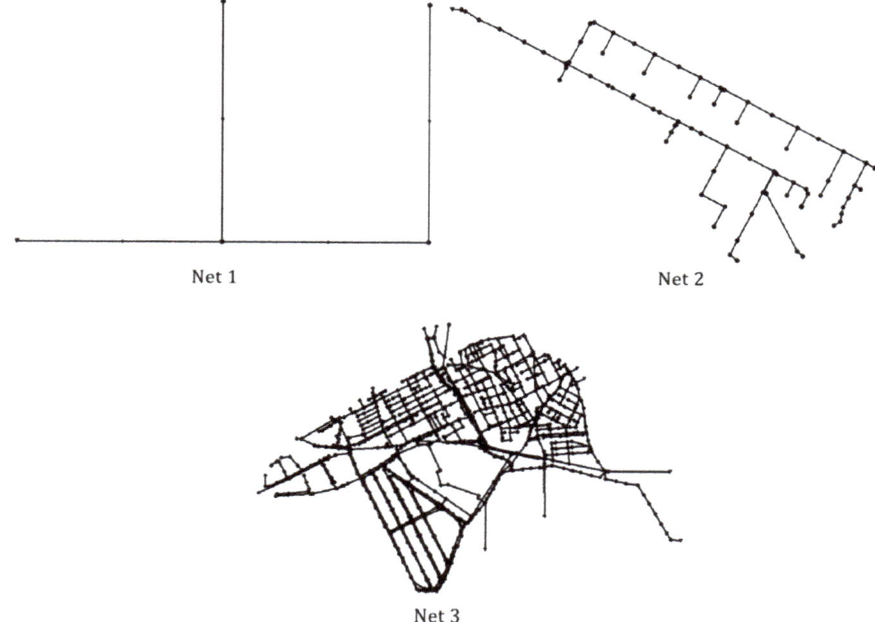

Figure 3. Networks used for calculation time evaluation.

On the other hand, three different rain situations have been tested. In the first situation, once again the expected rain that is provided in the SWMM user's manual sample is used. It consists of a 6-h long rainfall period, with intensity values that change every hour, and a 12-h long simulation. The second rain period corresponds to the data collected in a rain gage located in Almassora (Spain). All the records corresponding to October 2000 have been processed, with a 31-day long simulation and a rain time interval of one hour. Finally, for each of the three networks runoff, hydrographs obtained with the rain in case 1 have been collected. These hydrographs have been assigned as direct inflows in each of the nodes, and the hydraulic simulation has been done exclusively, thus avoiding hydrologic calculations.

In order to compare the differences in simulation times, the 9 cases (three networks × three rain events) have all been simulated using both the EPA SWMM native interface and the Toolkit. Execution times have been annotated for every case. For the comparison, a small application that allowed the measuring of the time employed in the calculation with millisecond precision was used.

Ten simulations were carried out for each case in which some parameters such as slopes or diameters were changed. Total spent time was also measured for every simulation. The results of this test are shown in Table 2.

Table 2. Execution time comparison (in milliseconds).

Case		Case 1 Rain (6 h)	Case 2 Rain (October 2000)	Case 3 Direct Inflows
Net 1	EPA SWMM	610	20,791	470
	Toolkit	460	20,775	430
Net 2	EPA SWMM	6731	313,803	4441
	Toolkit	4460	300,704	4280
Net 3	EPA SWMM	598,329	5,438,193	144,220
	Toolkit	135,549	4,935,524	126,308

It is possible to draw certain conclusions from the previous results. On the one hand, the greater the size of the network, the more time is required to calculate it. These data are obvious; however, it can also be observed that, the greater the size of the network, a larger percentage of time is spent in archive manipulation. When the Toolkit is used, there is a 25% time saving for network 1, 33% for network 2, and 77% for network 3 if the rain lasts 6 hours. It can also be observed that memory data manipulation slows down the process, that is, as the time that is destined for calculation increases, the relative importance of the time employed in archive data decreases. This is also an expected conclusion; however, when analyzing the results obtained in case 2, what is stated in the previous paragraph is ratified, since the Toolkit reduces the calculation time for a month-long rain by 0.2% for network 1, 4.2% for network 2 and 9.2% for network 3. Finally, the previous execution of the hydrologic calculation to write the results as direct inflow in a new archive supposes time savings of 10% in every case. Put differently, once the new archive without hydrologic data but with an entrance hydrograph for each node is obtained, the Toolkit may reduce calculation time by 10% when comparing this with the same operation using the EPA SWMM interface.

Short but intense rain periods are generally used in the optimization process, very similar to those seen in the case that was previously described. In these situations, one can observe how the Toolkit may significantly contribute to savings in calculation times. The larger the network, the more time saved. As has been observed, the saving may exceed 75% for very large networks.

3.2. Function Validation

Along its diverse history, SWMM has experienced numerous changes. The idea of the development team has always been that of maintaining reliability of the calculation, and this has been done through the implementation of a rigorous quality assurance testing program [18]. In this case, the validation process has followed a similar philosophy to that used in the aforementioned quality assurance program, thereby guaranteeing the perfect coincidence of the results obtained with the Toolkit. This way, and just like it was done during the SWMM 4-SWMM 5 transition, three tests have been performed [19]. These tests are based on two ideas.

First, in order to ensure the compatibility of the Toolkit with EPA SWMM, a series of tests have been carried out to compare the results of both tools. The tests were conducted using the example network suggested in the SWMM 5 User's Manual [14]. This network is presented in Figure 4, in which control sections S1, S2 and S3 have been established.

Second, a much simpler network was used to study specific control structures, as is the case of weirs, orifices and pumps. The reason for using a specific network for control structures derives from the way EPA SWMM calculates these types of devices. This way, different situations may be simulated (normal behavior, orifices behaving as a weir or weirs behaving as an orifice for surcharge conditions). Figure 5 shows a sketch of this network, which consists of a storage unit with a constant direct inflow so that a steady flow regime may be reached. Likewise, the outfall is simulated as a free discharge.

Conduits are long enough to allow them to reach the uniform flow in their middle points. Direct numerical calculations can thus be done, and they can be contrasted with those obtained with both EPA SWMM and Toolkit.

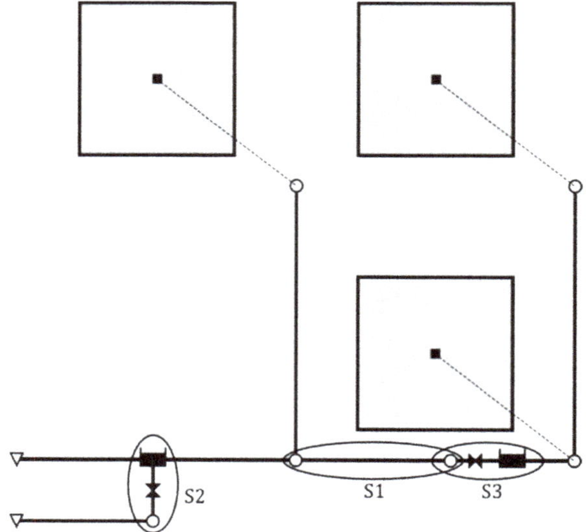

Figure 4. Network used for the study of the Toolkit's functions.

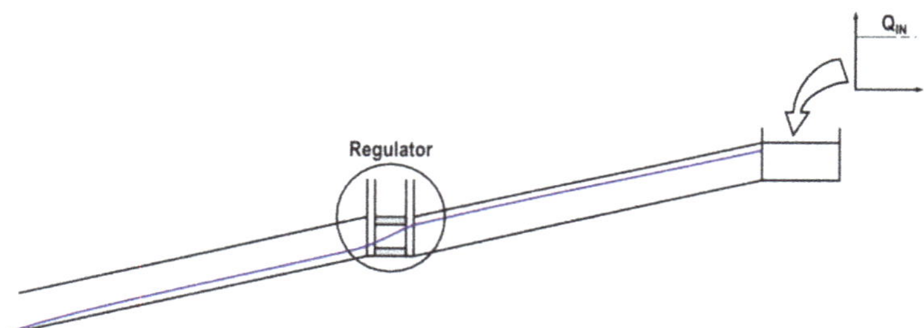

Figure 5. Network used for the exhaustive study of weirs and orifices.

The tests could be grouped in two phases, depending on the functions that each of these tested. In the first phase, only the Get functions were tested, and only section S1 in Figure 4 was used. The tests in this phase consisted of checking that the results obtained with EPA SWMM and the Toolkit are exactly coincidental without the need to manipulate information. The employed network was the one shown in Figure 4 with a one-hour long rainfall and duration of the simulation of 12 h. The results corresponding to both nodes and pipes were consulted. During the first test, it could be observed that the Toolkit collected all the calculation instants. EPA SWMM, instead, only collected those results corresponding to the report time step. This way, a 10-s hydraulic calculation and report time step was employed in order to ensure equality of conditions. Likewise, in order to ensure the precision of the results, EPA SWMM preferences were modified, so that the reported results were always presented with a numerical precision of four decimals. Regarding the nodes, all the variables that could be read with the Get functions were processed, except for those related to quality water parameters. The same procedure was applied for pipes.

In the second phase, two sets of tests were carried out involving Set functions. These functions allow the integration of EPA SWMM with any other external applications including more complex algorithms of iterative character, such as design, control, or calibration algorithms. Because of this, tests have been more elaborate for this group of functions. Next, a brief description of each one will be done. The first set was carried out in section S3 of Figure 4, and it included the basic elements such as storage units, junctions and conduits. The following parameters were modified:

- Storage Units: invert level, height (maximum depth) y and volume. Regarding the volume, it is defined by a function which relates depth and surface area through three parameters: A, B and C. The surface area of the storage unit varies with water depth following the function shown below:

$$S = A \cdot y^B + C \tag{1}$$

- Junctions: invert level and maximum depth.
- Conduits: diameter, invert levels at entrance and exit, and roughness coefficient. Furthermore, other tests were made in order to modify geometric shape and the corresponding geometric parameters. The latter tests were no so exhaustive as long as not all geometric shapes were tested.

This set's testing protocol will be briefly described. For each case within the set, three tests were carried out. The first included the analysis of the results obtained using EPA SWMM with a series of values of the analyzed parameter (10 different values ranging between a minimum and a maximum). Next, these same cases have been processed using the Toolkit and without parameter modification, that is, reading each of the 10 prepared archives. Finally, the obtained results have been analyzed by performing 10 calculations in one single process that used Set functions in order to modify parameters. As an example, Figure 6 compares the results obtained when modifying storage capacity. The data collected for each of the storage volume include maximum discharges at entrance and exit pipes as well as total flood values for the initial and the final nodes. The results obtained through the use of the library have been represented on the horizontal axis. In the vertical axis, instead, those results directly obtained using EPA SWMM are represented. As expected, the results are 100% coincidental.

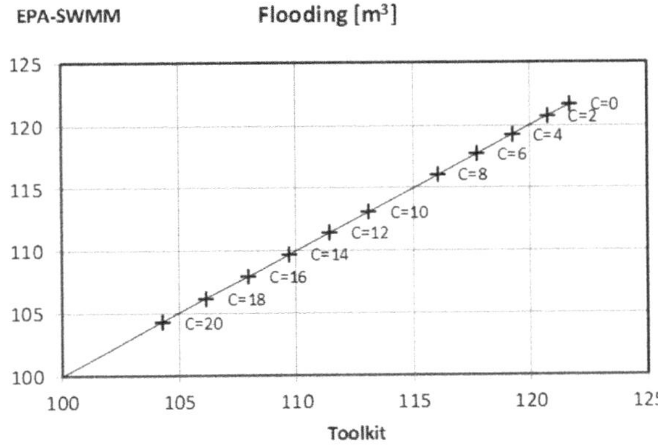

Figure 6. Flooding results comparison between EPA SWMM and a Toolkit-based application.

The second set was done for the exclusive study of control structures (orifices and weirs). Two networks were used on this occasion. Because it was important to use a section that would present discharge derivation, section S2 of Figure 4 was used. The characteristics of these elements were modified by using the developed functions. This way, the performance of both elements was proven to be the same regardless of whether EPA SWMM or the Toolkit was used. The same behavior was seen in both cases.

Considering that control structures may become very important for applications such as real time control, a more exhaustive analysis was performed. In the case of orifices, EPA SWMM discriminates the case of flooding conditions (behavior as orifice) from that of not-flooding conditions (behavior as weir). Therefore, all the plausible situations contemplated by EPA SWMM were processed as a function of depths upstream and downstream of control element. The transition between the two performance modes is produced from a specific depth denominated critical height. Depending on the case, EPA SWMM implements a different coefficient and exponent. The network shown in Figure 5 was used for this test. The detailed study that has been carried out helped to understand and perfectly reproduce the behavior shown by these control structures. As a conclusion, both EPA SWMM and the Toolkit presented the same results in all the cases that were analyzed.

4. Application

Usually, the design of a storm water network was solved by using the rational method. However, methods based on the concept of return period do not take into account economic issues. It might result in both overdesigned or underdesigned networks [20]. Cost-optimal design of hydraulic network is a complex problem for engineers. Meredith [6] applied dynamic programming to sewer studies including cost functions for pipes. Mays and Wenzel [21] also used dynamic programming to optimize the design of a gravity sewer network with known pipe flow direction. Recent studies try to solve the problem of designing sewage networks by heuristic algorithms [22,23]. Furthermore, costs associated with flooding are rarely considered. Defining flood damage functions are another challenge for storm water network design [24]. More recently, Guo *et al* [25] combined EPA SWMM with a cellular automata and genetic algorithm hybrid algorithm to design large sewer networks. In all previous studies, the inflows were considered constant. That is, the rainfall-runoff processes were not calculated. Furthermore, most of these studies considered a normal flow regime.

In this case study, the design of a simple sewage network is used as an example of application of the Toolkit. The dynamic wave approach is used for the hydraulic simulation. The network used for this application consists of 6 pipes, 6 manholes and an outfall. Node and pipe data are shown in Figure 7.

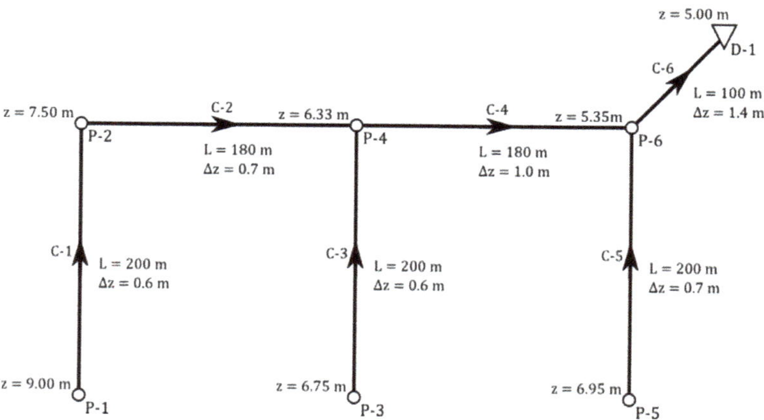

Figure 7. Sample network used as a case study.

Since the depth of the sewers should allow water flows by gravity, the design includes not only the dimensioning of the pipes, but also the depth at which these pipes must be installed. Hence, both the conduit diameters and their slopes must be calculated. It should be noted that diameters can be coded as discrete variables and they must be chosen among those listed in Table 3. On the other hand, slopes are derived from the elevation and depth of the conduit ends. Therefore, they are continuous variables that may assume any value. As it happens in every design problem, constraints have been used for maximum velocity, capacity of the conduit and minimum slope. Another restriction was used, based on a similar one proposed by Meredith [6]: the invert elevation of a pipe leaving a manhole cannot be higher than the lowest invert elevation of a pipe entering the manhole.

Table 3. Cost of conduits.

Diameter (mm)	Cost (€/m)
400	96.33 €
500	116.78 €
600	159.19 €
700	208.21 €
800	254.26 €
900	312.60 €
1000	319.36 €
1200	442.21 €
1400	575.64 €
1600	728.31 €
1800	858.91 €
2000	1,055.72 €

When calculating pipe installation costs, a formulation which involves both cost of the pipe according to Table 3 and the cost of installing such a pipe. In order to do so, a standardized trench was used (shown in Figure 8). It will depend on both the diameter (D) and the maximum depth the pipe can be installed (Δz). The trench model used is shown in Figure 8, as well as all the data that defines it. It is worth mentioning that the same type of trench has been used for all the pipes, regardless of the type of terrain in which pipes were located.

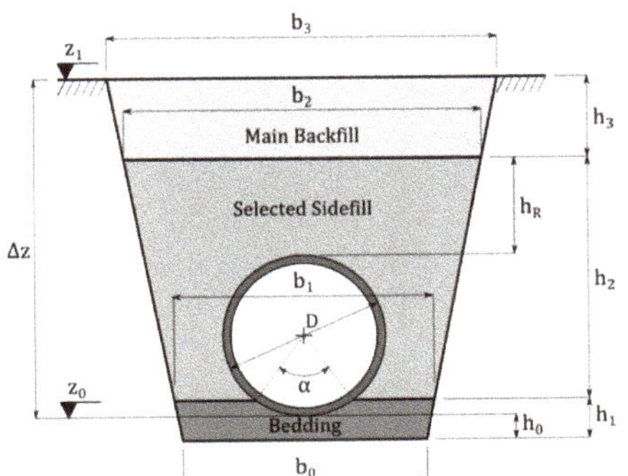

Figure 8. Trench model used.

This way, the data required to define the trench include:

- Pipeline depth, Δz. It will be defined by the design algorithm.
- Trench lateral slope, S_t. For this case study, this slope is considered constant and equal to 0.2 m/m.
- Base width (b_0). It is equal to pipe diameter plus 50 cm.
- Bedding thickness (h_0), also constant and equal to 15 cm.
- Selected sidefill height (h_R), measured from conduit crown level. Again, it is constant and equal to 30 cm.

- Pipeline charge angle α. It indicates pipeline portion in contact with bedding. In this case, it is equal to $90°$.
- Ground elevation, z_1
- Invert elevation, z_0

Likewise, z_0 and z_1 represent the invert and the ground elevations, respectively. The latter is calculated in the algorithm for all nodes with the exception of the outfall node D-1 for which $z_0 = 2.5$ m. Therefore, using these values and pipe diameters it is possible to calculate the rest of the geometrical parameters of the trench.

With the trench parameters, it is possible to calculate the volumes that correspond to each part of the trench construction. These calculations must then be associated to the costs of the different actuations that were performed on the terrain. A cost function associated with the trench is thus obtained, which is the sum of the terms shown in Table 4.

Table 4. Costs associated with the trench model.

Material/Operation	Cost ($€/m^3$)
Excavation costs C_1	14.21 €
Granular material costs C_2	13.97 €
Selected sidefill costs C_3	6.47 €
Non selected sidefill C_4	4.55 €
Transport costs C_5	3.19 €
Final costs (deposition) C_6	20.30 €

The geometrical calculation of the trench gives a cost function for pipe i that depends on the diameter (D_i), the depth (Δz_i) and the length (L_i) of the pipe:

$$C_i = f(D_i, \Delta z_i, L_i) \tag{2}$$

According to the object model of EPA SWMM, the elevation of both ends of each conduit will be computed and must be modified during the calculation process. Because of the different nature of the variables, two algorithms have been used for network design: a pseudo-genetic algorithm (PGA) and the particle swarm optimization (PSO) algorithm. PGA is a modified genetic algorithm that replaces the binary coding of each variable by an integer coding [26]. Therefore, PGA is meant to address problems of discrete character. PSO, instead, is an algorithm more suitable for continuous problems [27]. In both situations, modifications to the different generations have been performed with the help of the Toolkit functions.

In order to modify the slope of conduits, only the elevation of the downstream end of the pipe was modified, while the upstream end was fixed to the invert of the upstream node. The pipe slopes have been limited to be between 0% and 5%. The slope has been coded differently depending on the algorithm used. In the PSO method, due to its continuous nature, any value within this range has been allowed. In the PGA, instead, the range was divided into 100 intervals. This different coding is one of the reasons why PSO presents higher execution costs but better solutions than PGA.

Some restrictions were imposed. Maximum allowable velocity in pipes (v_{max}) was fixed at 4 m/s. As it is common in optimization problems with restrictions, if they are violated, penalizations will be applied to the cost functions. If the velocity in a pipe (v_i) is bigger than the maximum velocity, a binary variable $\delta_{1,i}$ will take the value of 1 and the penalty will be the exceedance times the penalty cost (λ_1). Finally, damage functions for flooding and surcharge conditions were assumed. As the target of the design was to avoid both situations, these functions were high enough to discard designs with flood or surcharge. Therefore, if the water level in node j (y_j) is higher than the level that provokes surcharging ($y_{j,max}$), another binary variable $\delta_{2,j}$ will take the value of 1 and a penalty cost (λ_2) will be added. Then, the fitness of a solution k will be given by an objective function F^k that accounts for the pipe costs and the penalties for constraint violations:

$$F^k = \sum_{i=1}^{NL} C_i^k + \lambda_1 \cdot \sum_{i=1}^{NL} \delta_{1,i} \cdot \left(v_i^k - v_{max}\right) + \lambda_2 \cdot \sum_{j=1}^{NM} \delta_{2,j} \cdot \left(y_j^k - y_{j,max}\right) \tag{3}$$

In this equation, C_i^k is the cost of pipe i in solution k according to Equation (2), NL is the number of links and NM is the number of manholes in the model.

In order to account for computational costs, nearly 7000 simulations were carried out for every algorithm. All the simulations were done with an initial population of 100 individuals, and the rest of the parameters were tuned for each method [28].

The optimal result obtained using the AG design model offers a solution that costs 290,441.68 €, while the solution obtained using PSO design model is 287,518.02 €. Both algorithms yielded the same solution for diameters, but since PSO is a continuous algorithm, its results are slightly better (1%) compared to those obtained by using APG. On the other hand, APG needs less than half the number of iterations than PSO. Furthermore, the rate of success of APG (as defined in [28]) is better. The summary of these results can be seen in Table 5.

Table 5. Summary of results for both algorithms.

Algorithm	PGA	PSO
Best solution	290,441.68 €	287,518.02 €
Average number of simulations	771	1670
Rate of success	37.1%	32.6%

5. Conclusions

The EPA SWMM program is a sewerage network hydraulic simulator that is widespread. However, its integration with other programs is quite limited [5]. Because of this fact, a function library (known as a Toolkit) that allows the use of EPA SWMM algorithm in more complex calculation programs has been presented.

The principal objective of the present study is the development, testing and application of a functions library that allows the execution of multiple simulations with the EPA SWMM. This library must allow data modification and results querying without reading or writing files.

The library is made up of three different groups of functions. First, the project management functions will be enclosed in the group of interfacing functions provided by the USEPA within the program. The second group consists of those functions that are associated with data and result gathering. These functions can be referred to as Get functions. Finally, a data modification functions (denominated as Set functions) were included in order to give the library a greater degree of flexibility.

This library has been tested in order to show an improvement in simulation times. With it, one can avoid the writing of the data archive for every simulation. Likewise, the functions that are included in this library allow the incorporation of the hydrologic calculation's results into the project's data, and the writing of the binary archive as well as the results report has been avoided. In exchange, the results are kept in the memory with the resulting savings in the calculation times. All this has led to saving in calculations times, which in some cases has exceeded 75%.

Finally, the library has been subjected to a result validation protocol that proves the obtained results are 100% similar to those obtained with the original program. The functions that are included in the library have been directly contrasted with the results obtained with the original program, thus obtaining a perfect adjustment for every case. Furthermore, because the library allows the reading of results directly from the calculating tool, a rounding error is avoided as result precision increases. This way, results obtained from the validation process confirm 100% compatibility between those results obtained using EPA SWMM and those obtained by the external programs using the Toolkit.

Once validated, the library has been applied to the design of a sewerage network in which the diameter calculation and the slope adjustment have been combined. This is done by using a cost function expression that takes into account not only pipe acquisition costs but also the costs associated with pipeline installation as well (excavation, trenching and backfilling). For the development of the cost function associated with each diameter, a parameterized-type trench has been defined, one that can be used in other applications. The principal conclusion that can be drawn from this study case is that it is possible to combine the Toolkit with an optimization program that modifies the network data based on the results obtained from the simulation without the need to deal with files, thereby allowing time savings during the simulation.

In this comparison, two different evolution algorithms have been used: one for solving discrete problems (PGA) and another for continuous problems (PSO). This application was meant to contrast the Toolkit utility in more complex problems that were based on repetitive successions of a single simulation. As an additional result, it has been proved that the PSO algorithm (conceived for continuous variable problems) produces better solutions compared to those produced by PGA. However, PGA gets the best solutions with a lower number of simulations, that is, in a smaller time and more frequently. This conclusion confirms what has already been observed with other authors in continuous optimization problems [29].

There is some field for future research regarding this work. The work was done in response to the problem of the design of a storm water network. Because of that, all the functions presented in this paper are related to hydraulic objects, that is, elements involved with hydraulic calculation. Hydrologic objects such as subcatchments or rain gages were not considered. Since the rainfall runoff process is another important part of the hydrological cycle, it would be interesting to extend the scope of this work to hydrological simulation. Finally, there is a growing interest in Sustainable Urban Drainage Systems (SUDS). EPA SWMM includes these systems and their design might be boosted with the help of similar functions for such a systems.

As a final conclusion, it can be said that the connection library with EPA SWMM represents a powerful tool for applications based on changing network parameters, as is the case of the formulation of sewerage network design. Besides, it allows saving of computational costs. All the aforementioned results allows the possibility of applying this library to other applications of different natures, which is precisely the case of the studies that are currently being implemented for sewerage network adaption to climate change.

Acknowledgments: The authors would like to thank the Colombian company PAVCO-MEXICHEM and the Colombian Administrative Department for Science, Technology and Innovation COLCIENCIAS, for financing the "Drenaje Urbano y Cambio Climático: Hacia los Sistemas de Drenaje Urbano del Futuro" investigation, under which the present paper was conceived. Likewise, the development of this paper has been possible thanks to the Spanish Ministry for Science and Innovation, who covered the "DPI2009-13674 OPERAGUA: Mejora de las técnicas de llenado y operación de redes de abastecimiento de agua" research project.

Author Contributions: All authors contributed extensively to the work presented in this paper. F.J. Martínez-Solano and Pedro L. Iglesias-Rey contributed to the subject of research, the development and the writing of the paper. Juan G. Saldarriaga and Daniel Vallejo contributed to the Toolkit debugging, supplied the simplified Bogotá sewer network and made the final revision of the manuscript.

Conflicts of Interest: The authors declare no conflict of interest.

References

1. James, W.; Huber, W.C.; Dickinson, R.E.; Pitt, R.E.; James, W.R.C.; Roesner, L.A.; Aldrich, J.A. *User's Guide to SWMM5-CHI Publications*, 13th ed.; CHI Water: Guelph, ON, Canada, 2011.
2. Mays, L.W.; Yen, B.C. Optimal Cost Designof Branched Sewer Systems. *Water Resour. Res.* **1975**, *11*, 37–47. [CrossRef]

3. Elimam, A.A.; Charalambous, C.; Ghobrial, F.H. Optimum design of large sewer networks. *J. Environ. Eng.* **1990**, *115*, 1171–1190. [CrossRef]

4. Van Nooijen, R.R.; Kolechkina, A. Speed of discrete optimization solvers for real time sewer control. *Urban Water J.* **2013**, *10*, 354–363. [CrossRef]

5. Banik, B.K.; Di Cristo, C.; Leopardi, A. SWMM5 Toolkit Development for Pollution Source Identification in Sewer Systems. *Procedia Eng.* **2014**, *89*, 750–757. [CrossRef]

6. Meredith, D.D. Chapter 9: Dynamic programming concepts and applications. In *Treatise on Urban Water Systems*; Colorado State University: Port Collins, CO, USA, 1971; pp. 599–652.

7. Moeini, R.; Afshar, M.H. Constrained Ant Colony Optimisation Algorithm for the layout and size optimisation of sanitary sewer networks. *Urban Water J.* **2013**, *10*, 154–173. [CrossRef]

8. Krebs, G.; Kokkonen, T.; Valtanen, M.; Koivusalo, H.; Setälä, H. A high resolution application of a stormwater management model (SWMM) using genetic parameter optimization. *Urban Water J.* **2013**, *10*, 394–410. [CrossRef]

9. Rossman, L.A. *Epanet 2 Users Manual*; U.S. Environmental Protection Agency: Cincinnaty, OH, USA, 2000.

10. Kandiah, V.; Jasper, M.; Drake, K.; Shafiee, M.E.; Barandouzi, M.; Berglund, D.; Brill, E.D.; Mahinthakumar, G.; Ranjithan, S.; Zechman, E. Population-Based Search Enabled by High Performance Computing for BWN-II Design. In Proceedings of the 14th Annual Conference on Water Distribution Systems Analysis, Adelaide, Australia, 24–27 September 2012.

11. Guidolin, M.; Burovskiy, P.; Kapelan, Z.; Savić, D.A. CWSNET: An Object-Oriented Toolkit for Water Distribution System Simulations. In Proceedings of the 12th Annual Conference on Water Distribution Systems Analysis, Tucson, AZ, USA, 12–15 September 2010.

12. Savić, D.; Bicik, J.; Morley, M.S. A DSS generator for multiobjective optimisation of spreadsheet-based models. *Environ. Model. Softw.* **2011**, *26*, 551–561. [CrossRef]

13. Marchi, A.; Simpson, A.R. Correction of the EPANET Inaccuracy in Computing the Efficiency of Variable Speed Pumps. *J. Water Resour. Plan. Manag.* **2013**, *139*, 456–459. [CrossRef]

14. Rossman, L.A. *Storm Water Management Model User's Manual*; U.S. Environmental Protection Agency: Cincinnaty, OH, USA, 2015.

15. Huber, W.C.; Dickinson, R.E. *Stormwater Management Model (SWMM) User's Manual, Version 4.0*; Environmental Research Laboratory, U.S. Environmental Protection Agency: Athens, GA, USA, 1988.

16. Chan, C.C.; Dickinson, R.E.; Burgess, E. Runtime Comparisons between SWMM 4 and SWMM 5 using Continuous Simulation Model Networks. *J. Water Manag. Model.* **2005**, *223*, 193–202.

17. Rossman, L.A. *Storm Water Management Model Quality Assurance Report: Dynamic Wave Flow Routing*; U.S. Environmental Protection Agency: Cincinnaty, OH, USA, 2006.

18. Schade, T.; Wagner, T. *Quality Assurance Project Plan SWMM Redevelopment*; National Risk Management Research Laboratory, U.S. Environmental Protection Agency: Cincinnaty, OH, USA, 2002.

19. Rossman, L.A.; Dickinson, R.; Schade, T.; Chan, C.; Burgess, E.H.; Sullivan, D.; Lai, D. SWMM 5-the Next Generation of EPA's Storm Water Management Model. *J. Water Manag. Model.* **2004**. [CrossRef]

20. Yazdi, J.; Sadollah, A.; Lee, E.H.; Yoo, D.G.; Kim, J.H. Application of multi-objective evolutionary algorithms for the rehabilitation of storm sewer pipe networks. *J. Flood Risk Manag.* **2015**. [CrossRef]

21. Mays, L.W.; Wenzel, H.G.J. Optimal Design of Multilevel Branching Sewer Systems. *Water Resour. Res.* **1976**, *12*, 913–917. [CrossRef]

22. Afshar, M.H.; Rohani, M. Optimal design of sewer networks using cellular automata-based hybrid methods: Discrete and continuous approaches. *Eng. Optim.* **2012**, *44*, 1–22. [CrossRef]

23. Cozzolino, L.; Cimorelli, L.; Covelli, C.; Mucherino, C.; Pianese, D. An Innovative Approach for Drainage Network Sizing. *Water* **2015**, *7*, 546–567. [CrossRef]

24. Messner, F.; Meyer, V. Flood damage, vulnerability and risk perception-challenges for flood research. In *Flood Risk Management: Hazards, Vulnerability and Mitigation Measures*; Schanze, J., Zeman, E., Marsalek, J., Eds.; Springer Netherlands: Dordrecht, The Netherlands, 2006; pp. 149–167.

25. Guo, Y.F.; Walters, G.A.; Khu, S.T.; Keedwell, E.C. Efficient Multiobjective Storm Sewer Design Using Cellular Automata and Genetic Algorithm Hybrid. *J. Water Resour. Plan. Manag.* **2008**, *134*, 511–515. [CrossRef]

26. Mora-Melia, D.; Iglesias-Rey, P.L.; Martinez-Solano, F.J.; Fuertes-Miquel, V.S. Design of Water Distribution Networks using a Pseudo-Genetic Algorithm and Sensitivity of Genetic Operators. *Water Resour. Manag.* **2013**, *27*, 4149–4162. [CrossRef]

27. Mora-Meliá, D.; Iglesias-Rey, P.L.; Fuertes-Miquel, V.S.; Martinez-Solano, F.J. Comparison of evolutionary algorithms for design of sewer systems. In *Environmental Hydraulics: Theoretical, Experimental and Computational Solutions*; CRC Press/Balkema: London, UK, 2010; pp. 261–263.
28. Mora-Melia, D.; Iglesias-Rey, P.L.; Martinez-Solano, F.J.; Ballesteros-Pérez, P. Efficiency of Evolutionary Algorithms in Water Network Pipe Sizing. *Water Resour. Manag.* **2015**, *29*, 4817–4831. [CrossRef]
29. Elbeltagi, E.; Hegazy, T.; Grierson, D. Comparison among five evolutionary-based optimization algorithms. *Adv. Eng. Inf.* **2005**, *19*, 43–53. [CrossRef]

Article

Network Capacity Assessment and Increase in Systems with Intermittent Water Supply

Amilkar E. Ilaya-Ayza *, Enrique Campbell, Rafael Pérez-García and Joaquín Izquierdo

FluIng-IMM, Universitat Politècnica de València, Camino de Vera s/n Edif. 5C, Spain; encamgo1@upv.es (E.C.); rperez@upv.es (R.P.-G.); jizquier@upv.es (J.I.)
* Correspondence: amilay@upv.es; Tel.: +34-693-918-718

Academic Editor: Helena Ramos
Received: 3 January 2016; Accepted: 28 March 2016; Published: 31 March 2016

Abstract: Water supply systems have been facing many challenges in recent decades due to the potential effects of climate change and rapid population growth. Water systems need to expand because of demographic growth. Therefore, evaluating and increasing system capacity is crucial. Specifically, we analyze network capacity as one of the main features of a system. When the network capacity starts to decrease, there is a risk that continuous supply will become intermittent. This paper discusses how network expansion carried out throughout the network life span typically reduces network capacity, thus transforming a system originally designed to work with continuous supply into a system with intermittent supply. A method is proposed to expand the network capacity in an environment of economic scarcity through a greedy algorithm that enables the definition of a schedule for pipe modification stages, and thus efficiently expands the network capacity. This method is, at the same time, an important step in the process of changing a water system from intermittent back to continuous supply—an achievement that remains one of the main challenges related to water and health in developing countries.

Keywords: water network expansion; water network capacity; intermittent water supply; theoretical maximum flow; system setting curve

1. Introduction

The world population is increasing at an exponential rate and available water resources are reducing due to pollution and the effects of the climate change that increase the severity of droughts and favor other extreme events [1]. This places stress on many public services and dramatically increases the gap between water supply and demand. In many countries, urban growth has exceeded, and continues exceeding, the growth of supply infrastructure [2]. This growth generates the conditions necessary for operators of water supply systems to embrace intermittent water supply (IWS) [3]. Many countries in Africa, Asia, and Latin America have IWS [4].

Reducing poverty and improving public health is an important part of the Millennium Development Goals [5]. Achieving this goal is threatened by water shortages, lack of supply guarantees [6] and poor quality of supplied water. Although it is considered that the drinking water supplied through pipeline systems is safe, many studies show that deficiencies in the network, caused by IWS, can create conditions so that water is not safe and reliable [7–13]. The lack of reliable water supply systems in developing countries can undermine much of the hope for improvements in public health [14], and have negative effects for drinking water system objectives [15].

IWS generally seeks to reduce the per capita water demand based on savings in capital and operating costs. However, instead of being smart, this strategy brings negative consequences that outweigh the positive factors [16,17]. Symptoms of system failure include very low levels of pressure, and insufficient supply in the remotest and highest points. Generally, intermittent supply is adopted by necessity rather than by design and results in serious system impairment [18].

The best way to protect water quality is by maintaining positive and continuous pressures throughout the network [13,19]. Thus, continuous water supply ensures security. The supply change from intermittent to continuous is one of the main challenges concerning water and health in developing countries [20].

The cost of risks to the health of users must also be considered (in terms of their incomes, medical treatments, *etc.*) as it is much greater than the cost of replacing deficient pipes [11] that are detrimental to continuous water supply.

According to several studies [21–23], IWS systems produce insufficient pressure in less favored sectors or areas (nodes located on high points and/or far away). Such conditions may be favorable for reducing water losses. However, they also produce inequity in the supply [18].

Insufficient funding and mismanagement [17] are two of the main causes of the origin of IWS. System improvements in these scenarios do not derive from increasing the water supply sources, but from improving system infrastructure and management. However, a shortage of funding does not allow operators to make large investments, so they should look for profitable long term planning strategies. In this sense, phased or gradual improvements can be a good option.

The growth of cities occurs horizontally and/or vertically, as a result of residential, industrial, and commercial developments, and community facilities, *etc.* [24]. This growth requires expanding the network capacity and the correction of anomalies or reduced system performance [25].

When undertaking the expansion of a water supply network, the goal is to supply a much larger demand. However, when this expansion does not take into account the network capacity and the influence of the new expansion, various scenarios may appear that reduce the capacity of the network and threaten the quantity and quality of the service. Reducing the capacity of the network may lead, for example, to intermittent supply.

Generally, the network of an IWS system has insufficient capacity. However, this situation can be imperceptible because the operators manage to cover the demanded flows through differentiation of supply schedules, sectorization, and use of household tanks. Increasing the network capacity in IWS systems that seek to reach continuous water supply (CWS) is a task that must be carefully analyzed.

In this paper, the *theoretical maximum flow* is proposed as an indicator of network capacity. This element is endowed with its true dimension as a quantitative element that is crucial in decision-making, assessment, management, maintenance, exploitation, and design of drinking water distribution systems.

The theoretical maximum flow is important to evaluate the behavior and evolution of a drinking water system and its relationship with intermittent supply. It also serves as a basis for proposing a greedy algorithm that enables the definition of a schedule for pipe modification stages, and thus efficiently expands network capacity.

The case of study has two parts. In the first part, we evaluate the growth of the southern subsystem of the city of Oruro (Bolivia), which was originally built to offer continuous supply. However, various network modifications imposed an ideal environment for intermittent supply. In the second part, we consider the possibility of increasing the current network capacity, as part of various actions to revert to CWS. Based on the IWS classification given by Totsuka *et al.* [17], the system only suffers economic scarcity and management problems, and not physical scarcity. Thus, only the actions related to infrastructure improvement are analyzed.

2. Methodology

This paper discusses how poorly planned network expansions can lead from CWS to IWS. We propose the use of the theoretical maximum flow as an indicator for evaluating the network capacity, and then analyzing its relationship with intermittent supply.

Generally, the capacity of a water distribution network is considered a qualitative concept, which is usually identified from user complaints of pressure reduction [26].

We consider that network capacity represents the maximum demand (or flow) that can be met while maintaining suitable pressures throughout the network, and strictly ensuring the minimum pressure required at the node with the lowest pressure. When this flow does not cover the demand of the population, the network has insufficient capacity. Given this scenario, the network responds by reducing pressure at nodes to achieve total user demand. This situation may threaten the continuity of water supply. Even though it is possible to use a PDD (pressure driven demand) analysis, in low pressure networks, a more conservative and, thus, safer design is obtained by making use of mathematical models using demand driven analysis (DDA), and representing its capacity deficiency as a flow magnitude, since negative pressures have no physical meaning in cases of deficient network capacity.

We propose the use of the theoretical maximum flow calculated with the pressure-restricted setting curve, explained later. To evaluate the capacity of a system network with IWS, the maximum theoretical flow is compared with the maximum flow required by the population in continuous supply. It is thus possible to establish the potential of converting CWS into IWS. Intermittent water supply is usually caused by the extension of the distribution network beyond its hydraulic capacity [27].

Therefore, a method is also proposed to increase the capacity of the network, as a part of a project for gradual transition towards CWS, while taking into account that the system suffers insufficient funding.

It is common to gradually carry out the process of expanding the network capacity with more or less localized interventions on its components, so as not to endanger the service and ensure a greater lifespan for the infrastructure [25].

Another important restriction on some systems with IWS is related to the economic constraints of the water company. Therefore, we propose a gradual expansion of capacity divided into stages, in which a schedule is defined for every stage (the optimal option being sought in each of these stages). Taking advantage of improvements in network capacity, CWS gradually spreads until the entire network is covered.

When a network requires expansion, it is common practice to use optimization techniques to find a solution with lower costs and satisfactory pressures. However, these processes tend to define the overall set of pipes regardless the necessary actions associated with stage-divided projects.

In this paper, the use of the theoretical maximum flow reduces the search space to an area equal to the number of pipes evaluated and multiplied by the number of candidate diameters and number of stages.

This advantage enables us to propose the strategic replacement of pipes in a context of economic scarcity by a greedy algorithm that enables the optimal option in each of the stages to be selected—in an attempt to reach an optimal general solution. A schedule of the stages for modifying the network is defined in this way, and the result is a gradual and more efficient transition to CWS.

As the influence of each of the pipes in the total capacity of the network is known, another advantage of the proposed method is the possibility of detecting bottlenecks in the network.

2.1. System Setting Curve

The setting curve is a very useful tool in the operation and management of a water supply system. This curve represents the need for energy production at the source in relation to the injected flow in the system, guaranteeing the minimum pressure at less favorable points. A distribution network does not have a well-defined resistance curve [28,29], because the curve slope changes—depending on the flow requirements and the resistance imposed on the network by user demand [30], from demand for minimum flows to demand for peak flows. However, the setting curve maintains a stable position, which is very advantageous and useful for solving problems related to water supply (see Figure 1).

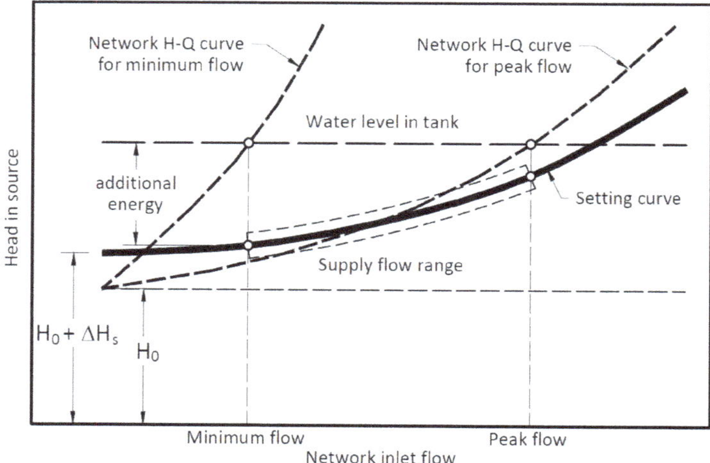

Figure 1. Representation of the network resistance and setting curves.

For the calculation of the setting curve, a reliable mathematical model of the network should be available. Thus, it is possible to evaluate the losses in terms of various load conditions in the network. Tracking the setting curve at all times ensures that the pressure injected into the households is that which is strictly necessary to provide a good service. This produces energy savings. Similarly, adhering to the curve moderates pressure fluctuations in the network, thus reducing the negative implications of such variations in the life span of the network [29].

The setting curve is used for control purposes in pumping systems [29] and in energy optimization of water supply systems [31–33]. An approximation to the setting curve is also used as a flow modulation curve or as a setting curve for pressure reducing valves in response to changes in system demands to optimize the operation of a district metered area [34,35].

The flowchart in Figure 2 summarizes the steps to determine the setting curve when there is only one feed point. In addition to the mathematical model of the network, the minimum pressure at the nodes (P_{min}) must also be given. This value will define the level of service to be achieved. Subsequently, scenarios for different load states, j, defined by peak factor K applied to the demand of all the nodes must be generated. Each load condition establishes an injected flow (Q_j) in the network that requires an available head at the source (H_a) that ensures the minimum pressure in less favorable nodes (P_{unf}). These pressures are compared with the minimum pressure until a desired very small margin of error is reached. The set of points thus obtained describes the setting curve. The hydraulic calculation for each load state is performed with DDA. It is recommended that, like in pressures, elevation and head units be meters; and the flow in liters per second.

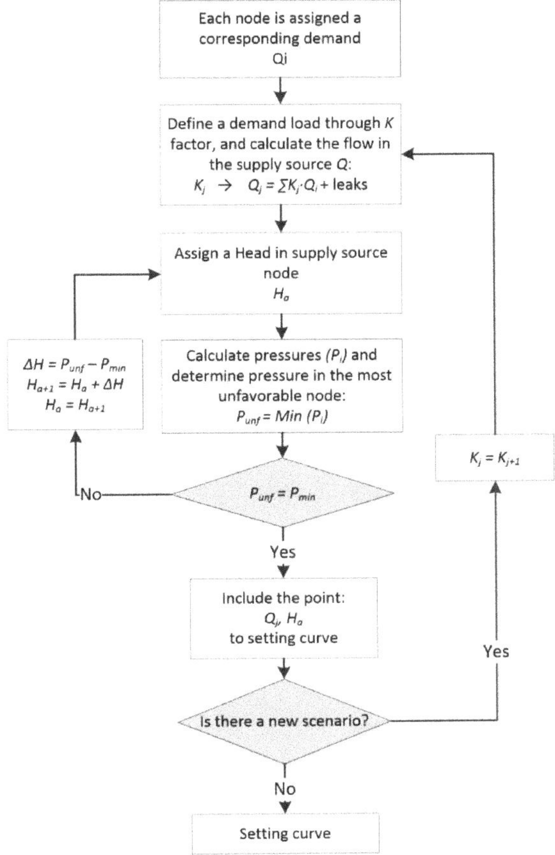

Figure 2. Flow diagram for the determination of the setting curve.

2.2. Theoretical Maximum Flow

We consider the theoretical maximum flow, as the maximum flow value that can be injected in the network by ensuring that the pressure is not lower than a minimum value established as a constraint. This flow represents the theoretical capacity of the distribution network.

The maximum flow through a simple pipe can be calculated by taking into account the upstream and downstream boundary conditions in terms of gauge pressure or head. In a network the calculation may be similarly performed. The downstream boundary conditions are set by the setting curve, which was built to meet a defined minimum pressure (P_{min}). The upstream condition, which provides the hydraulic potential of the network, is defined by the hydraulic head available at the source (H_s), which can be a reservoir (Figure 3) or a pump. The theoretical maximum flow (Q_{maxt}) is determined by the intersection of the setting curve and the supply curve (see Figure 4).

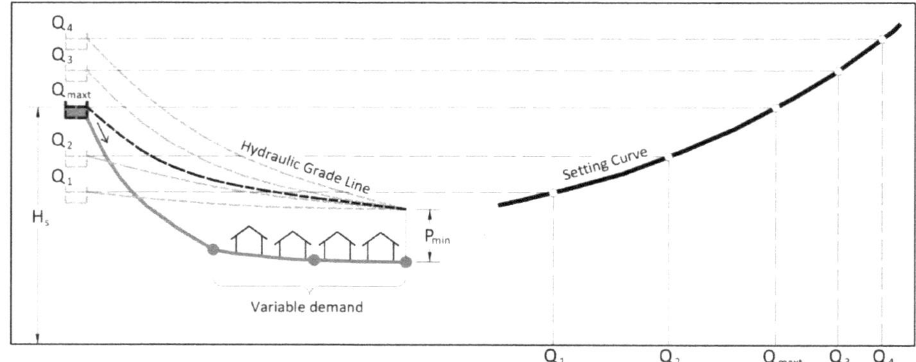

Figure 3. Representation of setting curve and theoretical maximum flow.

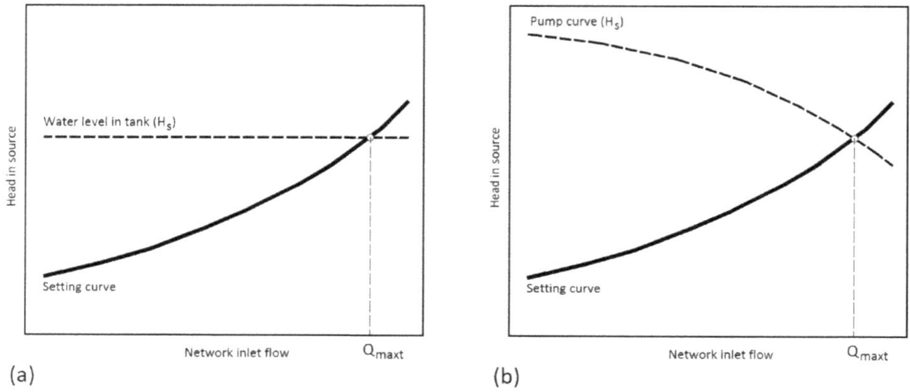

Figure 4. Theoretical maximum flow for: (**a**) a tank or reservoir; and (**b**) a pump.

The theoretical maximum flow or network capacity can be increased by changes in the supply source: by, for example, building new reservoirs or increasing the pump power or the number of pumps; reducing minimum pressure; and modifying the network (adding reinforcement, greater interconnectivity, making replacements, *etc.*). In our study, we address the latter case.

The theoretical maximum flow is strongly related to the setting curve which varies according to the network configuration, and the changes in the pipes or the service minimum pressure. One way to increase the network capacity without altering H_s is by changing the slope of the setting curve. This can be achieved by changing some physical characteristics of the network pipes.

2.3. Required Maximum Flow

For the process of increasing the capacity of the network it is necessary to have a point of reference, a target point defined by the required maximum flow (Q_{maxr}) and its corresponding source pressure head. In CWS, the required maximum flow corresponds to the maximum hourly flow.

The required maximum flow is calculated from the demand, using population growth forecasts, losses, and other future uses and water needs.

2.4. Procedure for Increasing the Capacity of the Network

The process of increasing the capacity of the network is very useful for the transformation of IWS into CWS. The process is based on changing the characteristics of the network so as to reduce the slope of the setting curve until the value of the required maximum flow is exceeded, see Figure 5.

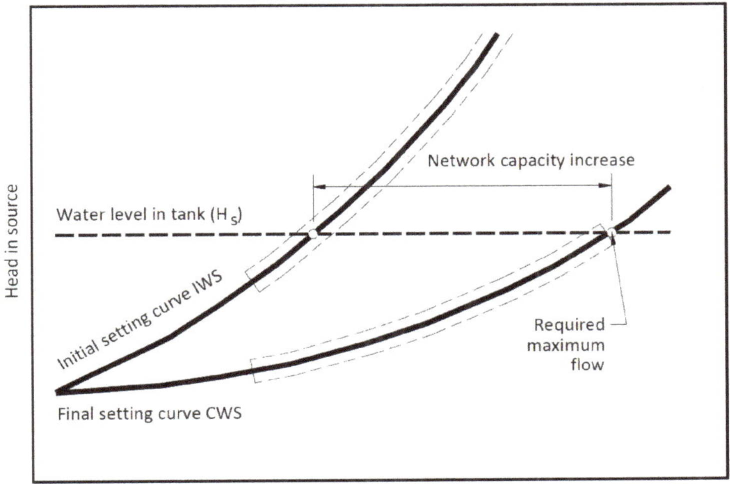

Figure 5. Network capacity increase process.

Because intermittent supply systems have hydraulic, structural [3,19,20], and water quality problems [21–25], we consider that, in the transition process to continuous supply, replacement is more effective than pipe reinforcement.

Starting with the current network, each pipe is replaced with another pipe with a different diameter. This change is evaluated by calculating the theoretical maximum flow. Thus, it is possible to calculate the increase in capacity involved in each of the changes. By incorporating the cost of replacing a pipe p an expansion rate is determined:

$$q_{p,d} = \frac{Q_{\max t}^{p,d} - Q_{\max t}^{o}}{C\left(d_p\right) \cdot L_p}, \tag{1}$$

where $q_{p,d}$ is expansion rate produced by substituting pipe p for a new pipe with diameter d; $Q_{\max t}^{o}$ is theoretical maximum flow of the original network or the network modified in the previous step; $Q_{\max t}^{p,d}$ is theoretical maximum flow after substituting pipe p for a new pipe with diameter d; $C(d_p)$ is unit length cost of pipe p replaced with a new pipe of diameter d; and L_p is length of modified pipe p.

The expansion rate (Equation (1)) is very useful in identifying pipes that constrain the network capacity.

Another approach for selecting the pipes to be replaced, which prioritizes the flow increase over the cost, is by raising the difference in the numerator of the expansion rate (Equation (1)) to an exponent, n, as in Equation (2). In this way, the pipes will initially be modified to larger diameters. This will subsequently offset the initial cost as smaller diameters will be required.

$$q_{p,d}^{n} = \frac{\left(Q_{\max t}^{p,d} - Q_{\max t}^{o} \right)^{n}}{C\left(d_{p} \right) \cdot L_{p}}, \tag{2}$$

By using the expansion rate in each of the expansion stages, the pipe to be modified or replaced is identified (as the pipe with the largest expansion rate value).

The process outlined in the flowchart in Figure 6 enables us to define a new network configuration by increasing its capacity after replacing the pipes that represent the lowest costs. The order of priority for the replacement of the pipes is also identified, and this enables us to plan the process of gradually increasing the network capacity. The total cost may be divided into stages based on this prioritization.

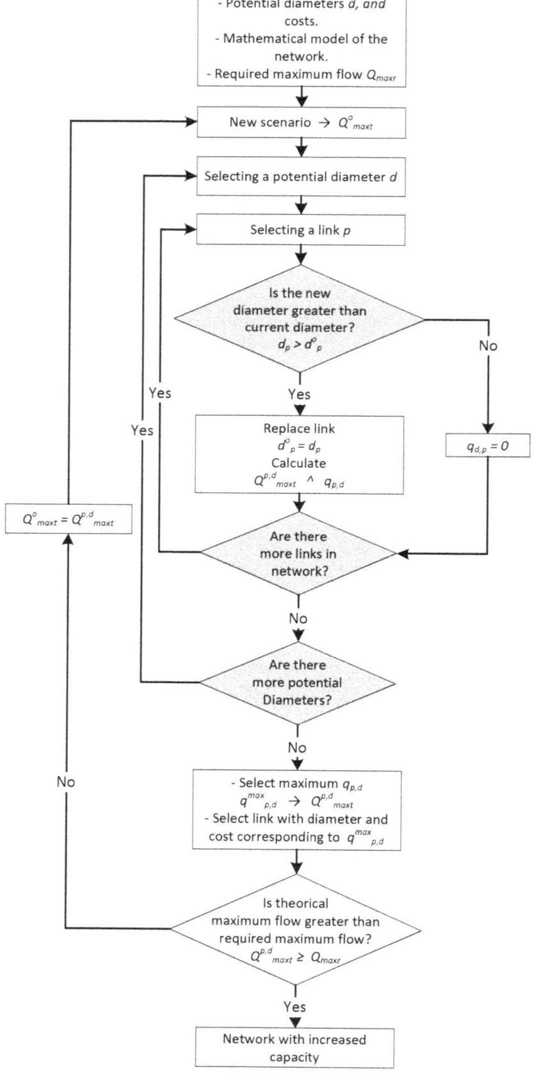

Figure 6. Flowchart for the process of increasing the network capacity.

3. Results and Discussion

The case of study has two parts:

The first step is to reconstruct the hydraulic conditions of the subsystem in the south of Oruro in the period 1968–2013. Considering that the initial network is the main part of the current network and that it has been maintained since the inauguration of the infrastructure, we will use the current network model. Due to a lack of information, the growth of the urban area is used as a reference to reconstruct the network evolution. As a result, the topology of the network during each of the study periods is estimated. Considering the population and water demand of each of the periods studied, it is possible to calculate the required maximum flow. However, to compare this value with the network capacity it is necessary that both elements have the same dimensions. For this purpose, we propose the use of the theoretical maximum flow indicator.

In the second part, based on the current network of the south Oruro area, we seek to identify and prioritize the order of the pipes that require replacement under economic and hydraulic criteria, in order to gradually increase the network capacity to achieve the necessary infrastructure to transform IWS into CWS (24/7). Besides the theoretical maximum flow, an expansion rate indicator will also be used.

3.1. First Part of the Study

The project "Drinking water for the city of Oruro" [36] was begun in 1968 and included the main infrastructure for water supply in the southern sub-area of the city—a tank and the mains. The subsystem offered continuous supply but is currently an IWS system. There are no records of when the system became intermittent. Based on the theoretical maximum flow, the transition process is analyzed in this part of the study.

Considering the size of the urban area [37], population growth [38], population density, and the number of current users, it is possible to estimate the population growth of the study area, defined by the urban growth (Table 1) during the given periods. Based on the current average supply of the subsystem (84.32 L/capita/d) and a peak factor of 2.5 (representative of other areas of the city that have CWS and similar population growth) the required maximum flow for each year can be calculated.

Table 1. Population and required maximum flow for the Oruro southern subsystem.

Year	Population	Q_{maxr} (L/s)
1968	13842	33.77
1972	16814	41.02
1985	33407	81.51
2007	37395	91.24
2013	37700	91.98

The urban growth of the city enables us to set growth scenarios for the subsystem supply network, for the changes in the mains and for the inlet to the sectors (see Figure 7). Based on these scenarios, the setting curve and the theoretical maximum flow or capacity of the network (Figure 8) is calculated. A minimum pressure (P_{min}) condition of 20 m is adopted; the minimum elevation of the water level in the reservoir (H_s) is 3771 masl (meters above sea level), and the average elevation of the network is 3723.8 masl.

Comparison between the required maximum flow and the theoretical maximum flow or network capacity is shown in Figure 9.

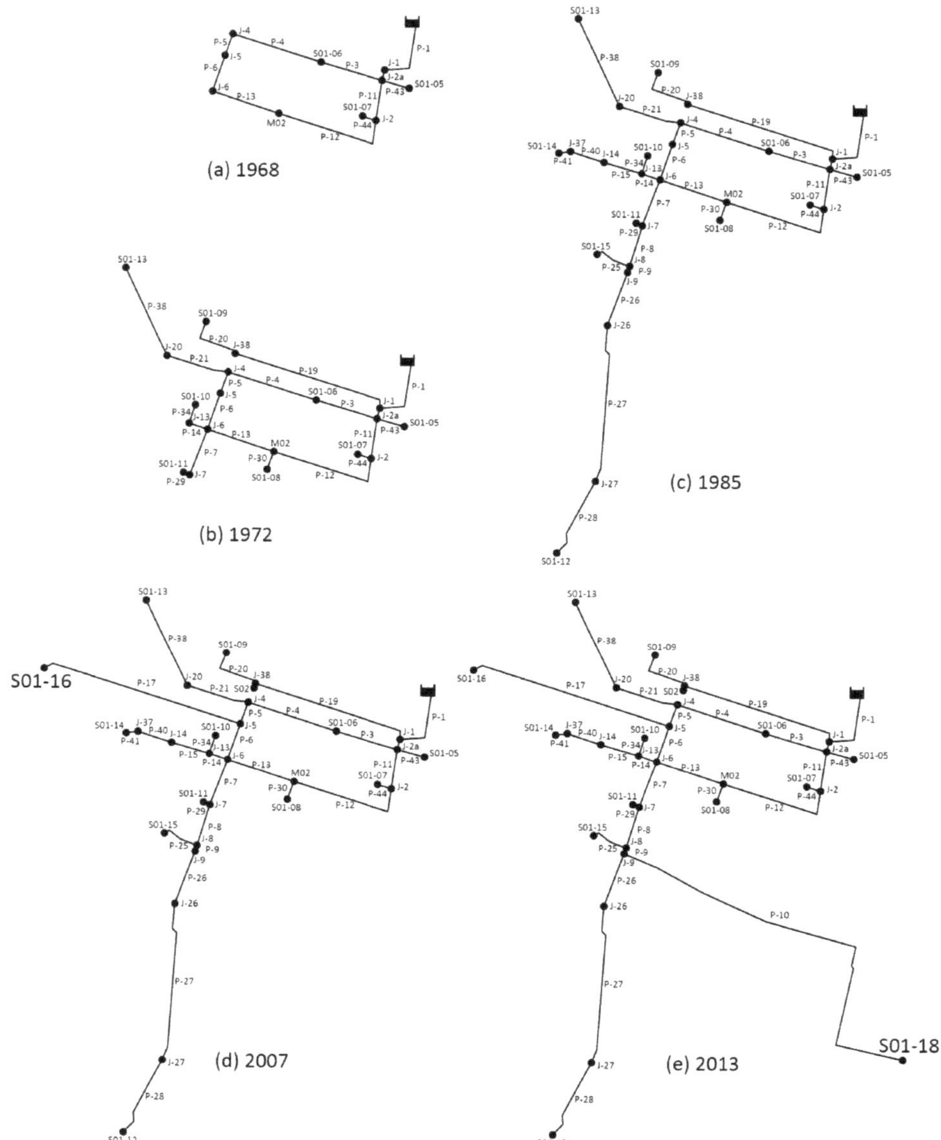

Figure 7. Growth of the Oruro southern subsystem network.

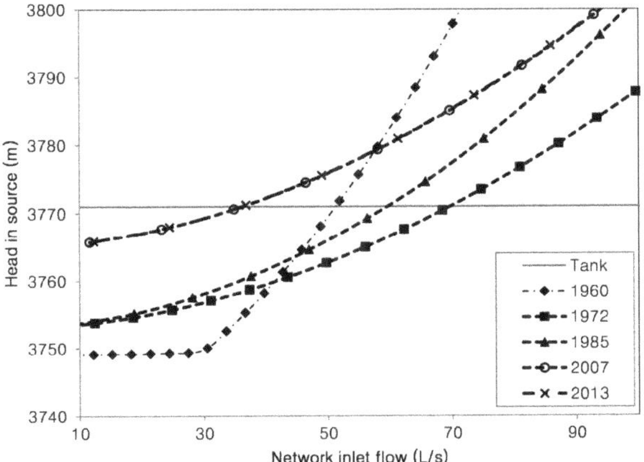

Figure 8. Setting curve in each of the years of study.

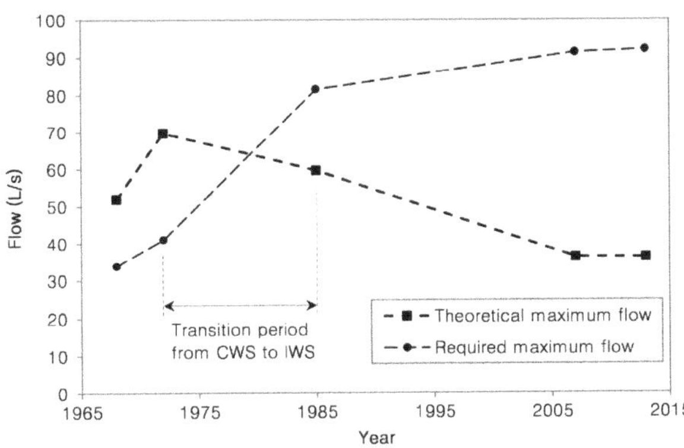

Figure 9. Development of the theoretical maximum flow and the required maximum flow in the study area.

The studied system was designed for continuous supply. This is evident because in 1968 the network capacity was greater than the required maximum flow. The network was designed to supply water to a larger population and had a reasonable slack for expansion. This scenario was maintained throughout the first period, 1968–1972; and expansions made during this period were developed in favorable areas. These expansions were developed properly and did not significantly weaken the network capacity.

In the period 1972–1985, the situation changed. The southern region of Oruro grew faster than the rest of the city [37]. The network expansions to cover the new sectors were very large. In addition, the selection of small diameters increased the pressure drop and network capacity was therefore reduced.

In 1985, the required maximum flow exceeded the theoretical maximum flow. This fact imposed a reduction of pressure in the network. The nodes located in unfavorable areas were prone to run out of water during peak consumption. People needed to protect their supply and opted for the use of household tanks. This scenario led to inequity in the supply: nodes located in favorable areas squandered water, due to lack of metering, while users located in unfavorable areas complained about a lack of water. Eventually, a perception of water scarcity prevailed among the population and the operator. A solution was sought.

To solve the water shortage in less favorable points, there are two potential solutions: the first option, though not very evident in scenarios of economic scarcity and poor management, is to expand the capacity of the network by replacing and reinforcing the mains; and the second option is to opt for intermittent supply (a widespread misconception derived from a lack of water and funding).

The adoption of intermittent supply limits the hours of supply in different areas, setting schedules that seek to reduce the required maximum flow. This action may be useful initially as the main sections transport lower flow rates during peak consumption and therefore sufficient flow reaches the points located in unfavorable areas. However, population growth will require expansion and this condemns the system to intermittent supply.

In this second period, the theoretical maximum flow and the required maximum flow curves intersect. A point was reached at which the capacity of the network no longer met the demand of the population with continuous supply (this situation has been maintained since then). Accordingly, we can say that a policy based on an IWS approach was implemented in the south of Oruro starting between 1970 and 1985 (see Figure 9).

In the third study period, 1985–2007, intermittent supply was consolidated. New expansion further reduced the capacity of the network. The expansion of the S01-16 sector used diameters that were too small and, in addition, supply was extended to an even higher point. These features made it a critical point, which conditioned the setting curve severely and, consequently, the value of the theoretical maximum flow.

Between 2007 and 2013, the subsystem was expanded with the S01-18 sector, which did not affect the capacity of the network because the sector is in a lower area and the installed diameters were suitable for the needs of the sector. It proved to be a good expansion, since it did not greatly affect the setting curve and the theoretical maximum flow.

The reduction of network capacity is a problem not perceived when the network works with intermittent supply, because water reaches all sectors, although at very low pressures; however, this situation will be primarily responsible for inequitable supply.

3.2. Second Part of the Study

The capacity of the subsystem network in the south of Oruro is currently insufficient and, as a result, supply is intermittent. It is necessary to increase this capacity if the aim is to achieve continuous supply. Due to a shortage of funds, the expansion is bound to be gradual and staged. Accordingly, it is necessary to know the order of pipe replacement. The budget available annually for the process of increasing the network capacity is Bs. 700,000 (seven hundred thousand Bolivian boliviano, equivalent to €89,172). The transition from IWS to CWS requires that the network has a capacity for continuous supply of 91.98 L/s.

Unit costs, which include all the elements necessary for the replacement of pipes, and which depend on the diameter, are presented in Table 2.

Table 2. Unit costs for replacing pipes as a function of diameter.

Diameter (mm)	Unit Cost (Bs/m)
75	385.02
100	399.66
150	442.46
200	502.93
250	608.39
300	703.17
350	832.61
400	1,030.54
500	1,371.51

When large diameter pipes are changed, the theoretical network maximum flow is increased (see Figure 10). This increase is more pronounced in pipes P-1, P-2, P-3, P-4, P-11, and P-13, and even more so in pipes P-12 and P-17. These last mentioned pipes are the most critical for the network and represent bottlenecks; so, any action to expand the network must take them into account. Augmenting the diameter of the remaining pipes causes minimal increases in the network capacity, so their importance in increasing network capacity is minimal. Generally, the first upgrade of diameter produces a significant increase in network capacity, while subsequent diameter upgrades do not produce such a large effect. The curve tends to reflect an asymptotic behavior (Figure 10).

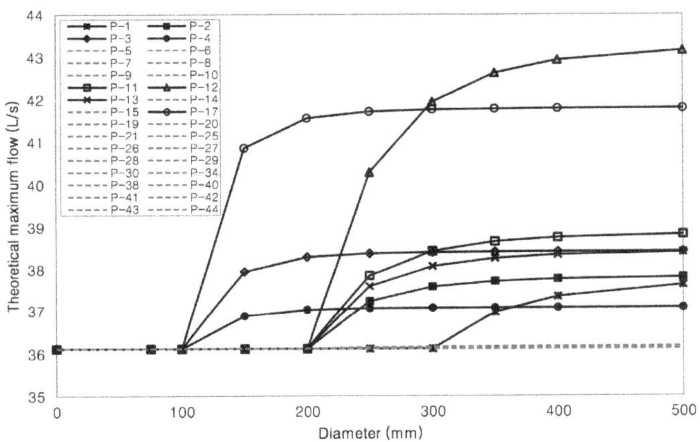

Figure 10. Network capacity increase based on diameter change.

The expansion process was developed using the expansion ratio q given in Equation (2) for $n = 1$, 2 and 3. The exponent was chosen depending on the theoretical maximum flow (Figure 11). The use of the indicator with an exponent greater than 1 enables lower costs to be reached when the expansion flow rate is larger; however, $n = 1$ can be used when the maximum flow requirement is lower. In any case, it is adequate for making a comparison with indicators.

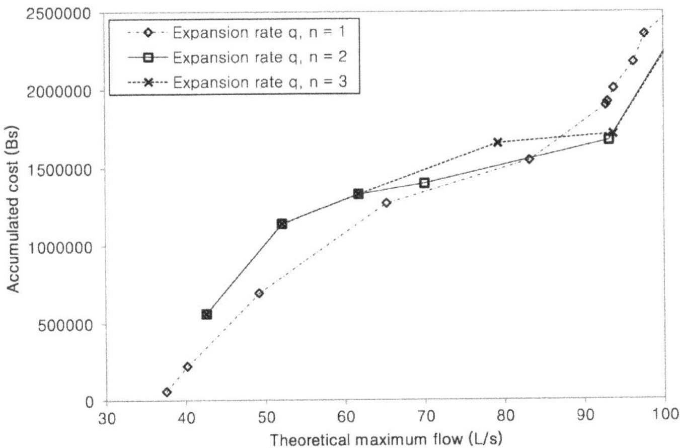

Figure 11. Comparison of cumulative costs depending on the expansion rate used.

To reach the required maximum flow of 91.98 L/s, the lowest cost is produced by the indicator q with $n = 2$. The setting curve is gradually changed in five steps that define the order of actions for the gradual improvement (Figure 12). With this prioritization, three stages of investment (last column of Table 3 and Figure 13) are defined.

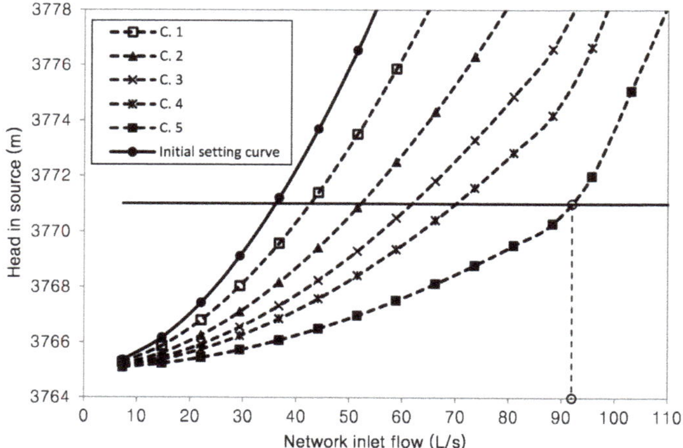

Figure 12. Setting curve evolution in the capacity increase process.

Table 3. Results of the network capacity increase process.

Iteration	q_{max} $n = 2$	Modified Pipe	New Diameter (mm)	Q_{maxt} (L/s)	Pipe Cost (Bs)	Accumulated Cost (Bs)	Cost per Stage (Bs)
1	0.07540	P-12	350	42.63	562,303.90	562,303.90	562,303.90
2	0.15582	P-17	200	52.14	581,264.92	1,143,568.83	581,264.92
3	0.48497	P-11	350	61.76	190,736.95	1,334,305.77	
4	1.00235	P-2	350	70.04	68,333.76	1,402,639.54	532,875.95
5	1.95625	P-13	300	93.18	273,805.24	1,676,444.78	

Due to increasing water-losses produced after transition from IWS to CWS [3], and which consequently increase the required maximum flow, it is necessary to implement an active leakage control [39] simultaneously with the network capacity expansion to avoid oversizing the network and wasting water.

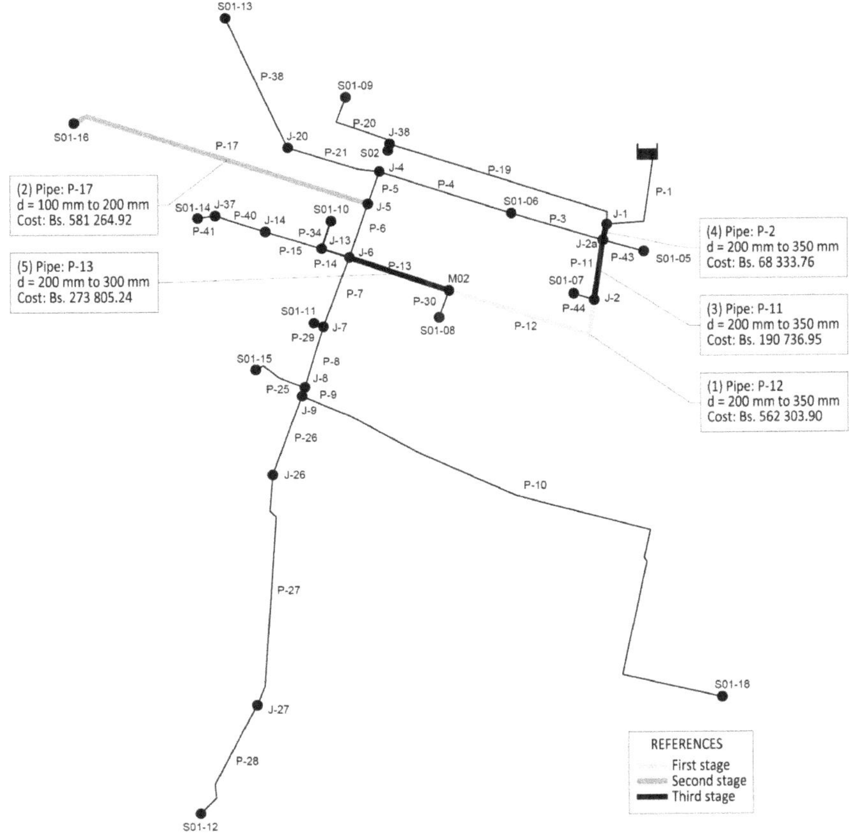

Figure 13. Stages of investment.

4. Conclusions

Although its use is generally applied to CWS, the setting curve is a useful tool for evaluating and improving systems with intermittent supply.

The approach using theoretical maximum flow enables network capacity to be given its true quantitative dimension. This is necessary for decision-making in the assessment, management, maintenance, operation, and design of a continuous or intermittent supply drinking water network. Specifically, in this paper, this approach enables us to evaluate a system with intermittent supply, track its evolution from its CWS origin (first part of the study) to the current IWS, and propose solutions to recover CWS (second part of the study). Thus, this study contributes to improve intermittent supply systems, which remains one of the main challenges related to water and health in developing countries.

In the first part, we can see how poorly planned expansion processes can systematically reduce network capacity. This reduction initially brought low pressure, flow failure in unfavorable areas, and user complaints. The situation was perceived as water scarcity rather than reduced network capacity. As a result, it was decided to opt for intermittent supply. Further expansion actions consolidated this approach.

It is important to consider that the rational extension of the network not only involves laying additional pipelines that reach the point of demand, but that this process should be accompanied by the reinforcement, rehabilitation, or replacement of the network water mains.

A process of expanding network capacity, as part of the transition from IWS to CWS with poor funding, is difficult to implement. Consequently, gradually prioritizing pipe replacements, based on achieving the greatest impact on the quality of service at the lowest cost, is a very useful criterion. The characteristics of the method allow us to reach a more efficient process in small networks than in large ones.

Acknowledgments: The authors are grateful to SeLA (water company in Oruro (Bolivia)) for providing information. The use of English has been revised with the help of John Rawlins.

Author Contributions: Ilaya-Ayza, A.E., Izquierdo, J., and Pérez-García, R. conceived and designed the study. Ilaya-Ayza, A.E. wrote the manuscript. All authors read and approved the final version of the manuscript.

Conflicts of Interest: The authors declare no conflict of interest.

References

1. Arnell, N.W. Climate change and water resources: A global perspective. In *Avoiding Dangerous Climate Change*; Schellnhuber, H.J., Cramer, W., Nakicenovic, N., Wigley, T., Yohe, G., Eds.; Cambridge University Press: Cambridge, UK, 2006; pp. 167–175.
2. Islam, S.; Alekal, P. Achieving 24 × 7 Water and Waterloss Management in Intermittent Supply Environment, 2009. Available online: http://www.iwawaterwiki.org/xwiki/bin/view/Articles/Achieving24x7waterand waterlossmanagementinintermittentsupplyenvironment (accessed on 15 October 2015).
3. Charalambous, B. The Effects of Intermittent Supply on Water Distribution Networks. Water Loss, 2012. Available online: http://www.leakssuite.com/wp-content/uploads/2012/09/2011_Charalambous.pdf (accessed on 15 October 2015).
4. Van den Berg, C.; Danilenko, A. *The IBNET Water Supply and Sanitation Performance Blue Book*; The World Bank: Washington, DC, USA, 2011.
5. United Nations. Resolution adopted by the General Assembly. In Proceedings of the 8th Plenary Meeting, New York, NY, USA, 8 September 2000; UN General Assembly: New York, NY, USA, 2000.
6. Fan, L.; Liu, G.; Wang, F.; Ritsema, C.J.; Geissen, V. Domestic water consumption under intermittent and continuous modes of water supply. *Water Resour. Manag.* **2014**, *28*, 853–865. [CrossRef]
7. Semenza, J.C.; Roberts, L.; Henderson, A.; Bogan, J.; Rubin, C.H. Water distribution system and diarrheal disease transmission: A case study in Uzbekistan. *Am. J. Trop. Med. Hyg.* **1998**, *59*, 941–946. [PubMed]
8. Mermin, J.H.; Villar, R.; Carpenter, J.; Roberts, L.; Gasanova, L.; Lomakina, S.; Hutwagner, L.; Mead, P.; Ross, B.; Mintz, E. A massive epidemic of multidrug-resistant typhoid fever in Tajikistan associated with consumption of municipal water. *J Infect. Dis.* **1999**, *179*, 1416–1422. [CrossRef] [PubMed]
9. Tokajian, S.; Hashwa, F. Water quality problems associated with intermittent water supply. *Water Sci. Technol.* **2003**, *47*, 229–234. [PubMed]
10. Tokajian, S.; Hashwa, F. Phenotypic and genotypic identification of Aeromonas spp. isolated from a chlorinated intermittent water distribution system in Lebanon. *J. Water Health.* **2004**, *2*, 115–122. [PubMed]
11. Sargaonkar, A.; Kamble, S.; Rao, R. Model study for rehabilitation planning of water supply network. *Comput. Environ. Urban. Syst.* **2013**, *39*, 172–181. [CrossRef]
12. Kumpel, E.; Nelson, K.L. Comparing microbial water quality in an intermittent and continuous piped water supply. *Water Res.* **2013**, *47*, 5176–5188. [CrossRef] [PubMed]
13. Kumpel, E.; Nelson, K.L. Mechanisms affecting water quality in an intermittent piped water supply. *Environ Sci. Technol.* **2014**, *48*, 2766–2775. [CrossRef] [PubMed]
14. Hunter, P.R.; Zmirou-Navier, D.; Hartemann, P. Estimating the impact on health of poor reliability of drinking water interventions in developing countries. *Sci. Total Environ.* **2009**, *407*, 2621–2624. [CrossRef] [PubMed]
15. Lee, E.; Schwab, K. Deficiencies in drinking water distribution systems in developing countries. *J. Water Health* **2005**, *3*, 109–127. [PubMed]
16. Woo, C.K. Managing water supply shortage: Interruption *vs.* pricing. *J. Public Econ.* **1994**, *54*, 145–160. [CrossRef]
17. Totsuka, N.; Trifunovic, N.; Vairavamoorthy, K. Intermittent urban water supply under water starving situations. In Proceedings of the 30th WEDC International Conference, Vientiane, Lao, 25–29 October 2004.

18. Vairavamoorthy, K.; Gorantiwar, S.D.; Pathiranaa, A. Managing urban water supplies in developing countries—Climate change and water scarcity scenarios. *Phys. Chem. Earth* **2008**, *33*, 330–339. [CrossRef]
19. Geldreich, E.E. *Microbial Quality of Water Supply in Distribution Systems*; CRC Lewis Publishers: Boca Raton, FL, USA, 1996.
20. Mohapatra, S.; Sargaonkar, A.; Labhasetwar, P.K. Distribution network assessment using EPANET for intermittent and continuous water supply. *Water Resour. Manag.* **2014**, *28*, 3745–3759. [CrossRef]
21. Vairavamoorthy, K.; Akinpelu, E.; Lin, Z.; Ali, M. Design of sustainable water distribution systems in developing countries. In Proceedings of the ASCE Conference, Houston, TX, USA, 10–13 October 2001.
22. De Marchis, M.; Fontanazza, C.M.; Freni, G.; La Loggia, G.; Napoli, E.; Notaro, V. A model of the filling process of an intermittent distribution network. *Urb. Water J.* **2010**, *7*, 321–333. [CrossRef]
23. Dahasahasra, S.V. A model for transforming an intermittent into a 24 × 7 water supply system. *Geospatial Today* **2007**, 34–39.
24. Neelakantan, T.; Rammurthy, D.; Smith, S.T.; Suribabu, C. Expansion and Upgradation of Intermittent Water Supply System. *Asian J. Appl. Sci.* **2014**, *7*, 470–485. [CrossRef]
25. Alegre, H.; Covas, D. Gestão patrimonial de infra estruturas de abastecimento de água. Uma abordagem centrada na reabilitação. Entidade Reguladora dos Serviços de Águas e Resíduos, Laboratório Nacional de Engenharia Civil, Instituto Superior Técnico, Lisbon, Portugal, 2010. Available online: https://poseur.portugal2020.pt/media/4039/guia_tecnico_16.pdf (accessed on 23 April 2015).
26. Alegre, H.; Baptista, J.M.; Cabrera, E., Jr.; Cubillo, F.; Duarte, P.; Hirner, W.; Merkel, W.; Parena, R. *Performance Indicators for Water Supply Services*; IWA publishing: London, UK, 2006.
27. McIntosh, A.C. *Asian Water Supplies Reaching the Urban Poor*; Asian Development Bank: Metro Manila, Philippines, 2003.
28. Bosserman, B.E. Pump system hydraulic design. In *Water Distribution Systems Handbook*; McGraw-Hill Professional Publishing: New York, NY, USA, 1999.
29. Martínez, F.; Vidal, R.; Andrés, M. La regulación de los sistemas hidráulicos. In *Ingeniería Hidráulica Aplicada a los Sistemas de Distribución de Agua*, 3rd ed.; ITA-UPV: Valencia, Spain, 2009; Volume 2.
30. Shammas, N.K.; Wang, L.K. *Water Engineering: Hydraulics, Distribution and Treatment*; John Wiley & Sons: Hoboken, NJ, USA, 2015.
31. Planells, P.; Carrión, P.; Ortega, J.; Moreno, M.; Tarjuelo, J. Pumping Selection and Regulation for Water-Distribution Networks. *J. Irrig. Drain. E-ASCE* **2005**, *131*, 273–281. [CrossRef]
32. Planells, P.; Tarjuelo, J.; Ortega, J. Optimización de estaciones de bombeo en riego a la demanda. *Ingeniería Agua* **2001**, *8*, 39–51. [CrossRef]
33. Oyarzún, S.A. Optimización energética de las redes de abastecimiento de Murcia. Master's Thesis, Universitat Politècnica de València, Valencia, Spain, 2011.
34. Prescott, S.L.; Ulanicki, B.; Shipley, N. Analysis of district metered area (DMA) performance. In Proceedings of the Advances in Water Supply Management: Proceedings of the International Conference on Computing and Control for the Water Industry, London, UK, 15–17 September 2003; pp. 59–67.
35. Ulanicki, B.; AbdelMeguid, H.; Bounds, P.; Patel, R. Pressure control in Dstrict Metering Areas with boundary and internal pressure reducing valves. In Proceedings of the 10th International Water Distribution System Analysis Conference (WDSA 2008), Kruger National Park, South Africa, 17–20 August 2008.
36. ANESAPA. SeLA—Oruro 50 Años de Servicio, 2014. Available online: http://www.anesapa.org/noticias/sela-oruro-50-anos-de-servicio/ (accessed on 18 March 2015).
37. Martinelly, E. Proyecto Cábala—Procesos de Integración Sociocultural y Económica en Ciudades Capitales de Bolivia, 2009. Available online: http://www.pieb.org/cabala/principal.html (accessed on 18 March 2015).
38. INE—Bolivia. Población por Censos Según Departamento, Área Geográfica y Sexo, Censos de 1950–1976–1992–2001. Available online: http://www.ine.gob.bo/indice/visualizador.aspx?ah=PC20111.HTM (accessed on 20 March 2015).
39. Lambert, A.O.; Fantozzi, M. Recent advances in calculating economic intervention frequency for active leakage control, and implications for calculation of economic leakage levels. *Water Sci. Technol.* **2005**, *5*, 263–271.

MDPI

St. Alban-Anlage 66

4052 Basel

Switzerland

Tel. +41 61 683 77 34

Fax +41 61 302 89 18

www.mdpi.com

Water Editorial Office

E-mail: water@mdpi.com

www.mdpi.com/journal/water